Enhancing Shelf Life of Food Products: Strategies, Challenges and Innovations

Enhancing Shelf Life of Food Products: Strategies, Challenges and Innovations

Guest Editors

Alessandro Bianchi
Francesca Venturi

MDPI

Basel • Beijing • Wuhan • Barcelona • Belgrade • Novi Sad • Cluj • Manchester

Guest Editors

Alessandro Bianchi
Department of Agriculture,
Food and Environment
University of Pisa
Pisa
Italy

Francesca Venturi
Department of Agriculture,
Food and Environment
University of Pisa
Pisa
Italy

Editorial Office
MDPI AG
Grosspeteranlage 5
4052 Basel, Switzerland

This is a reprint of the Special Issue, published open access by the journal *Foods* (ISSN 2304-8158), freely accessible at: https://www.mdpi.com/journal/foods/special_issues/XZ77LDIP0I.

For citation purposes, cite each article independently as indicated on the article page online and as indicated below:

Lastname, A.A.; Lastname, B.B. Article Title. *Journal Name* **Year**, *Volume Number*, Page Range.

ISBN 978-3-7258-6848-3 (Hbk)
ISBN 978-3-7258-6849-0 (PDF)
https://doi.org/10.3390/books978-3-7258-6849-0

Contents

About the Editors

Alessandro Bianchi

Alessandro Bianchi is a research fellow at the Department of Agriculture, Food, and Environment, University of Pisa, Italy. He holds a Ph.D. in Agricultural, Food, and Environmental Sciences. His research focuses on food technology, enology, and oleotechnology, with a particular emphasis on product quality, shelf-life evaluation, and the valorization of by-products. He has authored over 55 indexed scientific publications and serves as a reviewer for several international journals. His recent work explores the role of phenolic compounds and innovative processing strategies in enhancing the quality and stability of food and beverage products.

Francesca Venturi

Francesca Venturi is an Associate Professor at the Department of Agriculture, Food, and Environment, University of Pisa, Italy. Her scientific interest lies in food chemistry and technology, focusing on the characterization and preservation of sensory and nutritional quality in foods and beverages. She has coordinated numerous research projects and supervised graduate and doctoral students in the field of food science. Prof. Venturi's research integrates analytical chemistry and sensory analysis with innovative preservation techniques to ensure the quality, safety, and sustainability of food systems. She has authored over 120 indexed scientific publications and serves as a reviewer for several international journals.

Preface

The Special Issue 'Enhancing Shelf Life of Food Products: Strategies, Challenges and Innovations' was conceived to provide a comprehensive overview of the latest research efforts in the field of food preservation. The topic of shelf-life extension is increasingly relevant in the context of global sustainability goals and the reduction of food waste. This Reprint includes contributions that explore innovative processing methods, natural antimicrobial systems, and packaging technologies capable of maintaining the sensory and nutritional quality of foods. It highlights not only scientific advances but also industrial applications and consumer-oriented perspectives, demonstrating how interdisciplinary approaches can bridge the gap between research and real-world implementation. We would like to thank all the authors for their valuable contributions and the reviewers for their time and effort in improving the quality of the papers. We are also grateful to the editorial team of *Foods* for their support throughout the process.

Alessandro Bianchi and Francesca Venturi
Guest Editors

foods

MDPI

Editorial

Enhancing the Shelf Life of Food Products: Strategies, Challenges, and Innovations

Alessandro Bianchi [1,2,*] and Francesca Venturi [1,3]

1 Department of Agriculture, Food and Environment, University of Pisa, Via del Borghetto 80, 56124 Pisa, Italy; francesca.venturi@unipi.it
2 Department of Pharmacy, University of Pisa, Via Bonanno Pisano 12, 56126 Pisa, Italy
3 Interdepartmental Research Centre "Nutraceuticals and Food for Health", University of Pisa, Via del Borghetto 80, 56124 Pisa, Italy
* Correspondence: alessandro.bianchi@phd.unipi.it:

1. Introduction

The concept of shelf life is fascinating because it combines scientific issues with economic, governmental, and consumer considerations [1]. The latter factors, which can drastically differ from nation to nation, are dynamic and are subject to change in response to shifts in society, particularly in terms of consumer preferences and needs [2,3]. In the face of increasing global demand, shifting consumer expectations, and pressing sustainability goals, enhancing the shelf life of food products while maintaining their overall quality remains a central challenge in food science and technology [2–4].

Shelf life is not merely a label datum: it encapsulates complex interactions among microbial ecology, physicochemical stability, packaging systems, supply chain logistics, and consumer behavior [2]. In recent decades, the paradigm has shifted from relying on synthetic preservatives and high-intensity treatments toward milder, integrative, and sustainable strategies that better preserve nutritional and sensory attributes while reducing waste and aligning with the "clean label" movement [5,6]. Despite these advances, major gaps remain in bridging laboratory innovation and real-world applicability, in quantifying environmental tradeoffs, and in understanding dynamic microbial ecosystems under storage [7].

To obtain reliable shelf life data, it is essential to refer to a systematic and reliable approach. To achieve this aim, a comprehensive shelf life study can be divided into three fundamental steps. First, the identification of the most critical chemical, physical, or biological events leading to product quality depletion is required, followed by the definition of the relevant acceptability limits. Second, the selected quality indicators are monitored as a function of time under storage conditions that mimic foreseeable conditions (real-time shelf life testing) or under accelerated environments that enhance deteriorative reactions (accelerated shelf life testing, ASLT). Finally, experimental data are modeled to obtain an estimation or prediction of shelf life [1,7–9]. Real-time testing is especially suitable for perishable foods with short quality decay periods, while accelerated testing is more appropriate for microbiologically stable foods such as ambient and frozen products, where oxidative reactions dominate quality depletion [5,9,10].

Achieving the desired shelf life is a powerful driver for product and packaging innovation. In particular, packaging technologies have played a crucial role in extending product life and maintaining quality [11]. Modified atmosphere packaging (MAP) exemplifies this progress, relying on oxygen exclusion to limit oxidative rancidity and microbial spoilage [12–15]. Beyond MAP, a wide range of active and intelligent packaging systems have emerged to interact dynamically with the packaged food and its environment. 'Smart' packaging encompasses both active and intelligent forms: active materials may release or absorb substances (e.g., ethanol vapors to inhibit microbial growth), while intelligent devices—such as inexpensive colorimetric tags or RFID (Radio Frequency Identification) sensors—enable communication across the supply chain to ensure quality and safety [16,17]. Recent studies have also explored biobased smart materials incorporating natural pigments derived from plant products or food by-products—such as anthocyanins, curcumin, betalains, carotenoids, tannins, and chlorophyll—as visual indicators of spoilage [18–23]. These pigments, sensitive to volatile amines and organic acids released during deterioration, undergo visible color changes, allowing real-time, non-invasive monitoring of freshness and safety [20].

Besides packaging innovations, emerging food processing technologies have redefined shelf life enhancement paradigms. At the turn of the century, novel non-thermal technologies such as high hydrostatic pressure (HHP), pulsed electric fields (PEF), ultrasound (US), and high-pressure carbon dioxide (HPCD) processing have gained attention due to their ability to inactivate microorganisms while preserving the sensory and nutritional characteristics of foods [24–27]. These methods offer reduced thermal degradation, energy efficiency, and the possibility of combination with traditional treatments for optimized microbial and enzymatic control [28]. Integrating non-thermal methods with active packaging systems represents a promising direction for the future, potentially leading to synergistic effects in shelf life extension without compromising the "fresh-like" appeal of minimally processed foods [29–31].

Another rapidly growing field involves natural preservatives and antimicrobial compounds derived from plants. Chemical preservatives have long been used to control spoilage and pathogenic microorganisms, but growing consumer demand for minimally processed and additive-free products has shifted the focus toward natural alternatives [32–35]. Plant-derived bioactive compounds—including phenolics, alkaloids, flavonoids, steroids, and terpenes—exhibit significant antibacterial and antioxidant properties, providing a natural means of food protection during storage [36–40]. Essential oils, in particular, have gained traction for their strong antimicrobial activity, though challenges remain in their controlled release, sensory impact, and stability during processing and storage [39,41,42].

Looking ahead, the integration of predictive modeling, real-time sensing, and sustainable preservation strategies is expected to transform shelf life management [43]. Data-driven approaches, powered by artificial intelligence and digital twins, can simulate degradation kinetics under varying storage and distribution scenarios, improving the precision of shelf life prediction [39,44,45]. Moreover, advances in biodegradable and compostable packaging materials, coupled with circular economy principles, are pushing the field toward eco-designed shelf life systems that minimize waste and environmental impact. These multidimensional efforts are essential to support a holistic, science-based, and sustainability-oriented framework for shelf life extension and monitoring [43,44,46–48].

The present Special Issue was conceived to address these challenges. The editorial vision aimed to gather contributions that transcended incremental improvements, emphasizing interdisciplinary approaches that combine formulation, processing, packaging, microbiological control, and environmental sustainability. Across its contributions, the Issue encompasses diverse food matrices (meat, fish, fruits, berries, minimally processed vegetables, and dairy) and technological strategies (antimicrobials, coatings, mild processing, active and intelligent packaging, and life-cycle assessment). In the following sections, we synthesize the main insights, identify persisting research gaps, and propose a forward-looking agenda for future innovations in shelf life science.

2. Key Contributions and Their Implications

This Special Issue was conceived to gather original research and review papers addressing innovative strategies for the control, analysis, and preservation of microorganisms in foods, with a particular focus on natural antimicrobials, sustainable preservation technologies, valorization of food by-products, and quality–safety interactions during processing and storage. The eleven contributions collected in this Issue exemplify the breadth of current research, spanning from traditional food safety control to novel materials, environmental assessments, and sensory optimization. Below, a brief overview of each contribution is presented to encourage the reader to explore the full papers.

A central piece in this Issue is the review by Rabbani and collaborators (Contribution 8) on the effect of heat pasteurization and sterilization on milk. The authors comprehensively discuss how conventional thermal treatments—while ensuring microbial safety—unavoidably modify protein conformation, flavor, and micronutrient integrity, thus compromising quality. The review also highlights the growing interest in emerging non-thermal and combined technologies, such as pulsed electric fields and high-pressure processing, as promising tools to maintain both safety and nutritional value.

Building on this technological perspective, Dermesonlouoglou and colleagues (Contribution 6) present a hybrid preservation approach combining pulsed electric fields (PEF) and osmotic dehydration followed by modified-atmosphere packaging of fresh-cut and fried potatoes. This multi-step protocol effectively slows down microbial proliferation and oxidation, thereby illustrating the strength of hurdle technology, where mild complementary treatments act synergistically to preserve food quality without compromising texture or flavor.

In a related direction, Faisal and co-authors (Contribution 5) investigated an edible coating composed of turmeric extract and liquid smoke obtained from oil-palm empty fruit bunches, applied to mackerel filets. The coating significantly delayed microbial spoilage and maintained sensory acceptability for up to 48 h at room temperature, offering a low-cost preservation strategy particularly suited for warm-climate regions where cold storage is limited. Extending this concept to fruit products, Mari and collaborators (Contribution 7) integrated a Life Cycle Assessment (LCA) into a process combining osmotic dehydration and edible coatings for berries. Their work quantifies the environmental trade-offs of such innovative technologies, identifying energy use and coating formulation as major hotspots, and providing an evidence-based framework for the sustainable design of food preservation methods.

Natural antimicrobials and valorized ingredients are also at the core of several other contributions. Kačániová and co-authors (Contribution 1) demonstrated that the essential oil of *Eugenia caryophyllus* (clove) effectively inhibits *Salmonella enterica* and biofilm formation, improving the microbial safety and shelf life of sous-vide deer meat. Similarly, Tayel and collaborators (Contribution 9) evaluated dill (*Anethum graveolens*) essential oil as a preservative for fish filets stored under refrigeration, finding a substantial reduction in spoilage microorganisms and oxidative degradation, leading to longer freshness retention.

The incorporation of bioactive natural extracts into functional materials is another emerging trend showcased in this Issue. García-Juárez and colleagues (Contribution 3) developed gelatin nanoparticles loaded with bitter orange (*Citrus aurantium*) peel extract, revealing remarkable antioxidant and antibacterial activities, which highlight their potential as active components for edible coatings or packaging aimed at extending shelf life and reducing synthetic additives.

Complementing these innovations, Augustyńska-Prejsnar and co-authors (Contribution 4) investigated microbial and sensory quality changes in broiler chicken breast meat during refrigerated storage. Their findings provide valuable data on the relationship between microbial growth dynamics and sensory deterioration, supporting the establishment of realistic shelf life and quality criteria for fresh poultry.

The valorization of agro-industrial by-products was further addressed by D'Arrigo and collaborators (Contribution 2), who incorporated a red grape pomace ingredient—stabilized through blanching and high-pressure processing—into traditional dry-cured *salchichón* sausages. The inclusion enhanced antioxidant capacity and limited lipid and protein oxidation, offering a sustainable alternative to synthetic additives such as nitrites, while maintaining desirable sensory characteristics.

A different but complementary strategy was explored by Rodrigues and colleagues (Contribution 10), who examined dietary supplementation of Nile tilapia (*Oreochromis niloticus*) with the green alga *Chlorella pyrenoidosa*. The enriched diet improved the antioxidant stability and microbial resistance of the filets during refrigerated storage, demonstrating how nutritional interventions at the farming stage can enhance post-harvest quality and shelf life.

Finally, Panzani and co-authors (Contribution 11) compared traditional versus controlled drying methods for chestnuts (*Castanea sativa*), showing that controlled drying preserved more favorable chemical and aromatic profiles, resulting in superior sensory quality of chestnut flour. While not directly antimicrobial, this study reinforces the Issue's overarching theme—that innovative, controlled processing methods can simultaneously safeguard quality, extend shelf life, and align with sustainability goals.

Collectively, these contributions underscore the multifaceted nature of modern food preservation, spanning from molecular to environmental scales. They highlight how natural compounds, gentle physical technologies, valorized materials, and life cycle thinking converge toward the shared objective of achieving safe, high-quality, and sustainable foods, thus offering valuable guidance for both researchers and industry practitioners.

3. Future Directions and Research Agenda

Although remarkable progress has been achieved in the field, several aspects still warrant further investigation to ensure that advances translate effectively from research to real-world applications. Scaling up experimental results remains a key challenge, as most studies are performed under controlled laboratory conditions and must now be validated on a pilot or industrial scale. Similarly, shelf life optimization should be approached holistically, considering its interplay with the supply chain, cold-chain logistics, and consumer behavior. The development of reliable predictive models that integrate microbial kinetics, packaging performance, and storage conditions is another promising yet evolving area.

Moreover, future research would benefit from a more comprehensive evaluation of sustainability parameters—including energy, water, and carbon impacts—supported by life cycle and techno-economic analyses. Continued attention is also needed to ensure that novel materials, coatings, and active systems comply with regulatory frameworks and meet consumer expectations. Looking ahead, efforts should focus on hybrid smart packaging systems capable of dynamic monitoring and sensor feedback, as well as on material innovation toward biodegradable and functional solutions. The integration of digital tools such as data-driven models and digital twins could further enhance predictive accuracy and process optimization, paving the way for a more efficient, sustainable, and consumer-centered food packaging ecosystem.

4. Conclusions

This Special Issue has sought to bring together innovation, scientific rigor, and sustainability within the broad domain of food shelf life extension. The contributions collected herein clearly show that the future of the field lies not in single "silver bullet" technologies, but rather in smartly integrated, system-level strategies that combine mild processing, intelligent packaging, ecological assessment, and real-world validation.

Nevertheless, the journey toward truly sustainable and resilient food preservation is far from complete. It is hoped that this closing editorial will inspire further cross-disciplinary collaboration, motivating researchers to bridge the gap between laboratory experimentation and pilot- or industrial-scale implementation. Equally, it should encourage the exploration of new research pathways that connect microbiology, materials science, data analytics, and sustainability assessment.

Readers are invited to engage with the articles in this Special Issue, identify emerging synergies, and contribute to advancing the shared goal of safe, high-quality, and environmentally responsible food systems for the future.

Author Contributions: Conceptualization, A.B. and F.V.; methodology, A.B. and F.V.; validation, A.B. and F.V.; investigation, A.B. and F.V.; writing—original draft preparation, A.B. and F.V.; writing—review and editing, A.B. and F.V.; visualization, A.B. and F.V. All authors have read and agreed to the published version of the manuscript.

Funding: This work was funded by the European Union, within the NextGenerationEU project LOVE, PRIN 2022, grant number: 022M4WB3M and within NextGenerationEU project "CANTINA 5.0", PRIN PNRR 2022 call for Grants, grant number: P2022P5BZY.

Data Availability Statement: Not applicable.

Conflicts of Interest: The authors declare no conflicts of interest.

List of Contributions:

1. Kačániová, M.; Garzoli, S.; Ben Hsouna, A.; Ban, Z.; Elizondo-Luevano, J.; Kluz, M.; Ben Saad, R.; Haščík, P.; Čmiková, N.; Waskiewicz-Robak, B.; et al. Enhancing Deer Sous Vide Meat Shelf Life and Safety with Eugenia caryophyllus Essential Oil against *Salmonella enterica*. *Foods* **2024**, *13*, 2512. https://doi.org/10.3390/foods13162512.
2. D'Arrigo, M.; Petrón, M.; Delgado-Adámez, J.; García-Parra, J.; Martín-Mateos, M.; Ramírez-Bernabé, M. Dry-Cured Sausages "Salchichón" Manufactured with a Valorized Ingredient from Red Grape Pomace (Var. Tempranillo). *Foods* **2024**, *13*, 3133. https://doi.org/10.3390/foods13193133.
3. García-Juárez, A.; Garzón-García, A.; Ramos-Enríquez, J.; Tapia-Hernández, J.; Ruiz-Cruz, S.; Canizales-Rodríguez, D.; Del-Toro-Sánchez, C.; Rodríguez-Félix, F.; Ocaño-Higuera, V.; Ornelas-Paz, J. Evaluation of Antioxidant and Antibacterial Activity of Gelatin Nanoparticles with Bitter Orange Peel Extract for Food Applications. *Foods* **2024**, *13*, 3838. https://doi.org/10.3390/foods13233838.
4. Augustyńska-Prejsnar, A.; Kačániová, M.; Hanus, P.; Sokołowicz, Z.; Słowiński, M. Microbial and Sensory Quality Changes in Broiler Chicken Breast Meat During Refrigerated Storage. *Foods* **2024**, *13*, 4063. https://doi.org/10.3390/foods13244063.
5. Faisal, M.; Gani, A.; Muzaifa, M.; Heriansyah, M.; Desvita, H.; Kamaruzzaman, S.; Sauqi, A.; Ardiansa, D. Edible Coating Combining Liquid Smoke from Oil Palm Empty Fruit Bunches and Turmeric Extract to Prolong the Shelf Life of Mackerel. *Foods* **2025**, *14*, 139. https://doi.org/10.3390/foods14010139.

6. Dermesonlouoglou, E.; Seretis, G.; Katsouli, M.; Katsimichas, A.; Taoukis, P.; Giannakourou, M. Effect of Pulsed Electric Fields and Osmotic Dehydration on the Quality of Modified-Atmosphere-Packaged Fresh-Cut and Fried Potatoes. *Foods* **2025**, *14*, 420. https://doi.org/10.3390/foods14030420.
7. Mari, A.; Kekes, T.; Boukouvalas, C.; Krokida, M. Integrating Life Cycle Assessment in Innovative Berry Processing with Edible Coating and Osmotic Dehydration. *Foods* **2025**, *14*, 1167. https://doi.org/10.3390/foods14071167.
8. Rabbani, A.; Ayyash, M.; D'Costa, C.; Chen, G.; Xu, Y.; Kamal-Eldin, A. Effect of Heat Pasteurization and Sterilization on Milk Safety, Composition, Sensory Properties, and Nutritional Quality. *Foods* **2025**, *14*, 1342. https://doi.org/10.3390/foods14081342.
9. Tayel, A.; Hassanin, F.; Edris, S.; Hamad, A.; Sabeq, I. Preservative Potential of *Anethum graveolens* Essential Oil on Fish Fillet Quality and Shelf Life During Refrigerated Storage. *Foods* **2025**, *14*, 1591. https://doi.org/10.3390/foods14091591.
10. Rodrigues, L.; Aracati, M.; Luporini de Oliveira, S.; Carlino-Costa, C.; Alves Rodrigues, R.; Pereira, M.; Borba, H.; Menegasso Mansano, C.; Marques Rossi, G.; Galindo-Villegas, J.; et al. Dietary Supplementation with Green Alga (*Chlorella pyrenoidosa*) Enhances the Shelf Life of Refrigerated Nile Tilapia (*Oreochromis niloticus*) Fillets. *Foods* **2025**, *14*, 1642. https://doi.org/10.3390/foods14091642.
11. Panzani, S.; Venturi, F.; Bianchi, A.; Díaz-Guerrero, P.; Pieracci, Y.; Flamini, G.; Taglieri, I.; Sanmartin, C. Comparative Evaluation of Traditional and Controlled Drying Methods of Chestnuts (*Castanea sativa* Mill.): Impact on the Chemical Composition, Aromatic, and Sensory Profile of Flour. *Foods* **2025**, *14*, 1931. https://doi.org/10.3390/foods14111931.

References

1. Nicoli, M.C.; Calligaris, S. Secondary Shelf Life: An Underestimated Issue. *Food Eng. Rev.* **2018**, *10*, 57–65. [CrossRef]
2. Lohita, B.; Srijaya, M. Novel Technologies for Shelf-Life Extension of Food Products as a Competitive Advantage: A Review. In *Food Production, Diversity, and Safety Under Climate Change*; Chakraborty, R., Mathur, P., Roy, S., Eds.; Springer Nature Switzerland: Cham, Switzerland, 2024; pp. 285–306. ISBN 978-3-031-51647-4.
3. Dordevic, D.; Cazalilla, R.M.; Javurkova, Z.; Buchtova, H.; Jancikova, S. Consumers ' response to different shelf life food labelling. *Qual. Assur. Saf. Crop. Foods* **2020**, *12*, 24–34. [CrossRef]
4. Bianchi, A.; Venturi, F.; Palermo, C.; Taglieri, I.; Angelini, G.L.; Tavarini, S.; Sanmartin, C. Primary and secondary shelf-life of bread as a function of formulation and MAP conditions: Focus on physical-chemical and sensory markers. *Food Packag. Shelf Life* **2024**, *41*, 101241. [CrossRef]
5. Calligaris, S.; Manzocco, L.; Lagazio, C. Modeling shelf life using chemical, physical, and sensory indicators. In *Shelf Life Assessment of Food*; Nicoli, M.C., Ed.; CRC Press: Boca Raton, FL, USA; pp. 75–126. ISBN 9781439846032.
6. Djekić, I.; Velebit, B.; Pavlić, B.; Putnik, P.; Šojić Merkulov, D.; Bebek Markovinović, A.; Bursać Kovačević, D. Food Quality 4.0: Sustainable Food Manufacturing for the Twenty-First Century. *Food Eng. Rev.* **2023**, *15*, 577–608. [CrossRef]
7. Manzocco, L.; Calligaris, S.; Nicoli, M.C. 9—Methods for food shelf life determination and prediction. In *Oxidation in Foods and Beverages and Antioxidant Applications*; Decker, E.A., Ed.; Woodhead Publishing: Sawston, UK, 2010; pp. 196–222. ISBN 978-1-84569-648-1.
8. Calligaris, S.; Manzocco, L.; Anese, M.; Nicoli, M.C. Shelf-life Assessment of Food Undergoing Oxidation–A Review. *Crit. Rev. Food Sci. Nutr.* **2016**, *56*, 1903–1912. [CrossRef]
9. Manzocco, L. The Acceptability Limit in Food Shelf Life Studies. *Crit. Rev. Food Sci. Nutr.* **2016**, *56*, 1640–1646. [CrossRef] [PubMed]
10. Calligaris, S.; Manzocco, L. Critical Indicators in Shelf Life Assessment. In *Shelf Life Assessment of Food*; Nicoli, M.C., Ed.; CRC Press: Boca Raton, FL, USA, 2012; pp. 61–74. ISBN 9781439846032.
11. Sanmartin, C.; Taglieri, I.; Bianchi, A.; Parichanon, P.; Puccinelli, M.; Pardossi, A.; Venturi, F. Effects of Temperature and Packaging Atmosphere on Shelf Life, Biochemical, and Sensory Attributes of Glasswort (*Salicornia europaea* L.) Grown Hydroponically at Different Salinity Levels. *Foods* **2024**, *13*, 3260. [CrossRef] [PubMed]
12. Kirwan, M.; Brown, H.; Williams, J. Packaged Product Quality and Shelf Life. In *Food and Beverage Packaging Technology*, 2nd ed.; Blackwell Publishing Ltd.: Oxford, UK, 2011; pp. 59–83. [CrossRef]
13. Narasimha Rao, D.; Sachindra, N.M. Modified atmosphere and vacuum packaging of meat and poultry products. *Food Rev. Int.* **2002**, *18*, 263–293. [CrossRef]

14. Bianchi, A.; Sanmartin, C.; Taglieri, I.; Macaluso, M.; Venturi, F.; Napoli, M.; Mancini, M.; Fabbri, C.; Zinnai, A. Effect of Fertilization Regime of Common Wheat (*Triticum aestivum*) on Flour Quality and Shelf-Life of PDO Tuscan Bread. *Foods* **2023**, *12*, 2672. [CrossRef]
15. Monacci, E.; Sanmartin, C.; Bianchi, A.; Pettinelli, S.; Taglieri, I.; Mencarelli, F. Plastic film packaging for the postharvest quality of fresh hop inflorescence (*Humulus lupulus*) cv. Cascade. *Postharvest Biol. Technol.* **2023**, *206*, 112575. [CrossRef]
16. Hempel, A.W.; O'Sullivan, M.G.; Papkovsky, D.B.; Kerry, J.P. Use of smart packaging technologies for monitoring and extending the shelf-life quality of modified atmosphere packaged (MAP) bread: Application of intelligent oxygen sensors and active ethanol emitters. *Eur. Food Res. Technol.* **2013**, *237*, 117–124. [CrossRef]
17. Costa, F.; Genovesi, S.; Borgese, M.; Michel, A.; Dicandia, F.A.; Manara, G. A Review of RFID Sensors, the New Frontier of Internet of Things. *Sensors* **2021**, *21*, 3138. [CrossRef]
18. Bhargava, N.; Sharanagat, V.S.; Mor, R.S.; Kumar, K. Active and intelligent biodegradable packaging films using food and food waste-derived bioactive compounds: A review. *Trends Food Sci. Technol.* **2020**, *105*, 385–401. [CrossRef]
19. Kuswandi, B.; Wicaksono, Y.; Jayus; Abdullah, A.; Heng, L.Y.; Ahmad, M. Smart packaging: Sensors for monitoring of food quality and safety. *Sens. Instrum. Food Qual. Saf.* **2011**, *5*, 137–146. [CrossRef]
20. Zhang, K.; Huang, T.-S.; Yan, H.; Hu, X.; Ren, T. Novel pH-sensitive films based on starch/polyvinyl alcohol and food anthocyanins as a visual indicator of shrimp deterioration. *Int. J. Biol. Macromol.* **2020**, *145*, 768–776. [CrossRef]
21. Ardiyansyah; Kurnianto, M.F.; Poerwanto, B.; Wahyono, A.; Apriliyanti, M.W.; Lestari, I.P. Monitoring of banana deteriorations using intelligent-packaging containing brazilien extract (*Caesalpina sappan* L.). *IOP Conf. Ser. Earth Environ. Sci.* **2020**, *411*, 12043. [CrossRef]
22. Kanatt, S.R.; Rao, M.S.; Chawla, S.P.; Sharma, A. Active chitosan–polyvinyl alcohol films with natural extracts. *Food Hydrocoll.* **2012**, *29*, 290–297. [CrossRef]
23. Chi, W.; Cao, L.; Sun, G.; Meng, F.; Zhang, C.; Li, J.; Wang, L. Developing a highly pH-sensitive κ-carrageenan-based intelligent film incorporating grape skin powder via a cleaner process. *J. Clean. Prod.* **2020**, *244*, 118862. [CrossRef]
24. van Boekel, M.; Fogliano, V.; Pellegrini, N.; Stanton, C.; Scholz, G.; Lalljie, S.; Somoza, V.; Knorr, D.; Jasti, P.R.; Eisenbrand, G. A review on the beneficial aspects of food processing. *Mol. Nutr. Food Res.* **2010**, *54*, 1215–1247. [CrossRef] [PubMed]
25. Pereira, R.N.; Vicente, A.A. Environmental impact of novel thermal and non-thermal technologies in food processing. *Food Res. Int.* **2010**, *43*, 1936–1943. [CrossRef]
26. Younis, K.; Jahan, K.; Qadri, O.S.; Osama, K.; Yousuf, O. Non-thermal processing techniques for sustainable food supply chains. *Sustain. Futur.* **2025**, *9*, 100658. [CrossRef]
27. Olaniyi, M.B.; Olaniyi, A.A.; Alawode, R.A.; Ogunbamowo, P.O. Roles of some Innovative Non-Thermal Processing Techniques on Food Quality and Safety. *e-Proc. Fac. Agric. Int. Conf.* **2024**, *12*, 423–428.
28. Khan, M.K.; Ahmad, K.; Hassan, S.; Imran, M.; Ahmad, N.; Xu, C. Effect of novel technologies on polyphenols during food processing. *Innov. Food Sci. Emerg. Technol.* **2018**, *45*, 361–381. [CrossRef]
29. Nabi, B.G.; Mukhtar, K.; Arshad, R.N.; Radicetti, E.; Tedeschi, P.; Shahbaz, M.U.; Walayat, N.; Nawaz, A.; Inam-Ur-Raheem, M.; Aadil, R.M. High-Pressure Processing for Sustainable Food Supply. *Sustainability* **2021**, *13*, 13908. [CrossRef]
30. Puebla-Duarte, A.L.; Santos-Sauceda, I.; Rodríguez-Félix, F.; Iturralde-García, R.D.; Fernández-Quiroz, D.; Pérez-Cabral, I.D.; Del-Toro-Sánchez, C.L. Active and Intelligent Packaging: A Review of the Possible Application of Cyclodextrins in Food Storage and Safety Indicators. *Polymers* **2023**, *15*, 4317. [CrossRef]
31. Chávez-mejía, A.C.; Magaña-lόpez, R.; Durán-álvarez, J.C.; Jiménez-cisneros, B.E. The synergistic effect of non-thermal techniques and modified atmosphere packaging in food preservation. *Int. J. Environ. Agric. Biotechnol.* **2019**, *8*, 246–259. [CrossRef]
32. Sweet, R.; Kroon, P.A.; Webber, M.A. Activity of antibacterial phytochemicals and their potential use as natural food preservatives. *Crit. Rev. Food Sci. Nutr.* **2024**, *64*, 2076–2087. [CrossRef]
33. El Alami El Hassani, N.; Baraket, A.; Alem, C. Recent advances in natural food preservatives: A sustainable solution for food safety and shelf life extension. *J. Food Meas. Charact.* **2025**, *19*, 293–315. [CrossRef]
34. Čmiková, N.; Vukic, M.D.; Vukovic, N.L.; Verešová, A.; Bianchi, A.; Garzoli, S.; Ben Saad, R.; Ben Hsouna, A.; Ban, Z.; Kačániová, M. Phytochemical investigation, evaluation of the biological activities and preservative effect of the essential oil of *Juniperus communis* L. dried berries on the vacuum-packed carrot after the application of *Salmonella enterica*. *Sci. Hortic.* **2024**, *336*, 113442. [CrossRef]
35. Al-Maqtari, Q.A.; Rehman, A.; Mahdi, A.A.; Al-Ansi, W.; Wei, M.; Yanyu, Z.; Phyo, H.M.; Galeboe, O.; Yao, W. Application of essential oils as preservatives in food systems: Challenges and future prospectives—A review. *Phytochem. Rev.* **2022**, *21*, 1209–1246. [CrossRef]

36. Kačániová, M.; Čmiková, N.; Ban, Z.; Garzoli, S.; Elizondo-Luevano, J.H.; Ben Hsouna, A.; Ben Saad, R.; Bianchi, A.; Venturi, F.; Kluz, M.I.; et al. Enhancing the Shelf Life of Sous-Vide Red Deer Meat with *Piper nigrum* Essential Oil: A Study on Antimicrobial Efficacy against Listeria monocytogenes. *Molecules* **2024**, *29*, 4179. [CrossRef]
37. Gouveia, A.R.; Alves, M.; Silva, J.A.; Saraiva, C. The Antimicrobial Effect of Rosemary and Thyme Essential Oils Against *Listeria monocytogenes* in Sous Vide Cook-chill Beef During Storage. *Procedia Food Sci.* **2016**, *7*, 173–176. [CrossRef]
38. Hyldgaard, M.; Mygind, T.; Meyer, R.L. Essential Oils in Food Preservation: Mode of Action, Synergies, and Interactions with Food Matrix Components. *Front. Microbiol.* **2012**, *3*, 12. [CrossRef]
39. Fadiji, T.; Rashvand, M.; Daramola, M.O.; Iwarere, S.A. A Review on Antimicrobial Packaging for Extending the Shelf Life of Food. *Processes* **2023**, *11*, 590. [CrossRef]
40. Sağlam, K.; Mzoughi, M. Chapter 20—Phytochemicals in food preservation: Antimicrobial and antioxidant properties. In *Phytoceuticals in Food for Health and Wellness*; Sarkar, T., Smaoui, S., Lai, W.-F., Eds.; Academic Press: Cambridge, MA, USA, 2026; pp. 397–429. ISBN 978-0-443-26494-8.
41. Bianchi, A.; Farina, P.; Venturi, F.; Trusendi, F.; Flamini, G.; Ascrizzi, R.; Sarrocco, S.; Ortega-Andrade, S.; Echeverria, M.C.; Conti, B.; et al. Reduced Doses of Diatomaceous Earth and Basil Essential Oil on Stored Grain Against the Wheat-Damaging *Sitophilus oryzae*: Influence on Bread Quality and Sensory Profile. *Foods* **2025**, *14*, 572. [CrossRef]
42. Wang, J.; Zhao, F.; Huang, J.; Li, Q.; Yang, Q.; Ju, J. Application of essential oils as slow-release antimicrobial agents in food preservation: Preparation strategies, release mechanisms and application cases. *Crit. Rev. Food Sci. Nutr.* **2024**, *64*, 6272–6297. [CrossRef] [PubMed]
43. Peres, F.A.; Bondarczuk, B.A.; Gomes, L.D.; Jardim, L.D.; Corrêa, R.G.; Baierle, I.C. Advances in Food Quality Management Driven by Industry 4.0: A Systematic Review-Based Framework. *Foods* **2025**, *14*, 2429. [CrossRef] [PubMed]
44. Rashvand, M.; Ren, Y.; Sun, D.-W.; Senge, J.; Krupitzer, C.; Fadiji, T.; Miró, M.S.; Shenfield, A.; Watson, N.J.; Zhang, H. Artificial intelligence for prediction of shelf-life of various food products: Recent advances and ongoing challenges. *Trends Food Sci. Technol.* **2025**, *159*, 104989. [CrossRef]
45. Zou, Y.; Wu, J.; Meng, X.; Wang, X.; Manzardo, A. Digital twin integration for dynamic quality loss control in fruit supply chains. *J. Food Eng.* **2025**, *397*, 112577. [CrossRef]
46. Ali Eltabey, R. Utilizing Digital Technologies to Ensure Food Safety. *Int. J. Artif. Intell. Emerg. Technol.* **2023**, *6*, 42–65. [CrossRef]
47. Shehzad, K. Predictive AI Models for Food Spoilage and Shelf-Life Estimation. *Glob. Trends Sci. Technol.* **2025**, *1*, 75–94. [CrossRef]
48. Hussain, S.; Akhter, R.; Maktedar, S.S. Advancements in sustainable food packaging: From eco-friendly materials to innovative technologies. *Sustain. Food Technol.* **2024**, *2*, 1297–1364. [CrossRef]

foods

MDPI

Article

Comparative Evaluation of Traditional and Controlled Drying Methods of Chestnuts (*Castanea sativa* Mill.): Impact on the Chemical Composition, Aromatic, and Sensory Profile of Flour

Sofia Panzani [1], Francesca Venturi [1,2], Alessandro Bianchi [1,*], Pierina Díaz-Guerrero [1], Ylenia Pieracci [1,3], Guido Flamini [2,3], Isabella Taglieri [1,2] and Chiara Sanmartin [1,2]

1 Department of Agriculture, Food and Environment, University of Pisa, Via del Borghetto 80, 56124 Pisa, Italy; sofia.panzani@phd.unipi.it (S.P.); francesca.venturi@unipi.it (F.V.); pierina.guerrero@phd.unipi.it (P.D.-G.); ylenia.pieracci@agr.unipi.it (Y.P.); isabella.taglieri@unipi.it (I.T.); chiara.sanmartin@unipi.it (C.S.)
2 Interdepartmental Research Centre "Nutraceuticals and Food for Health", University of Pisa, Via del Borghetto 80, 56124 Pisa, Italy; guido.flamini@unipi.it
3 Department of Pharmacy, University of Pisa, Via Bonanno Pisano 12, 56126 Pisa, Italy
* Correspondence: alessandro.bianchi@phd.unipi.it

Abstract: Chestnut flour, obtained through drying and milling of *Castanea sativa* fruits, has evolved from a subsistence food into a sought-after niche product, appreciated for its naturally gluten-free profile, high starch content, and richness in micronutrients. Over the past decade, its demand has steadily increased due to consumer perception of the health benefits associated with chestnut consumption. As the market for chestnut flour expanded from small-scale to large-scale production, alternative methods to the traditional process were developed. Its distinctive aroma and flavor are strongly influenced by processing methods, which are the focus of this study. Two drying approaches were compared: a traditional smoke-based method (drying house named *metato*) characterized by a wood-drying method and a controlled laboratory process using a forced-air dryer that maintained a constant temperature of 40 °C. The impact of these methods on the physico-chemical composition, volatile organic compounds (VOCs), and sensory properties of the flour was evaluated using chemical, instrumental, and sensory analyses. The traditional method enhanced the flour's aromatic complexity and typicity through the application of smoke, which has been demonstrated to generate volatile organic compounds (VOCs), such as guaiacol, furfural, and o-creosol, that are associated with the smoked aroma. Nevertheless, if not properly managed, it can lead to undesirable sensory notes due to excessive smoke exposure. In contrast, the laboratory-controlled process ensured better preservation of bioactive compounds—such as polyphenols (351 mg GAE/100 g dm) and ascorbic acid (322 mg/kg dm)—while retaining the aroma notes associated with fresh chestnuts. Optimizing processing methods may support the valorization of chestnut flour as a high-quality ingredient in the modern gluten-free and functional food market.

Keywords: gluten-free flour; volatile organic compound; bioactive compounds; sensory analysis; forced-air drying; optimizing process

1. Introduction

The chestnut tree (*Castanea* spp.), belonging to the Fagaceae family, is distributed across three main geographical areas and represented by four species: *Castanea sativa* Mill.

in Europe, *Castanea crenata* Sieb. & Zucc. in Japan, *Castanea mollissima* Bl. in China and Korea, and *Castanea dentata* Borkh. in North America [1,2].

Among these, *Castanea sativa* Mill., native to Europe, Asia Minor, and North Africa, is the only species that grows wild in Europe and has been cultivated and valued since the time of the Roman Empire for its fruit and wood due to its adaptability to temperate climates [2–4]. Historically, chestnuts (*Castanea sativa* Mill.) have played a crucial role in the diets of many rural communities of mountainous marginal areas unsuitable for wheat cultivation [4,5].

Chestnuts are nutritionally rich, providing vitamins (B1, B2, C, E), minerals (Ca, P, K, Mg, S, Fe, Cu, Zn, Mn), dietary fiber, starch (~36% fresh weight), and phenolic compounds, while being low in fat (~3%) and cholesterol free [1,3,6–9]. However, their composition varies significantly based on cultivar and harvest year [6,10,11].

To extend their shelf life, given their high water content (~50%) [3], chestnuts were traditionally processed into flour [8,12], a naturally gluten-free product increasingly valued for both traditional and modern culinary applications [13–17].

Chestnut flour, being a nutritious, naturally gluten-free product, is particularly suited for consumers with gluten intolerance or sensitivity [6,13–16].

Italy counts fifteen chestnut-based products that have obtained the European Union's Protected Designation of Origin (PDO) status. Among these, five are produced in the Tuscany region, including the renowned *"Farina di Neccio della Garfagnana PDO"* produced in the Garfagnana area and *"Farina di Castagne della Lunigiana PDO"* produced in the Lunigiana area [18].

Traditional chestnut flour production relies on the use of the *metato*, a two-level structure typically built from stone, lime and sand, where chestnuts are placed on upper racks made of chestnut sticks, or wire mesh in more recently restored structures immediately after harvest [19]. Drying is achieved through a slow, continuous fire maintained on the lower level, fueled exclusively with chestnut wood and kept steaming, covering it with residual peels from the previous year's harvest [18].

This technique allows a gradual drying process, accompanied by mild smoking, without burning the fruit. As a result, fresh chestnuts are transformed into dried fruit with modified nutritional and aromatic properties, while moisture content is reduced to about 10% [20–22]. The dried chestnuts are then threshed to remove the outer shell (about 20% of the fruit), separate the edible nut, and finally ground in traditional mills using stone millstones, often water powered. The traditional drying process typically lasts a minimum of 40 days, and during the drying period, chestnuts are turned at least once to promote uniform moisture reduction.

Global chestnut production has increased steadily over the last decade due to consumers' salient beliefs in the health benefits of chestnut consumption, and as the chestnut flour market has shifted from small to large scale, alternative methods to the traditional one have developed [8].

Industrial processes employ higher drying temperatures (40–80 °C) and mechanization, shortening drying times to as little as 48 h but potentially altering the nutritional and sensory profiles of the flour these are intended to ensure greater standardization, a feature which the traditional method currently lacks [12,22–25].

While traditional methods are valued for enhancing flavor, they often lack standardized process control, resulting in greater variability compared to modern controlled drying techniques.

The evolving demands of the agri-food sector in contemporary society are closely linked to ensuring human health safety and promoting sustainable development in production and processing methods [26].

While reviving traditional methods can be appealing, it is essential to assess how effectively these ancient practices align with modern expectations for product quality and health benefits. The traditional and industrial drying methods yield distinctly different products, as each technique alters the characteristics of the product in specific ways.

The quality of the chestnut dried in the "*metato*" might be harmed by the continuous change in temperature due to the manual managing of the wood fire [18]. Traditional drying could even lead to a high formation of toxic volatile compounds, such as furfural, guaiacol, and o-cresol. The latter two make the product taste smokey, or even burned if present in large amounts, and are formed precisely during the thermal degradation of lignin.

The traditional method, prized for imparting a distinctive smoky flavor to the product, should not be abandoned. However, it is essential to regulate temperature levels and maintain consistency throughout the process to ensure optimal quality. Clearly, improvements are needed in controlling the drying conditions inside the *metato*, as well as in managing the different stages of chestnut drying [22].

Moreover, due to climate change, the raw materials can vary significantly from year to year, for example showing lower moisture content, making it even more important to carefully monitor and adjust the drying process accordingly [27]. Moreover, the enhancement of the conventional technique would result in a reduction of waste, for instance, due to the development of mold and burns caused by inadequate drying management.

In this context, a critical evaluation of the impact of different processing methods, both traditional and non-traditional, on the chemical and sensory properties of chestnut flour becomes particularly relevant.

The present study provides a comparative evaluation of chestnut flours obtained using traditional drying techniques and a simulated industrial low-temperature process carried out under controlled laboratory conditions using a ventilated dryer. By characterizing chestnuts and the resulting flours using physical and chemical analyses as well as aromatic and descriptive sensory analysis, combined with the monitoring of processing conditions, this study aimed to evaluate the impact of different drying processes on the physicochemical and sensory properties of the final product.

2. Materials and Methods

2.1. Raw Materials and Production Process

The chestnuts and flour samples were provided by local growers of Garfagnana (Lucca, Italy) and coded as follows:

- M1t0: Fresh chestnuts
- M2t0: Fresh chestnuts
- FM1: Traditional flour obtained in *metato* 1 from M1t0 chestnuts
- FM2: Traditional flour obtained in *metato* 2 from M2t0 chestnuts
- FL: Laboratory-processed flour from M2t0 chestnuts

The chestnuts (*Castanea sativa* Mill.) used belong to the Carpinese, Pontecosi, and Rossola varieties. However, the percentage varietal composition of the different batches is not known.

Fresh chestnut samples were stabilized as soon as received. To ensure the preservation of the bioactive components, the samples were chopped (Monsieur Cuisine food processor,

Silvercrest, Bad Wimpfen, Germany) and freeze-dried (LyoQuest lyophilizer, Azbil Telstar, S.L.U., Terrassa, Spain). The freeze-drying process lasted a total of 48 h, with a condenser temperature of −52.4 °C, a shelf temperature of 25 °C, and a pressure of 0.072 mBar.

The analyzed flours were produced using two distinct methods.

Samples FM1 and FM2 were produced using the traditional method. The two *metati* (drying houses), M1 and M2, were obtained from the municipality of Molazzana (Lucca, Italy), at the following coordinates:

- M1: 44.05939° N, 10.354385° E; approximately 800 m above sea level.
- M2: 44.066249° N, 10.371313° E; approximately 900 m above sea level.

The total drying time was different for the two *metati*. For M1, the drying process started on 16 October 2023 and ended on 29 November 2023, for a total of 44 days. In the case of M2, the process was started on 22 October 2023 and ended on 24 December 2023, for a total of 63 days.

Temperature and relative humidity trends (Figure S1a,b), within the traditional drying structures (*metati*), were continuously monitored using PKDLA1 data loggers (Parkside®, Neckarsulm, Germany). The devices were positioned on the top layer of chestnuts on each drying floor to ensure representative measurements of the environmental conditions affecting the drying process.

The FL flour was produced at the Food Science and Technology Laboratory (Department of Agriculture, Food and Environment, University of Pisa) simulating a small-scale industrial process using a Fruit Jerky Plus 6 dryer (Klarstein, Berlin, Germany). Fresh chestnuts from the same batch as M2t0 (later used for FM2 production) were dried at 40 °C with frequent weight monitoring to assess the drying progression for about 3 days (71 h; Figure S2).

After drying, the chestnuts were manually peeled (removal of both the outer shell and inner pellicle), chopped, and ground using a Monsieur Cuisine food processor (Silvercrest brand). Grinding was performed in ON/OFF cycles to avoid overheating. A temperature probe was used throughout the process to ensure that the product temperature never exceeded the drying temperature.

2.2. Chemical Analysis

2.2.1. Determination of Dry Matter and Water Activity

The dry matter (dm) of samples was determined on about 5 g, dried at 105 °C until constant weight and expressed as percentage [28].

Water activity (a_w) was measured [29] using a HygroPalm HP23-AW-A (Rotronic, Bassersdorf, Switzerland) hygrometer equipped with an HC2-AW probe, calibrated with the different range standards supplied with the instruments.

2.2.2. Determination of Total Lipids

The total lipid (TL) content in the flour and fresh chestnut samples was determined using a Soxhlet extractor (SER148, Velp Scientifica Srl, Usmate Velate, Italy) following the methods previously reported [30]. Approximately 5 g of sample were placed in pre-dried and pre-weighed cellulose thimbles, which were previously dried in an oven at 105 °C. After drying, lipid extraction was carried out using *n*-hexane (ACS reagent, Sigma-Aldrich, Steinheim, Germany) as solvent. Following the extraction, the thimbles containing the de-oiled sample and the corresponding vessels containing the extracted oil were dried to remove any residual water and then weighed.

2.2.3. Determination of Free Fatty Acidity

The determination of free fatty acidity (FFA) was conducted on 4 g of flour samples and chestnuts as previously reported [31]. Each sample was subjected to an extraction with 100 mL of a 50% hydroalcoholic solution neutralized with 0.1 N HCl (ACS reagent, Sigma-Aldrich, Steinheim, Germany) and maintained under magnetic stirring for 3 h.

Afterward, the sample was centrifuged at 966× *g* for 5 min, and a 50 mL aliquot of the supernatant was titrated with 0.02 N NaOH (ACS reagent, Sigma-Aldrich, Steinheim, Germany), using phenolphthalein as an indicator. By performing this procedure, the number of mL of 0.02 N alkali used for titration corresponds to the degree of acidity, which is referred to 100 parts of dry matter. The result was expressed as g oleic acid/100 g dm.

2.2.4. Total Starch Sugars and Ascorbic Acid

Determinations of total starch, sugars, and ascorbic acid were performed using Megazyme enzyme kits (Megazyme®, Dublin, Ireland) following the instructions for each kit.

For the total starch content, the Megazyme® K-TSHK was used, following the procedure described for samples containing resistant starch, simple sugars, and/or maltodextrins.

Glucose, fructose, and sucrose content were performed with the kit Megazyme® K-SUFRG.

Ascorbic acid content was determined with Megazyme® K-ASCO.

2.2.5. Bioactive Compounds and Antioxidant Activity Analysis

In order to carry out the assays, a methanol extract was first prepared [32]. Briefly, 4 g were weighed in a Falcon tube, and 40 mL of 80% (*v*/*v*) methanol (Sigma-Aldrich, Steinheim, Germany) was added. The mixture was vortexed for approximately 15–20 s and subsequently subjected to sonication in an ultrasonic water bath for 30 min (temperature: 20 ± 2 °C, power: 100%, frequency: 40 kHz). After sonication, the samples were centrifuged at 10,000 rpm (10,733× *g*) for 10 min. The supernatant was filtered with a 0.45 µm cellulose acetate syringe filter. Extracts were stored at −20 °C for a minimum of 2 h before use in the assays. Hereafter, the methanolic extract is referred to as the "sample".

The determination of total polyphenols content (TPC) was performed following the Folin–Ciocalteu method [33]. In 1 cm pathlength cuvettes, 100 µL of sample was added, followed by 1500 µL of 1:10 diluted Folin–Ciocalteu solution (Sigma-Aldrich, Steinheim, Germany) and 1875 µL of 7% (*w*/*v*) Na_2CO_3 solution (ACS reagent, Sigma-Aldrich, Steinheim, Germany). For the blank, the same volume of distilled water was used in place of the sample.

The cuvettes were then shaken and incubated in the dark for 1 h and 30 min. Finally, the absorbance was measured at 765 nm spectrophotometrically using a Cary 60 UV spectrophotometer (Agilent Technologies Inc., Santa Clara, CA, USA). The TPC was expressed as mg of gallic acid equivalents (GAE) per 100 g of dry matter (dm), based on calibration curve in the range 0–2 g/L of gallic acid (>98%, Sigma-Aldrich, Steinheim, Germany).

The determination of the total tannin content (TTC) was carried out according to the Bate–Smith method [34], based on the property of catechins, both monomeric and condensed, to oxidize in an acidic and alcoholic medium at high temperature (100 °C), releasing colored proanthocyanidins (Bate–Smith reaction). A volume of 2 mL of methanolic extract was placed in a test tube (A) with 6 mL of reaction mixture previously prepared as follows: 500 mL of 12 N HCl (ACS reagent, Sigma-Aldrich, Steinheim, Germany) + 500 mL of *n*-butanol (ACS reagent, Sigma-Aldrich, Steinheim, Germany) + 150 mg of $Fe_2(SO_4)_3$ (ACS reagent, Sigma-Aldrich, Steinheim, Germany). The half of the mixture of the tube A was then transferred in glass test tube (B), closed with a screw cap and a Teflon seal, and placed in a hot oil bath (T = 105 °C) for 30 min. Then, the B tube was cooled at room temperature. At high temperatures, the acid hydrolysis of proanthocyanidins takes place.

Finally, the mixture in both A and B tubes was read spectrophotometrically at 550 nm with a 1 cm quartz cuvettes. The difference between the two readings was compared to the calibration curve in the range of 0–1 g/L of catechins (>98%, Sigma-Aldrich, Steinheim, Germany), allowing determination of the TTC value, expressed as mg of catechin equivalents (CE) per 100 g dm.

The antioxidant activity was assessed with three different assays: the ABTS method by reading the absorbance at 734 nm, the DPPH method by reading the absorbance at 515 nm, and the FRAP method by reading the absorbance at 593 nm [35]. The results were expressed as μmol Trolox equivalents (TE) per g dm of sample, based on different standard curves of Trolox (Sigma-Aldrich, Steinheim, Germany): 0–200 μmol L^{-1} for the DPPH assay, 0.2–1.5 mM range for ABTS, and 0–2.0 mM for the FRAP assay.

2.3. Color Determination

Color determination was performed using a tristimulus bench colorimeter (Eoptis® CLM-196; Benchtop, Trento, Italy) with a white reference standard according to the CIE L* a* b* system (applying standard illuminant D65 with a 2° observer angle). The color of the samples was defined based on the three chromatic coordinates: lightness (L*), green-red (a*), and blue–yellow (b*) components.

The cylindric coordinates, namely, chroma (C*) and hue (h*); the white index (WI); and yellow index (YI) were calculated as previously reported [31] using the following equations:

$$C^* = \sqrt{(a^*)^2 + (b^*)^2} \quad (1)$$

$$h^* = \arctan\left(\frac{b^*}{a^*}\right) \quad (2)$$

$$WI = 100 - \sqrt{(100 - L^*)^2 + (a^*)^2 + (b^*)^2} \quad (3)$$

$$YI = 142.86 \times \left(\frac{b^*}{L^*}\right) \quad (4)$$

The color difference between the samples was evaluated with the ΔE^*_{ab} value calculated as previously reported [35] using the following equation:

$$\Delta E^*_{ab} = \sqrt{\Delta(L^*_1 - L^*_2)^2 + \Delta(a^*_1 - a^*_2)^2 + \Delta(b^*_1 - b^*_2)^2} \quad (5)$$

2.4. Volatile Organic Compound (VOC) Analysis

The aromatic profile of flours and fresh chestnuts were determined using the headspace solid-phase microextraction (HS-SPME) method, as reported by Pieracci et al. [36]. All samples were placed in glass vials, sealed with aluminum foil, and left to equilibrate for 30 min. After equilibration, a Supelco SPME fiber (100 μm, PDMS) (Supelco analytical, Bellefonte, PA, USA), conditioned according to the manufacturer's guidelines, was exposed to the headspace for 30 min at room temperature. Then, the fiber was withdrawn into the needle and injected into a GC-MS system. The GC-EIMS (gas chromatography-electron impact mass spectrometry) analyses were conducted with an Agilent 7890B gas chromatograph (Agilent Technologies Inc., Santa Clara, CA, USA) equipped with an Agilent HP-5MS capillary column (30 m × 0.25 mm; coating thickness 0.25 mm) and an Agilent 5977B single quadrupole mass detector. The analytical conditions were set as follows: oven temperature ramp from 60 to 240 °C at 3 °C/min; injector temperature of 250 °C; transfer line temperature of 240 °C; helium used as the carrier gas at a flow rate of 1 mL/min. The analysis of the HS is performed with the splitless method. The acquisition parameters of the single quadrupole mass spectrometer were: full scan; scan range, 30–300 m/z; scan time, 1.0 s.

Peak identification was performed by comparing the retention times with those of authentic standards, evaluating their linear retention indices relative to a series of n-hydrocarbons (C8–C27), and matching the mass spectra against both commercial and laboratory-developed spectral libraries compiled from pure compounds, known essential oil components, and published MS data [37–39].

Odor description of VOCs was obtained from FLAVORNET [40] and The Good Scents Company Information System [41].

2.5. Sensory Evaluation

The flour sensory profile was evaluated using quantitative descriptive analysis (QDA) by a panel of eight trained judges (four females and four males, aged between 23 and 63 years) who are members of the "Experts Panel" of Department of Agriculture, Food and Environment of the University of Pisa. A sub-group of trained panelists participated in a consensus panel specifically designed generated descriptors and their definitions, prior to the tasting sessions. The research was conducted according to the ethical guidelines, and informed consent was obtained from all participants and obtained the approval of the Ethics Committee of the University of Pisa (protocol no. 0088081/2024).

The QDA sensory test was conducted with a final set of 30 descriptive parameters (Figure 1) composed of both quantitative descriptors (sub-divided into visual, olfactory, tactile, and taste perceptions) and hedonic descriptors. In addition, judges were permitted to input free descriptors for each descriptor category.

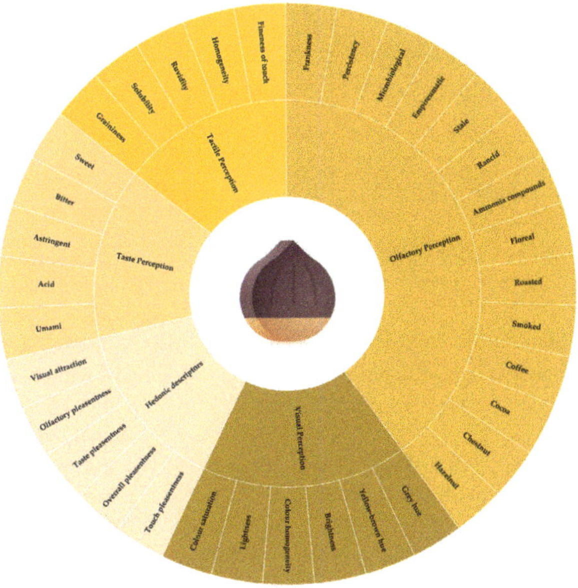

Figure 1. Descriptive parameters (quantitative and hedonic descriptors) used in the QDA test.

Each judge assigned a score from 0 to 9 for each attribute (Figure S3), where 0 represents the absence of perception and 9 the maximum intensity, digitally acquired by the Input Sensory Soft 2.0 (ISS, Centro Studi Assaggiatori, Brescia, Italy). The samples were presented in a different order at each tasting session, and 5-minute intervals between each sample were set. Furthermore, a sample was randomly replicated to verify the performance of the panel at each tasting session

Finally, the overall hedonic index (HI), which represents the overall acceptability of the product, was calculated based on the mean of the hedonic parameters, which was converted to a scale from 0 to 10, as previously reported [42].

2.6. Statistical Analysis

Results were statistically analyzed with one-way analysis of variance (ANOVA) using CoStat version 6.451 software (CoHort Software, Pacific Grove, CA, USA) applying Tukey's Honestly Significant Difference test ($p < 0.05$).

Sensory profile data were processed using Big Sensory Soft version 2.0 software (Centro Studi Assaggiatori, Brescia, Italy), and statistical analyses were performed by two-way interquartile ANOVA, with samples and panelists as main factors [43].

Finally, hierarchical cluster analysis (HCA), applying the Ward method and using two-way clustering, and other figures were performed using JMP Student version 18 software (SAS Institute, Cary, NC, USA).

3. Results and Discussion

3.1. Physico-Chemical Characterization

Table 1 reports the physico-chemical characterization of the analyzed chestnut and chestnut flour samples.

Table 1. Physico-chemical characterization chestnut and chestnut flour samples.

Parameter	Units	p-Value [1]	FL	FM1	FM2	M1t0	M2t0
Dry matter (dm)	%	***	89.10 ± 0.09^{b}	91.72 ± 0.07^{a}	91.40 ± 0.05^{a}	53.56 ± 0.20^{c}	49.56 ± 0.13^{d}
a_w		ns	0.50 ± 0.07	0.41 ± 0.01	0.43 ± 0.02	n.d.	n.d.
TL	g/100 g dm	***	4.14 ± 0.01^{a}	2.13 ± 0.01^{b}	4.50 ± 0.01^{a}	1.30 ± 0.01^{c}	1.58 ± 0.01^{bc}
FFA	g oleic acid/100 g dm	***	0.13 ± 0.01^{a}	0.14 ± 0.01^{a}	0.14 ± 0.01^{a}	0.07 ± 0.01^{b}	0.10 ± 0.01^{b}

The values represent the mean (±SD). Different letters in each row correspond to statistically different values (Turkey's HSD, $p < 0.05$). [1] Significant level: ns = $p \geq 0.05$; *** = $p < 0.001$. n.d. = not detected.

The two flours obtained using the traditional process (FM1 and FM2) showed slightly higher dry matter (dm) and lower water activity (a_w) compared to the flour produced under controlled conditions (FL). Variability was also observed in the total lipid content (TL), with statistically significant difference among FL, FM2, and FM1. This may be attributed to the heterogeneity of the raw material (e.g., different varietal composition and harvest location).

The TL of fresh chestnuts (samples M1t0 and M2t0) may have been affected by the freeze-drying process used as a pre-treatment to stabilize the fresh sample; this process may have altered the structural characteristics, affecting the extraction yield of the lipids.

Lipid composition, along with dm content and a_w, plays a critical role in determining the product's shelf life.

Due to its low a_w and high dm content, chestnut flour is relatively resistant to microbial growth. However, although chestnut flour has a relatively low-fat content, its lipid fraction is characterized by a high percentage of unsaturated fatty acids (UFA), primarily linoleic acid (C18:2 ω6) and oleic acid (C18:1 ω9), as reported in the literature for different chestnut varieties [2,30,44]. UFAs are prone to oxidation, which can compromise the quality of the final product, especially under suboptimal storage conditions such as elevated temperature, light exposure, or inadequate atmospheric composition. The free fatty acid (FFA) value is a measure of lipid degradation. An increase in the value reflects a higher degree of lipid hydrolysis, leading to the availability of molecules more prone to subsequent oxidative reactions, which also impacts sensory quality. In the samples analyzed (Table 1), FFA values were very low. High levels of free fatty acids would negatively impact oxidative stability and, consequently, the product's shelf life, increasing the risk of lipid rancidity with an impact on the sensorial quality. It is evident that elevated concentrations of free fatty acids can result in a decline in oxidative stability, thereby diminishing the product's shelf life and heightening the probability of lipid rancidity. This phenomenon can also exert an influence on the sensorial quality of the product.

The data related to the carbohydrate fraction reveal an unexpected trend, particularly with respect to free sugars. Notably, sucrose (Table 2) exhibited a statistically significant increase after processing it into flour, contrary to expectations.

Table 2. Sucrose, D-glucose, and D-fructose content expressed as g/100 g of dm.

Parameters	Units	*p*-Value [1]	FL	FM1	FM2	M1t0	M2t0
Sucrose	g/100 g dm	***	19.04 ± 2.14^{a}	21.57 ± 0.51^{a}	15.93 ± 0.69^{b}	9.88 ± 1.26^{c}	9.31 ± 0.33^{c}
D-Glucose	g/100 g dm	***	0.70 ± 0.04^{c}	1.70 ± 0.11^{a}	0.61 ± 0.06^{c}	0.91 ± 0.09^{b}	0.75 ± 0.06^{bc}
D-Fructose	g/100 g dm	***	0.58 ± 0.03^{b}	1.44 ± 0.04^{a}	0.48 ± 0.03^{c}	0.44 ± 0.03^{c}	0.40 ± 0.04^{c}

The values represent the mean (±SD). Different letters in each row correspond to statistically different values (Turkey's HSD, $p < 0.05$). [1] Significant level: *** = $p < 0.001$.

This phenomenon has also been observed by other research groups; however, to the best of our knowledge, no explanation can be provided at this time [6,45].

The data also show that during the transformation from fresh chestnuts (M1t0, M2t0) to flour (FL, FM1, FM2), the free sugar content undergoes a statistically significant increase due to the drying processes (Table 2). These processes, through water removal and heating of the raw material, cause the breakdown of part of the starch into simpler sugars.

The different sugar compositions could be a result of the varying compositions of the raw chestnut used, which may belong to different varieties [46]. As previously reported by Piccolo et al. [6], in a study concerning the varieties from the same geographical area, the different composition could impact the response to heat treatment.

Sugar transformation is affected by numerous variables, with heat treatment being the main factor [45,47], but the phytosanitary condition of chestnuts also plays a role.

The time and conditions of storage prior to heat treatment, mainly temperature and humidity, are also critical factors [48,49]. If these conditions are inadequate, they may induce physicochemical changes in the fruit, including the degradation of carbohydrate reserves necessary for cellular respiration, a process that begins when the fruit naturally falls and continues until enzymatic activity and water activity allow germination to occur [3].

Figure 2 shows the total starch content of fresh chestnuts and their derived flours. Although significant decreases in starch content are observed when comparing M1t0 and M2t0 with their corresponding flours (M1t0: FM1; M2t0: FM2, FL), these differences are not substantial enough to account for the increase in free sugar content reported in the flours (Table 1).

Correia et al. [50] report that the starch granules of *C. sativa* are composed of type C crystalline units, a mixture of type A and type B structures. Correlating with the findings of Wang et al. [51], who reports that during freeze-drying there is a modification of the starch component of *C. mollissima*, it can be hypothesized that the freeze-drying treatment applied to the chestnuts for analysis may have altered their components. Freeze-drying may influence the type B crystals, which are destroyed due to the removal of the bound water located between the helices.

Buléon et al. [52] reported that the B-type structure is composed of repeated maltoside units; consequently, breaking the bonds between the helical chains leads to the release of maltodextrins (linear chains of D-glucose units ranging from 2 to 20). Following the removal of internal bound water, if the length of the resulting chains falls within the range of maltodextrins, these could have been eliminated during the starch extraction process used for the enzymatic kit, which includes pre-washing steps aimed at removing free sugars and maltodextrins that would interfere with the quantification of total starch.

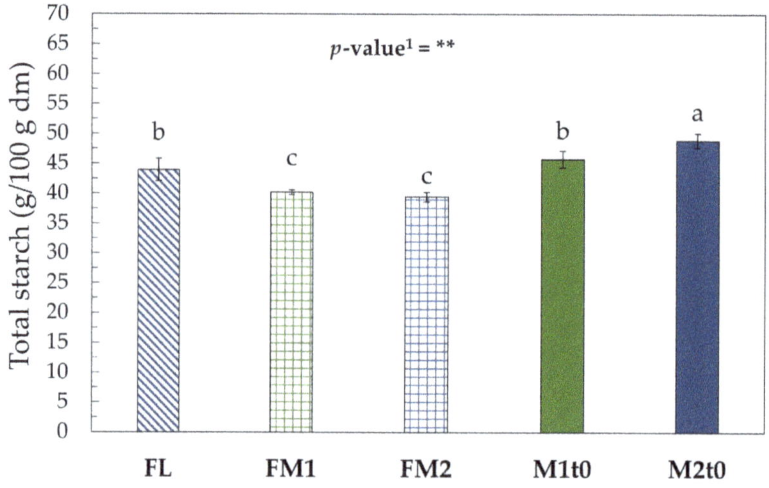

Figure 2. Total starch (g/100 g dm). Values represent the mean ± standard deviation (SD) of three samples. Different letters indicate significant difference among values (Turkey's HSD, $p < 0.05$). [1] Significance level: ** = $p < 0.01$.

3.2. Bioactive Compounds and Antioxidant Activity

The trend of TPC and TTC in the flours samples is shown in Figure 3a,b, which highlights the presence of significant differences.

The differences observed between the samples of chestnut flour can be attributed to several factors. The FL sample underwent a shorter heat treatment at controlled temperatures, which favored rapid dehydration of the product (Figure S2). This may help to preserve the phenolic compounds present in the raw chestnuts, which are more likely to degrade in traditionally produced flours that were exposed to heat for longer periods under less controlled conditions (Figure S1a,b). These observations are consistent with findings reported by Conti et al. [22].

The assays performed to evaluate the antioxidant activity of the flour samples (Figure 4a–c) produced results consistent with the trends observed in the TPC and TTC assays. These findings confirm that the processing conditions adopted for the FL sample led to reduced degradation of bioactive compounds, further supporting the hypothesis that shorter and more controlled heat treatments help preserve the nutritional and functional properties of chestnut flour [15,53].

This trend could also be correlated with the behavior of ascorbic acid as reported in Figure 5.

The behavior of chestnut bioactive compounds should also be considered in relation to the different chestnut varieties, which respond differently [6].

In this case (FM1 and FM2 comparison), the composition of the batch used for flour production was not defined in terms of the proportion of the different varieties, which could have influenced the results.

Moreover, the reduced processing time combined with constant temperature conditions likely contributed to limiting the degradation of thermolabile compounds.

As reported in the literature, ascorbic acid is highly sensitive to thermal treatments. According to Nguyen [54], lower drying temperatures and shorter drying times favor better retention of vitamin C in dried chestnuts. The vitamin C content remained comparable in FL and in FM2, both derived from the same batch of chestnuts.

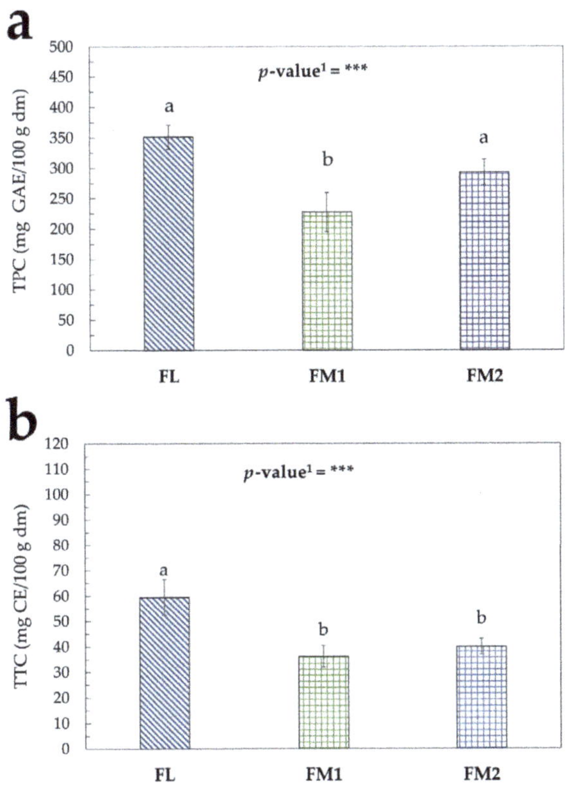

Figure 3. Bioactive compounds in flour samples: (**a**) TPC (mg GAE/100 g dm); (**b**) TTC (mg CE/100 g dm). Values represent the mean ± standard deviation (SD) of three samples. Different letters indicate significant difference among values (Turkey's HSD, $p < 0.05$). [1] Significance level: *** = $p < 0.001$.

The FM1 sample, despite undergoing a shorter drying process than FM2, showed a significantly lower ascorbic acid concentration, likely due to compositional differences in the raw material. Barros et al. [55] reported that the different behavior of ascorbic acid during processing could also be influenced by variation in shell and peel properties, which can alter thermal transfer and the diffusivity coefficient of the chestnut.

It can therefore be concluded that the drying process conducted under controlled laboratory conditions and stable temperatures—ensuring shorter exposure times to heat and minimizing thermal fluctuations—resulted in better preservation of the bioactive compounds (Figures S1a,b and S2).

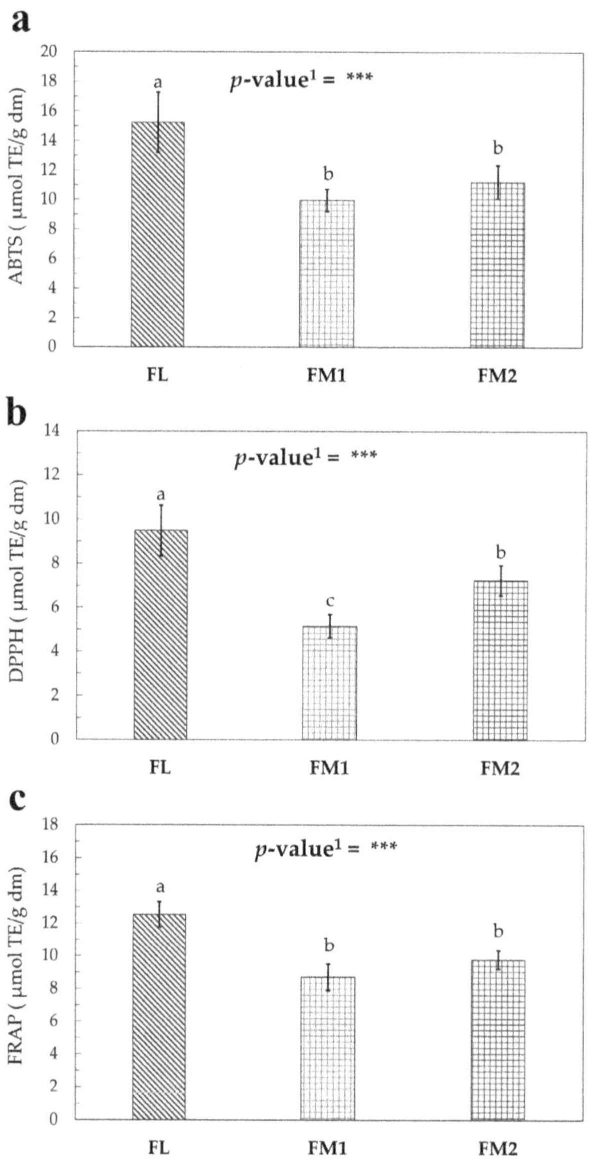

Figure 4. Antioxidant activity assay of flour samples expressed as µmol TE/g dm: (**a**) ABTS; (**b**) DPPH; (**c**) FRAP. Values represent the mean ± standard deviation (SD) of three samples. Different letters indicate significant difference among values (Turkey's HSD, $p < 0.05$). [1] Significance level: *** = $p < 0.001$.

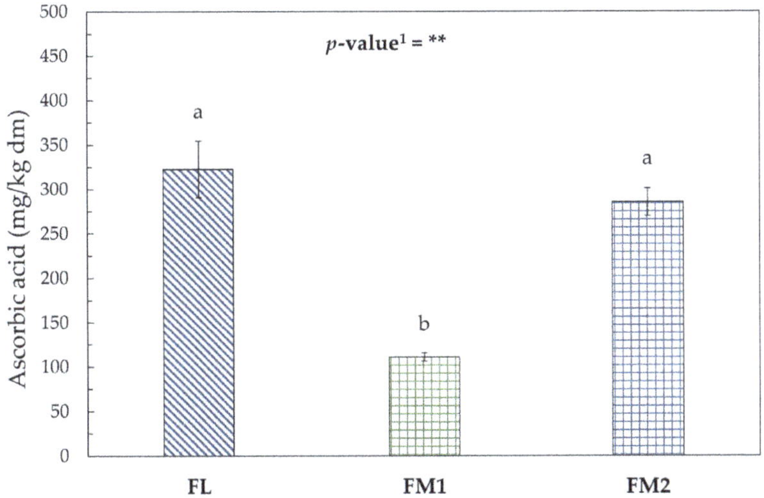

Figure 5. Ascorbic acid (mg/kg dm) in flour samples. Values represent the mean ± standard deviation (SD) of three samples. Different letters indicate significant difference among values (Turkey's HSD, $p < 0.05$). [1] Significance level: ** = $p < 0.01$.

The temperature trend inside the *metati*, as illustrated in Figure S1a,b, fluctuated between approximately 5 °C and 35 °C, causing an instability during the drying process. Instead, relative humidity (%) varied between approximately 30% and 90%. For the M1 and M2 *metati*, the weighted averages of temperature (°C) and relative humidity (%) were 19.0 °C and 81.5% as well as 17.4 °C and 71.2%, respectively. These factors were probably affected by external environmental conditions, unlike the laboratory dryer where these variables are kept constant over time allowing for more constant dehydration (Figure S2).

3.3. Color Evaluation

The color of chestnuts changes during the processing into flour. Undoubtedly, the compounds present in the raw fruit undergo modifications due to spontaneous reactions.

The results of the colorimetric evaluation are reported in Table 3. The results show an increase in lightness (L*) from raw chestnuts (M1t0, M2t0) to flour samples (FL, FM1, FM2), accompanied by an increase in the white index (WI) and a decrease in the yellow index (YI), following the thermal degradation and the oxidative phenomena that affecting the carotenoids [56,57].

Table 3. Color parameters of different samples.

Parameter	*p*-Value [1]	FL	FM1	FM2	M1t0	M2t0
L*	***	88.43 ± 0.84 [a]	84.14 ± 1.23 [b]	87.14 ± 0.97 [a]	68.66 ± 0.81 [c]	71.19 ± 0.84 [c]
a*	***	0.25 ± 0.03 [d]	1.26 ± 0.05 [a]	0.67 ± 0.03 [c]	0.98 ± 0.06 [b]	0.98 ± 0.06 [b]
b*	***	11.86 ± 0.72 [c]	14.13 ± 0.52 [b]	12.80 ± 0.11 [bc]	22.32 ± 0.79 [a]	22.32 ± 0.79 [a]
C*	***	11.86 ± 0.73 [c]	12.56 ± 1.91 [b]	13.41 ± 1.52 [bc]	22.34 ± 0.79 [a]	22.34 ± 0.79 [a]
h*	***	1.55 ± 0.01 [a]	1.48 ± 0.01 [c]	1.51 ± 0.04 [b]	1.53 ± 0.01 [b]	1.53 ± 0.01 [b]
WI	***	83.41 ± 0.40 [a]	78.72 ± 1.20 [b]	79.86 ± 2.87 [a]	61.51 ± 0.56 [d]	63.54 ± 0.58 [c]
YI	***	19.16 ± 1.03 [c]	24.01 ± 1.18 [b]	22.54 ± 3.14 [bc]	46.44 ± 1.67 [a]	44.79 ± 1.61 [a]

Values are presented as the mean ± standard deviation (SD) of three samples. In the same row, different letters indicate significant difference among values (Turkey's HSD, $p < 0.05$). [1] Significance level: *** = $p < 0.001$.

The calculation of ΔE^*_{ab} (Table 4) provides values representing the color differences between the flour samples.

Table 4. Color differences (ΔE^*_{ab}) among the different flour samples.

ΔE^*_{ab}	FL	FM1	FM2
FL		2.32	3.83
FM1			1.55
FM2			

In this case, the analyzed samples showed only slight differences, but these fell into two distinct ranges. Specifically, the difference between the FL sample and the FM1 sample falls within the ΔE^*_{ab} below 2.5, indicating a slight but perceptible color difference [58].

On the other hand, the ΔE^*_{ab} between the FL and FL2 samples, obtained from the same batch of chestnuts, shows a value of 3.83, indicating a noticeable but distinguishable color variation [31].

Since both flours were produced from the same batch of chestnuts, the differences can mainly be attributed to the different processing methods applied.

3.4. *Volatile Organic Compound (VOC) Profile*

The complete chemical composition of the volatile emissions of the samples analyzed is reported in Table 5.

Table 5. Composition of the VOC emissions (relative abundance (%)) of the fresh chestnuts and derived flours.

Compound	l.r.i [2]	Odor	*p*-Value [1]	Relative Abundance (%)				
				FL	FM1	FM2	M1t0	M2t0
				Monoterpene hydrocarbons				
α-Pinene	941	Balsamic	***	0.8 ± 0.02^{a}	$-^{3,b}$	$-^{b}$	$-^{b}$	$-^{b}$
Sabinene	977	Woody	***	1.6 ± 0.03^{a}	$-^{b}$	$-^{b}$	$-^{b}$	$-^{b}$
β-Pinene	982	Herbal	***	3.2 ± 0.06^{a}	$-^{b}$	$-^{b}$	$-^{b}$	$-^{b}$
Myrcene	993	Spicy	***	13.6 ± 0.53^{a}	$-^{b}$	$-^{b}$	$-^{b}$	$-^{b}$
δ-3-Carene	1012	Citrus	***	2.8 ± 0.34^{a}	$-^{b}$	$-^{b}$	$-^{b}$	$-^{b}$
p-Cymene	1028	Citrus	***	3.2 ± 0.06^{a}	$-^{b}$	$-^{b}$	$-^{b}$	$-^{b}$
Limonene	1032	Citrus	***	$-^{c}$	2.0 ± 0.06^{a}	$-^{c}$	$-^{c}$	0.1 ± 0^{b}
γ-Terpinene	1062	Woody	***	20.4 ± 0.81^{a}	$-^{b}$	$-^{b}$	$-^{b}$	$-^{b}$
Terpinolene	1090	Herbal	***	1.6 ± 0.03^{a}	$-^{b}$	$-^{b}$	$-^{b}$	$-^{b}$
				Alcohols				
Isobutyl alcohol	627	Musty	*	$-^{b}$	$-^{b}$	$-^{b}$	1.2 ± 0.05^{ab}	2.3 ± 1.92^{a}
1-Butanol	657	Fusel	***	$-^{b}$	$-^{b}$	$-^{b}$	0.2 ± 0.01^{a}	$-^{b}$
Isopentyl alcohol	736	Musty	***	$-^{c}$	$-^{c}$	$-^{c}$	13 ± 0.45^{a}	6.5 ± 2.47^{b}
2-Methylbutanol	737	Fusel	***	$-^{b}$	$-^{b}$	$-^{b}$	6.3 ± 0.39^{a}	5.3 ± 2.32^{a}
1-Pentanol	766	Balsamic	***	$-^{c}$	0.9 ± 0.01^{a}	0.4 ± 0.2^{b}	$-^{c}$	$-^{c}$
1,3-Butanediol	788		***	5.6 ± 0.11^{bc}	7.7 ± 0.1^{b}	2.7 ± 0.06^{c}	5.6 ± 2.74^{bc}	22.9 ± 1.1^{a}
2,3-Butanediol	789		***	4.8 ± 0.1^{bc}	9.9 ± 0.27^{b}	4 ± 0.01^{c}	8.9 ± 3.64^{bc}	33.8 ± 2.38^{a}
Furfuryl alcohol	858	Burning	***	$-^{b}$	$-^{b}$	1 ± 0.02^{a}	$-^{b}$	$-^{b}$
1-Hexanol	871	Herbal	***	2.8 ± 0.34^{b}	2.2 ± 0.19^{b}	1.1 ± 0.08^{cd}	5.6 ± 0.93^{a}	0.2 ± 0^{d}
1-Heptanol	970	Herbal	***	$-^{b}$	$-^{b}$	$-^{b}$	0.4 ± 0.13^{a}	0.1 ± 0^{b}
1-Octen-3-ol	982	Musty	***	$-^{b}$	1 ± 0.11^{a}	1.1 ± 0.08^{a}	$-^{b}$	$-^{b}$
3-Ethyl-1-hexanol	1031	Floral	***	$-^{b}$	$-^{b}$	8.8 ± 0.41^{a}	$-^{b}$	$-^{b}$
1-Octanol	1071	Herbal	***	$-^{c}$	0.7 ± 0.06^{a}	$-^{c}$	0.2 ± 0.01^{b}	$-^{c}$
Phenylethyl alcohol	1111	Floral	***	2 ± 0.36^{b}	2.1 ± 0.09^{b}	0.7 ± 0.09^{c}	3.7 ± 0.39^{a}	2.7 ± 0.3^{b}

Table 5. *Cont.*

Compound	l.r.i [2]	Odor	p-Value [1]	Relative Abundance (%)				
				FL	FM1	FM2	M1t0	M2t0
Ethers								
2-Acetylfuran	913	Fruity	***	-^{b}	-^{b}	0.6 ± 0.01^{a}	-^{b}	-^{b}
2-Pentyl furan	992	Fruity	***	-^{b}	3.5 ± 0.69^{a}	-^{b}	-^{b}	-^{b}
γ-Caprolactone	1056	Herbal	***	-^{b}	0.8 ± 0.01^{a}	-^{b}	-^{b}	-^{b}
2,3-Dihydrobenzofuran	1221	Harsh	***	-^{b}	-^{b}	-^{b}	-^{b}	0.3 ± 0.05^{a}
Phenols								
Phenol	983		***	-^{d}	1.6 ± 0.01^{b}	4.5 ± 0.1^{a}	0.5 ± 0.25^{c}	0.1 ± 0^{d}
o-Cresol	1057	Musty	***	-^{c}	-^{c}	3.9 ± 0.3^{a}	0.9 ± 0.27^{b}	-^{c}
p-Cresol	1078	Floral	***	-^{c}	-^{c}	1.3 ± 0.13^{a}	0.4 ± 0.13^{b}	-^{c}
o-Guaiacol	1091	Smoky	***	-^{d}	2.6 ± 0.07^{b}	14.0 ± 0.02^{a}	1.6 ± 0.76^{c}	0.1 ± 0^{d}
Veratrole	1149	Musty	***	-^{b}	0.7 ± 0.01^{a}	-^{b}	-^{b}	-^{b}
Creosol	1193	Smoky	***	-^{c}	1.2 ± 0.15^{b}	4.9 ± 0.09^{a}	-^{c}	-^{c}
3-Phenylpropanol	1232	Spicy, floral	***	-^{b}	-^{b}	-^{b}	20.2 ± 3.41^{a}	-^{b}
p-Ethylguaiacol	1280	Smoky	**	-^{b}	-^{b}	-^{b}	4.6 ± 1.92^{a}	-^{b}
p-Vinylguaiacol	1314	Spicy		-^{b}	-^{b}	-^{b}	-^{b}	1.1 ± 0.1^{a}
Esters								
Ethyl acetate	611	Fruity	**	-^{b}	1 ± 0.06^{b}	0.7 ± 0.12^{b}	0.6 ± 0.09^{b}	4.5 ± 1.98^{a}
Isopentyl acetate	876	Fruity	***	-^{b}	-^{b}	-^{b}	-^{b}	0.7 ± 0.05^{a}
Butyrolactone	914		***	4.8 ± 0.1^{a}	2.8 ± 0.12^{b}	1.4 ± 0.03^{c}	-^{d}	-^{d}
Ethyl hexanoate	998	Fruity	***	-^{b}	-^{b}	-^{b}	-^{b}	0.1 ± 0.00^{a}
Ethyl 2-phenylacetate	1246	Fruity, honey	***	-^{b}	-^{b}	-^{b}	-^{b}	0.1 ± 0.00^{a}
2-Phenylethyl acetate	1259	Fruity, floral	***	-^{b}	-^{b}	-^{b}	-^{b}	0.1 ± 0.00^{a}
Aldehydes								
Hexanal	802	Herbal	***	3.6 ± 0.33^{c}	10.6 ± 0.09^{a}	8.8 ± 0.61^{b}	-^{d}	-^{d}
Furfural	839	Baked	***	-^{c}	2.6 ± 0.04^{a}	1.8 ± 0.16^{b}	-^{c}	-^{c}
Heptanal	901	Herbal	***	-^{c}	0.8 ± 0.01^{a}	0.5 ± 0.09^{b}	-^{c}	-^{c}
Octanal	1001		***	2.4 ± 0.05^{b}	3 ± 0.19^{a}	1.4 ± 0.03^{c}	-^{d}	-^{d}
5-Ethylcyclopent-1-enecarboxaldehyde	1035		***	-^{b}	0.4 ± 0^{a}	-^{b}	-^{b}	-^{b}
(E)-2-Octenal	1063	Fruity	***	-^{b}	0.6 ± 0^{a}	0.7 ± 0.12^{a}	-^{b}	-^{b}
Nonanal	1102	Fruity	***	4.8 ± 0.1^{a}	4.9 ± 0.29^{a}	2.9 ± 0.34^{b}	-^{c}	-^{c}
Decanal	1204	Fruity	***	1.2 ± 0.38^{a}	-^{b}	-^{b}	-^{b}	-^{b}
(E,E)-2,4-Nonadienal	1215	Herbal	***	0.8 ± 0.02^{a}	-^{b}	-^{b}	-^{b}	-^{b}
Ketones								
2-Pentanone	696	Fruity	***	-^{b}	-^{b}	-^{b}	2.8 ± 0.12^{a}	-^{b}
Acetoin	709	Buttery	***	-^{c}	-^{c}	-^{c}	17.8 ± 3.36^{a}	8.1 ± 3.17^{b}
2-Heptanone	894	Fruity/herbal	***	-^{c}	0.6 ± 0^{b}	0.5 ± 0.09^{b}	2.7 ± 0.23^{b}	-^{c}
6-Methyl-5-hepten-2-one	987	Herbal	***	-^{b}	-^{b}	-^{b}	0.5 ± 0.02^{a}	-^{b}
3-Octen-2-one [4]	1042	Spicy, herbal	***	-^{c}	0.9 ± 0.05^{a}	0.6 ± 0.01^{b}	-^{c}	-^{c}
2-Nonanone	1093	Herbal	***	-^{b}	-^{b}	-^{b}	1.5 ± 0.18^{a}	-^{b}
Acids								
Acetic acid	599	Acidic	***	18.8 ± 0.78^{b}	32.0 ± 0.99^{a}	21.2 ± 1.35^{b}	-^{d}	9.3 ± 3.96^{c}
Other non-terpene derivatives								
Styrene	896	Balsamic/plastic	*	-^{b}	-^{b}	-^{b}	-^{b}	1.2 ± 0.91^{a}
n-Undecane	1100		***	-^{b}	-^{b}	0.4 ± 0.01^{a}	-^{b}	-^{b}
Naphthalene	1181	Pungent	***	-^{c}	2.5 ± 0.02^{b}	4 ± 0.01^{a}	-^{c}	-^{c}
n-Dodecane	1200		***	0.8 ± 0.02^{a}	-^{b}	0.8 ± 0.02^{a}	-^{b}	-^{b}
Chemical classes								
Terpenes								
Monoterpene hydrocarbons			***	47.2 ± 0.15^{a}	2.00 ± 0.06^{b}	-^{c}	-^{c}	0.1 ± 0^{c}

Table 5. *Cont.*

Compound	l.r.i [2]	Odor	*p*-Value [1]	Relative Abundance (%)				
				FL	FM1	FM2	M1t0	M2t0
			Non-terpene derivatives					
Alcohols/ethers/phenols			***	15.2 ± 0.49 d	34.9 ± 0.89 c	49.1 ± 0.53 b	73.3 ± 2.74 a	75.3 ± 3.68 a
Esters			**	4.8 ± 0.1 a	3.8 ± 0.18 ab	2.2 ± 0.15 bc	0.6 ± 0.09 c	5.5 ± 2.03 a
Aldehydes/ketones			***	12.8 ± 0.54 bc	24.5 ± 0.06 a	17.3 ± 0.10 b	25.2 ± 2.81 a	8.1 ± 3.17 c
Acids			***	18.8 ± 0.78 b	32.0 ± 0.99 a	21.2 ± 1.35 b	- d	9.3 ± 3.96 c
Others			**	0.8 ± 0.02 c	2.5 ± 0.02 b	5.2 ± 0.02 a	- d	1.2 ± 0.91 c
Total identified				**99.6 ± 0.01**	**99.7 ± 0.1**	**95 ± 0.55**	**99 ± 0.16**	**99.6 ± 0.05**

Values are presented as the mean ± standard deviation (SD) of three samples. The superscript lowercase letters in the row indicate statistical differences among the samples (Turkey's HSD, $p < 0.05$). [1] Significance level: *** $p < 0.001$, ** $p < 0.01$, * $p < 0.05$; [2] Linear retention indices on a HP5-MS capillary column; [3] Not detected; [4] Correct isomer not determined.

The GC-MS analysis permitted the identification of a total of 62 compounds, accounting for 95.0–99.7% of the whole volatile profiles. The identified chemical constituents mainly belonged to the chemical classes of monoterpene hydrocarbons and non-terpenes derivatives. In particular, monoterpene hydrocarbons were detected in high percentages (47.2%) in the laboratory-processed flour (FL), mainly characterized by γ-terpinene (20.4%) and myrcene (13.6%), responsible for woody and spicy notes, respectively. Nevertheless, the volatile profile of FL flour also showed noteworthy relative amounts of the non-terpenic acids (18.8%), alcohols (15.2%), and aldehydes (12.8%), with acetic acid, 1,3- and 2,3-butanediol, and nonanal being the chief compounds, respectively.

Acids, alcohols, and aldehydes also represented the main chemical classes of the volatile emission of the samples FM1 and FM2. Acetic acid was the only compound belonging to the class of non-terpenic acids and was detected at the highest relative abundance in the FM1 flour (32.0%), although FM2 (21.2%) and FL (18.8%) also showed high percentages. The presence of acetic acid could be due to bacteria and yeasts typically present on fresh raw materials, as demonstrated by studies conducted on other products [59]. The higher presence of this molecule in the FM1 sample may be related to the post-harvest conditions of the fresh chestnuts before processing. As these conditions are not standardized, significant differences among samples may occur.

Among non-terpenic alcohols, the 1,3- and 2,3-butanediol isomers represented important components of the volatile emission of FM1 and FM2 as well, and their presence could be also related to microbial metabolism, particularly to yeast activity [60,61]. Nevertheless, the traditional flour *metato* 2 exhibited remarkable relative amounts of 3-ethyl-1-hexanol, associated with floral scent notes, which were not detected in any of the other samples. Finally, regarding aldehydes, although considerable percentages of nonanal were also found in FM1 and FM2, these samples were particularly rich in hexanal (10.6 and 8.8%, respectively) responsible for green aroma notes.

Analyzing the data, it was observed that the aromatic complexity of the flours (FL, FM1, and FM2) particularly differ from that of the fresh chestnuts (M1t0 and M2t0), probably as the result of the thermal treatment and the traditional smoking process applied to traditional flours. In detail, both fresh chestnut samples were characterized by alcohols as the most abundant chemical class. However, while M1t0 showed isopentyl alcohol as the main compound of this class, accounting for 13.0%, M2t0 exhibited 2,3-butanediol (33.8%) followed by 1,3-butanediol (22.9%) as major constituents.

Interestingly, the volatile emission of fresh chestnuts did not show the presence of aldehydes, which instead have been characterized in significant percentages, were greatly found in the flours, as previously reported. Indeed, aldehyde compounds, as evidenced in the literature by various research groups [15,20], can derive from the drying process, as these compounds were not found in the fresh chestnut samples (M1t0 and M2t0). The temperatures used both in the traditional drying process and the laboratory processes, although not high, could be sufficient to start lipid peroxidation, leading to the formation of aldehydes and ketones. In our case, compounds like hexanal, heptanal, octanal, (*E*)-2-octenal, nonanal, and decanal were detected, which might originate from the degradation of unsaturated lipids present in the chestnuts [3,15,20,21]. Nevertheless, fresh chestnuts spontaneously emitted remarkable percentages of ketones, with great differences between the samples analyzed. Indeed, in addition to showing higher relative amounts of the buttery scented acetoin, which in turn was the only ketone detected in M2t0, M1t0 also featured not negligible percentages of 2-pentanone and 2-heptanone. Fresh chestnuts *metato* 1 (M1t0) headspace, different from that of *metato* 2 (M2t0), was also characterized by relevant amounts of phenolic compounds, mainly represented by 3-phenylpropanol (20.2%), responsible for spicy and floral notes, and p-ethylguaiacol (4.6%), associated with smoky flavors [20,21]. Considerable percentages of phenols were detected in the volatile emission of the flours, even though, in this case, they were mainly represented by o-guaiacol, creosol, and phenol, found in higher relative quantities in FM2 than in FM1. The presence of these compounds could be related to the degradation of lignin into simpler phenolic compounds during the flour thermal treatment and smoking process of the traditional method, in agreement with Cantini et al. [19]. Thermal treatment is also related to the presence of furanic compounds that can originate from the degradation and the rearrangement of carbohydrates via the Maillard reaction [62,63]. Among these compounds, furfural, furfuryl alcohol, 2-acetylfuran, and 2-pentylfuran were detected exclusively in the flours. Conversely, 2,3-dihydrobenzofuran was found in low percentages in M0t2 [20,22].

Hierarchical cluster analysis (HCA) graphically summarized the above discussed differences in the volatile emissions of the analyzed samples (Figure S4). Indeed, the obtained dendrogram was constituted by two first-level clusters, dividing fresh chestnuts from the flours. In turn, the traditional flours FM1 and FM2 constituted a sub-cluster by themselves, within the upper cluster also comprising the sample FL, highlighting greatest similarities in their volatile emissions.

3.5. Sensory Profile

The sensory analysis conducted on the three chestnut flour samples permitted the characterization of their respective sensory profiles. Figure 6a,b compares the profiles obtained using descriptive parameters from the sensory evaluation. The descriptors for which statistically significant differences were observed among the samples are marked with an asterisk.

The main differences were observed between sample FL and samples FM1 and FM2. As expected, based on colorimetric data, the samples exhibited significant visual differences in terms of color saturation and yellow–brown hues. From an olfactory perspective, sample FL, which was not subjected to smoking, showed a value of zero for the empyreumatic descriptor, unlike the other two samples.

Furthermore, sample FL displayed a higher score for the chestnut descriptor, whereas FM1 and FM2 were more strongly characterized by smoky, coffee, and cocoa notes (Figure 6a), typically associated with traditional smoking processes [20]. These results are also consistent with the values of VOCs such as o-guaiacol, furfural, and creosol, which are associated with odors like smoky and baked, as reported in Table 5 and present in samples FM1 and FM2.

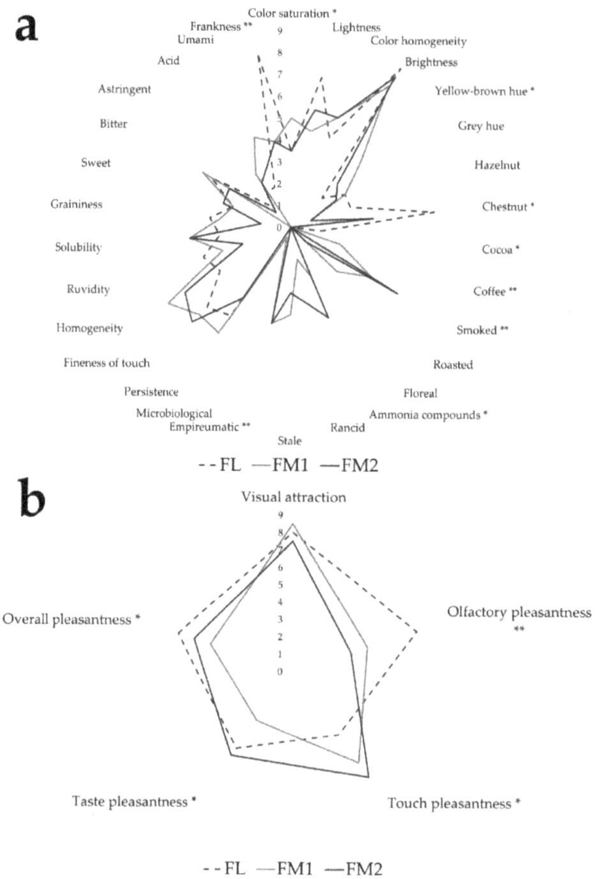

Figure 6. Median of the descriptor of the sensory profile of chestnut flour: (**a**) quantitative; (**b**) hedonic significance level: * = $p < 0.05$; ** = $p < 0.01$; without asterisk = $p \geq 0.05$.

Regarding frankness (absence of defects), sample FL also reported higher positive values than the other samples.

Sample FM2 exhibited higher intensity in the smoky and bitter descriptors (Figure 6a), likely attributable to the extended duration of the traditional drying process and prolonged exposure to thermal degradation phenomena. In contrast, sample FL, processed under controlled conditions, displayed a more delicate profile characterized by sweet and fruity notes, suggesting better preservation of the original aromatic components.

In addition, the panel evaluated the free descriptors associated with the olfactory profile of the samples, as illustrated in Figure 7a–c. The graphs report the average values of the free attribute and the frequency of their use.

In sample FL (Figure 7a), descriptors related to fruity and vegetal families were identified, with the predominant descriptor being raw chestnut. In contrast, samples FM1 (Figure 7b) and FM2 (Figure 7c) were predominantly characterized by descriptors associated with the empyreumatic category (e.g., pungent, metallic) most likely as a result of the traditional smoking process, which was not applied to sample FL. Sample FM2 also exhibited greater aromatic complexity compared to the other samples, a finding that is consistent with the volatile compound (VOC) analysis previously discussed. Both

FM1 and FM2 flours were produced using traditional smoking techniques for 43 and 66 days, respectively; however, sample FM2 underwent a markedly prolonged smoking treatment, which may have led to excessive exposure to smoke and contributed to its distinct aromatic profile.

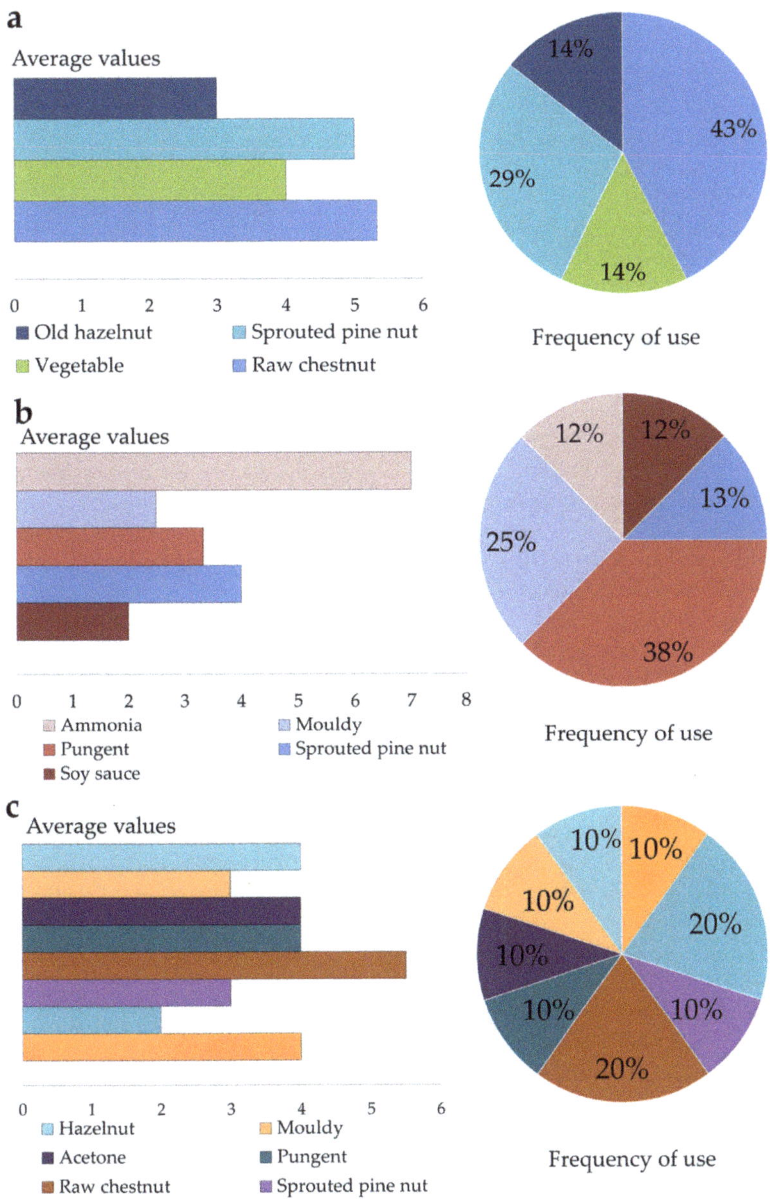

Figure 7. Average of free descriptors and their frequency of use for each flour sample: (**a**) FL; (**b**) FM1; (**c**) FM2.

Concerning the hedonic descriptors, FM1 and FM2 received the lowest scores for olfactory pleasantness, which negatively affected their overall evaluation. In contrast, FL flour obtained the highest ratings for both olfactory and overall pleasantness, although it scored lowest for tactile pleasantness (Figure 6b).

To summarize, the highest hedonic index (HI) was achieved by the sample FL (7.2), followed by FM2 (6.7), while FM1 recorded the lowest value (6.2) (Figure S5).

These results are consistent with the instrumental analysis of volatile compounds and confirm that different processing methods significantly affect the sensory perception of the final product. Sensory analysis thus proved to be an essential tool for integrating and completing the qualitative evaluation of the flours under investigation.

4. Conclusions

Traditional chestnut flour is a product obtained from the drying and milling of fruits from various cultivars of *Castanea sativa*. Historically, this foodstuff has been of fundamental importance as a staple food.

Today, it is considered a niche product, increasingly valued for its distinctive sensory properties and its notable nutritional qualities, being rich in carbohydrates, minerals, and vitamins, and naturally gluten free.

The present study aimed to evaluate the impact of traditional and controlled processing methods on the quality of chestnut flour, representing a preliminary investigation into the main critical aspects of chestnut flour production. To this end, a chemical-physical and volatile and sensory characterization was performed on both fresh chestnuts and the flours obtained from them.

The laboratory-scale drying process, conducted at controlled temperatures over shorter times, allowed better preservation of bioactive compounds, as noted for the total phenolic content and antioxidant activity, and overall compositional integrity. In addition, the primary smell descriptors associated with the fresh raw material were also retained. Traditional processing, while imparting a distinctive aromatic profile due to smoke exposure during drying, proved to be less controllable compared to laboratory-scale drying using a ventilated system (stable at 40 °C), and the process is highly affected by weather conditions, as evident from the observed temperature range, which spans from 5 °C to 35 °C. Although this traditional method has the potential to enrich the flour with complex and unique aromatic notes, inadequate control of its duration or intensity may lead to an excessively marked empyreumatic profile, due to the presence of some VOCs, like as o-guaiacol, furfural, and creosol, which, if present in excessively high concentrations, could result in a less pleasant or unbalanced sensory perception.

To enhance the quality of traditionally produced flour, it is crucial to gain a deeper understanding of the drying phase. The adoption of technologies capable of monitoring the key parameters (e.g., temperature, humidity, and drying time) could minimize the formation of undesirable compounds while safeguarding the product's traditional sensory identity, helping manufacturers to understand when the drying process needs to be remodeled and when the optimum level of drying has been reached. Such technological advancements would not only optimize the nutritional and sensory quality of the flour but would also contribute to more efficient and sustainable production practices, particularly for producers aiming to scale up without compromising product excellence.

As this is a preliminary study, future research will focus on a broader nutritional characterization of chestnut flour, including protein and mineral content, to deepen the understanding of how different drying methods influence its quality. These insights will support the development of optimized drying protocols that preserve the nutritional and sensory attributes of the final product while maintaining the identity and authenticity of traditional processing.

Supplementary Materials: The following supporting information can be downloaded at: https://www.mdpi.com/article/10.3390/foods14111931/s1. Figure S1: Trend of the temperature (°C) and relative humidity (%): (a) inside the *metato* 1; (b) inside the *metato* 2. Figure S2: Weight loss in the drying of the FL sample. Figure S3: Sensory sheet used by the judges on the ISS portal. Q = quantitative descriptors; H = Hedonic descriptor. Figure S4: HCA of VOCs. Figure S5: Hedonic index (HI) of the chestnut flour samples. Different letters indicate significant differences among values (Turkey's HSD, $p < 0.05$).

Author Contributions: Conceptualization, S.P., F.V. and C.S.; methodology, A.B., G.F. and C.S.; software, A.B.; validation, A.B., F.V. and C.S.; formal analysis, S.P., A.B., P.D.-G., Y.P. and I.T.; investigation, S.P., A.B., P.D.-G., Y.P., G.F., F.V. and C.S.; resources, F.V., G.F. and C.S.; data curation, S.P., A.B., P.D.-G., Y.P. and I.T.; writing—original draft preparation, S.P., A.B., P.D.-G., Y.P., I.T., G.F., F.V. and C.S.; writing—review and editing, S.P., A.B., P.D.-G., Y.P., I.T., G.F., F.V. and C.S.; visualization, A.B., F.V. and C.S.; supervision, F.V. and C.S. All authors have read and agreed to the published version of the manuscript.

Funding: This research received no external funding.

Institutional Review Board Statement: The study was conducted in accordance with the Declaration of Helsinki, and approved by the Ethics Committee of the University of Pisa (protocol no. 0088081/2024) on 3 March 2024.

Informed Consent Statement: Informed consent was obtained from all subjects involved in the study.

Data Availability Statement: The original contributions presented in the study are included in the article/Supplementary Materials; further inquiries can be directed to the corresponding author.

Conflicts of Interest: The authors declare no conflicts of interest.

References

1. Aglietti, C.; Cappelli, A.; Andreani, A. From Chestnut Tree (*Castanea sativa*) to Flour and Foods: A Systematic Review of the Main Criticalities and Control Strategies towards the Relaunch of Chestnut Production Chain. *Sustainability* **2022**, *14*, 12181. [CrossRef]
2. De Vasconcelos, M.C.B.M.; Bennett, R.N.; Rosa, E.A.S.; Ferreira-Cardoso, J. V Composition of European chestnut (*Castanea sativa* Mill.) and association with health effects: Fresh and processed products. *J. Sci. Food Agric.* **2010**, *90*, 1578–1589. [CrossRef]
3. De Vasconcelos, M.d.C.B.M.; Nunes, F.; Viguera, C.G.; Bennett, R.N.; Rosa, E.A.S.; Ferreira-Cardoso, J.V. Industrial processing effects on chestnut fruits (*Castanea sativa* Mill.) 3. Minerals, free sugars, carotenoids and antioxidant vitamins. *Int. J. Food Sci. Technol.* **2010**, *45*, 496–505. [CrossRef]
4. Rutter, P.A.; Miller, G.; Payne, J.A. Chestnuts (*Castanea*). In *Acta Horticulturae*; International Society for Horticultural Science (ISHS): Leuven, Belgium, 1991; pp. 761–790.
5. Pezzi, G.; Gambini, S.; Buldrini, F.; Ferretti, F.; Muzzi, E.; Maresi, G.; Nascimbene, J. Contrasting patterns of tree features, lichen, and plant diversity in managed and abandoned old-growth chestnut orchards of the northern Apennines (Italy). *For. Ecol. Manag.* **2020**, *470–471*, 118207. [CrossRef]
6. Piccolo, E.L.; Landi, M.; Ceccanti, C.; Mininni, A.N.; Marchetti, L.; Massai, R.; Guidi, L.; Remorini, D. Nutritional and nutraceutical properties of raw and traditionally obtained flour from chestnut fruit grown in Tuscany. *Eur. Food Res. Technol.* **2020**, *246*, 1867–1876. [CrossRef]
7. Santos, M.J.; Pinto, T.; Vilela, A. Sweet Chestnut (*Castanea sativa* Mill.) Nutritional and Phenolic Composition Interactions with Chestnut Flavor Physiology. *Foods* **2022**, *11*, 4052. [CrossRef]
8. Massantini, R.; Moscetti, R.; Frangipane, M.T. Evaluating progress of chestnut quality: A review of recent developments. *Trends Food Sci. Technol.* **2021**, *113*, 245–254. [CrossRef]
9. Li, R.; Sharma, A.K.; Zhu, J.; Zheng, B.; Xiao, G.; Chen, L. Nutritional biology of chestnuts: A perspective review. *Food Chem.* **2022**, *395*, 133575. [CrossRef]

10. Sacchetti, G.; Neri, L.; Dimitri, G.; Mastrocola, D. Chemical composition and functional properties of three sweet chestnut (*Castanea sativa* Mill.) Ecotypes from Italy. In *Acta Horticulturae*; International Society for Horticultural Science (ISHS): Leuven, Belgium, 2009; pp. 41–46.
11. Neri, L.; Dimitri, G.; Sacchetti, G. Chemical composition and antioxidant activity of cured chestnuts from three sweet chestnut (*Castanea sativa* Mill.) ecotypes from Italy. *J. Food Compos. Anal.* **2010**, *23*, 23–29. [CrossRef]
12. Attanasio, G.; Cinquanta, L.; Albanese, D.; Matteo, M. Di Effects of drying temperatures on physico-chemical properties of dried and rehydrated chestnuts (*Castanea sativa*). *Food Chem.* **2004**, *88*, 583–590. [CrossRef]
13. Demirkesen, I.; Mert, B.; Sumnu, G.; Sahin, S. Utilization of chestnut flour in gluten-free bread formulations. *J. Food Eng.* **2010**, *101*, 329–336. [CrossRef]
14. Melo, B.G.d.; Tagliapietra, B.L.; Clerici, M.T.P.S. Evolution of the technological, sensory, and nutritional quality of gluten-free cookies: A critical review. *Food Sci. Technol.* **2023**, *43*, e75822. [CrossRef]
15. Dall'Asta, C.; Cirlini, M.; Morini, E.; Rinaldi, M.; Ganino, T.; Chiavaro, E. Effect of chestnut flour supplementation on physico-chemical properties and volatiles in bread making. *LWT-Food Sci. Technol.* **2013**, *53*, 233–239. [CrossRef]
16. Frati, A.; Landi, D.; Marinelli, C.; Gianni, G.; Fontana, L.; Migliorini, M.; Pierucci, F.; Garcia-Gil, M.; Meacci, E. Nutraceutical properties of chestnut flours: Beneficial effects on skeletal muscle atrophy. *Food Funct.* **2014**, *5*, 2870–2882. [CrossRef]
17. Mete, M.; Altiner, L. Chestnut Flour and Applications of Utilization. *Int. J. Food Eng. Res.* **2017**, *2017*, 9–16.
18. Bellini, E. The Chestnut And Its Resources: Images And Considerations. In *Acta Horticulturae*; International Society for Horticultural Science (ISHS): Leuven, Belgium, 2005; pp. 85–96.
19. Cantini, C.; Salusti, P.; Poggioni, L.; Romi, M. Analisi del profilo aromatico delle farine di castagna e relazioni con le proprietà sensoriali. In Proceedings of the VII Convegno Nazionale sul Castagno, ACTA Italus Hortus 25: Pergine Valsugana, Trento, Italy, 11–14 June 2019; pp. 1–3.
20. Cirlini, M.; Dall'Asta, C.; Silvanini, A.; Beghè, D.; Fabbri, A.; Galaverna, G.; Ganino, T. Volatile fingerprinting of chestnut flours from traditional Emilia Romagna (Italy) cultivars. *Food Chem.* **2012**, *134*, 662–668. [CrossRef]
21. Krist, S.; Unterweger, H.; Bandion, F.; Buchbauer, G. Volatile compound analysis of SPME headspace and extract samples from roasted Italian chestnuts (*Castanea sativa* Mill.) using GC-MS. *Eur. Food Res. Technol.* **2004**, *219*, 470–473. [CrossRef]
22. Conti, V.; Salusti, P.; Romi, M.; Cantini, C. Effects of Drying Methods and Temperatures on the Quality of Chestnut Flours. *Foods* **2022**, *11*, 1364. [CrossRef]
23. Ahmed, J.; Al-Attar, H. Effect of drying method on rheological, thermal, and structural properties of chestnut flour doughs. *Food Hydrocoll.* **2015**, *51*, 76–87. [CrossRef]
24. Moreira, R.; Chenlo, F.; Torres, M.D.; Rama, B.; Arufe, S. Air drying of chopped chestnuts at several conditions: Effect on colour and chemical characteristics of chestnut flour. *Int. Food Res. J.* **2015**, *22*, 407–413.
25. Zhang, L.; Wang, Z.; Shi, G.; Yang, H.; Wang, X.; Zhao, H.; Zhao, S. Effects of drying methods on the nutritional aspects, flavor, and processing properties of Chinese chestnuts. *J. Food Sci. Technol.* **2018**, *55*, 3391–3398. [CrossRef] [PubMed]
26. Conidi, C.; Donato, L.; Algieri, C.; Cassano, A. Valorization of chestnut processing by-products: A membrane-assisted green strategy for purifying valuable compounds from shells. *J. Clean. Prod.* **2022**, *378*, 134564. [CrossRef]
27. Salgueiro, L.; Martins, A.P.; Correia, H. Raw materials: The importance of quality and safety. A review. *Flavour. Fragr. J.* **2010**, *25*, 253–271. [CrossRef]
28. Monacci, E.; Sanmartin, C.; Bianchi, A.; Pettinelli, S.; Najar, B.; Mencarelli, F.; Taglieri, I. Chemical Quality and Characterization of Essential Oils in Postharvest Hop cv. Cascade: Ventilated Room Temperature as a Sustainable Alternative to Hot-Stove and Freeze-Drying Processes. *Beverages* **2025**, *11*, 54. [CrossRef]
29. Bianchi, A.; Taglieri, I.; Zinnai, A.; Macaluso, M.; Sanmartin, C.; Venturi, F. Effect of Argon as Filling Gas of the Storage Atmosphere on the Shelf-Life of Sourdough Bread—Case Study on PDO Tuscan Bread. *Foods* **2022**, *11*, 3470. [CrossRef]
30. Borges, O.; Gonçalves, B.; de Carvalho, J.L.S.; Correia, P.; Silva, A.P. Nutritional quality of chestnut (*Castanea sativa* Mill.) cultivars from Portugal. *Food Chem.* **2008**, *106*, 976–984. [CrossRef]
31. Bianchi, A.; Capparelli, S.; Taglieri, I.; Sanmartin, C.; Pistelli, L.; Venturi, F. Salty Biscuits Enriched with Fresh and Dried Bee Pollen: Chemical, Technological, and Sensory Characterization. *Foods* **2025**, *14*, 527. [CrossRef] [PubMed]
32. Bianchi, A.; Sanmartin, C.; Taglieri, I.; Macaluso, M.; Venturi, F.; Napoli, M.; Mancini, M.; Fabbri, C.; Zinnai, A. Effect of Fertilization Regime of Common Wheat (*Triticum aestivum*) on Flour Quality and Shelf-Life of PDO Tuscan Bread. *Foods* **2023**, *12*, 2672. [CrossRef]
33. Monacci, E.; Sanmartin, C.; Bianchi, A.; Pettinelli, S.; Taglieri, I.; Mencarelli, F. Plastic film packaging for the postharvest quality of fresh hop inflorescence (*Humulus lupulus*) cv. Cascade. *Postharvest Biol. Technol.* **2023**, *206*, 112575. [CrossRef]
34. Bate-Smith, E.C. Tannins of herbaceous leguminosae. *Phytochemistry* **1973**, *12*, 1809–1812. [CrossRef]

35. Bianchi, A.; Venturi, F.; Zinnai, A.; Taglieri, I.; Najar, B.; Macaluso, M.; Merlani, G.; Angelini, L.G.; Tavarini, S.; Clemente, C.; et al. Valorization of an Old Variety of *Triticum aestivum*: A Study of Its Suitability for Breadmaking Focusing on Sensory and Nutritional Quality. *Foods* **2023**, *12*, 1351. [CrossRef]
36. Pieracci, Y.; Vento, M.; Pistelli, L.; Lombardi, T.; Pistelli, L. Halophyte *Artemisia caerulescens* L.: Metabolites from In Vitro Shoots and Wild Plants. *Plants* **2022**, *11*, 1081. [CrossRef]
37. Adams, R.P. *Identification of Essential Oil Components by Gas Chromatography/Mass Spectroscopy*; Allured Pub. Corp.: Carol Stream, IL, USA, 2007; ISBN 0-931710-42-1.
38. Davies, N.W. Gas chromatographic retention indices of monoterpenes and sesquiterpenes on methyl silicon and Carbowax 20M phases. *J. Chromatogr. A* **1990**, *503*, 1–24. [CrossRef]
39. Masada, Y. *Analysis of Essential Oils by Gas Chromatography and Mass Spectrometry*; Halsted Press (Division of John Wiley & Sons): New York, NY, USA, 1976; Volume 14.
40. FLAVORNET. Available online: https://www.flavornet.org/index.html (accessed on 1 May 2025).
41. The Good Scents Company Information System. Available online: https://www.thegoodscentscompany.com/index.html (accessed on 1 May 2025).
42. Bianchi, A.; Venturi, F.; Palermo, C.; Taglieri, I.; Angelini, G.L.; Tavarini, S.; Sanmartin, C. Primary and secondary shelf-life of bread as a function of formulation and MAP conditions: Focus on physical-chemical and sensory markers. *Food Packag. Shelf Life* **2024**, *41*, 101241. [CrossRef]
43. Hammer, Ø.; Harper, D.A.T.; Ryan, P.D. Past: Paleontological Statistics Software Package for Education and Data Analysis. *Palaeontologia. Palaeontol. Electron.* **2001**, *4*, 9.
44. Zhu, F. Properties and Food Uses of Chestnut Flour and Starch. *Food Bioprocess. Technol.* **2017**, *10*, 1173–1191. [CrossRef]
45. Delgado, T.; Pereira, J.A.; Ramalhosa, E.; Casal, S. Effect of hot air convective drying on sugar composition of chestnut (*Castanea sativa* Mill.) slices. *J. Food Process. Preserv.* **2018**, *42*, e13567. [CrossRef]
46. Barreira, J.C.M.; Pereira, J.A.; Oliveira, M.B.P.P.; Ferreira, I.C.F.R. Sugars Profiles of Different Chestnut (*Castanea sativa* Mill.) and Almond (*Prunus dulcis*) Cultivars by HPLC-RI. *Plant Foods Hum. Nutr.* **2010**, *65*, 38–43. [CrossRef]
47. Correia, P.; Leitão, A.; Beirão-da-Costa, M.L. The effect of drying temperatures on morphological and chemical properties of dried chestnuts flours. *J. Food Eng.* **2009**, *90*, 325–332. [CrossRef]
48. Ertan, E.; Erdal, E.; Alkan, G.; Algül, B.E. Effects of Different Postharvest Storage Methods on the Quality Parameters of Chestnuts (*Castanea sativa* Mill.). *HortScience Horts* **2015**, *50*, 577–581. [CrossRef]
49. Chenlo, F.; Moreira, R.; Chaguri, L.; Torres, M.D. Effects of storage conditions on sugars and moisture content of whole chestnut fruits. *J. Food Process. Preserv.* **2010**, *34*, 609–620. [CrossRef]
50. Correia, P.; Cruz-Lopes, L.; Beirão-da-Costa, L. Morphology and structure of chestnut starch isolated by alkali and enzymatic methods. *Food Hydrocoll.* **2012**, *28*, 313–319. [CrossRef]
51. Wang, S.; Liu, C.; Wang, S. Drying methods used in starch isolation change properties of C-type chestnut (*Castanea mollissima*) starches. *LWT* **2016**, *73*, 663–669. [CrossRef]
52. Buléon, A.; Véronèse, G.; Putaux, J.-L. Self-Association and Crystallization of Amylose. *Aust. J. Chem.* **2007**, *60*, 706–718. [CrossRef]
53. Fedorková, S.; Musilová, J.; Ňorbová, M.; Vollmannová, A.; Čeryová, N.; Lidiková, J. Impact of Heat Treatments on The Antioxidant Activity and Total Phenolic Content of Sweet Chestnuts (*Castanea sativa* Mill.). *Sci. Bull. Ser. F. Biotechnol.* **2024**, *XXVIII*, 31–36.
54. Nguyen, M.P. Ascorbic Acid and Total Phenolic Contents of Dried Roasted Chestnut (*Castanea sativa*) Affected by Drying, Roasting and Preservation. *Biosci. Biotechnol. Res. Commun.* **2020**, *13*, 1–9. [CrossRef]
55. Barros, A.I.R.N.A.; Nunes, F.M.; Gonçalves, B.; Bennett, R.N.; Silva, A.P. Effect of cooking on total vitamin C contents and antioxidant activity of sweet chestnuts (*Castanea sativa* Mill.). *Food Chem.* **2011**, *128*, 165–172. [CrossRef]
56. Mladenović Drinić, S.D.; Vukadinović, J.Z.; Srdić, J.; Milašinović Šeremešić, M.S.; Anđelković, V.B. Effect of Cooking on the Content of Carotenoids and Tocopherols in Sweet Corn. *Food Feed Res.* **2021**, *48*, 119–129. [CrossRef]
57. Boon, C.S.; McClements, D.J.; Weiss, J.; Decker, E.A. Factors Influencing the Chemical Stability of Carotenoids in Foods. *Crit. Rev. Food Sci. Nutr.* **2010**, *50*, 515–532. [CrossRef]
58. Martínez, J.A.; Melgosa, M.; Pérez, M.M.; Hita, E.; Negueruela, A.I. Note. Visual and Instrumental Color Evaluation in Red Wines. *Food Sci. Technol. Int.* **2001**, *7*, 439–444. [CrossRef]
59. Pasqualone, A.; Paradiso, V.M.; Summo, C.; Caponio, F.; Gomes, T. Influence of Drying Conditions on Volatile Compounds of Pasta. *Food Bioprocess. Technol.* **2014**, *7*, 719–731. [CrossRef]
60. Romano, P.; Suzzi, G.; Brandolini, V.; Menziani, E.; Domizio, P. Determination of 2,3-butanediol in high and low acetoin producers of *Saccharomyces cerevisiae* wine yeasts by automated multiple development (AMD). *Lett. Appl. Microbiol.* **1996**, *22*, 299–302. [CrossRef] [PubMed]

61. Garg, S.K.; Jain, A. Fermentative production of 2,3-butanediol: A review. *Bioresour. Technol.* **1995**, *51*, 103–109. [CrossRef]
62. Gong, X.; Huang, J.; Xu, Y.; Li, Z.; Li, L.; Li, D.; Belwal, T.; Jeandet, P.; Luo, Z.; Xu, Y. Deterioration of plant volatile organic compounds in food: Consequence, mechanism, detection, and control. *Trends Food Sci. Technol.* **2023**, *131*, 61–76. [CrossRef]
63. Martins, S.I.F.S.; Jongen, W.M.F.; van Boekel, M.A.J.S. A review of Maillard reaction in food and implications to kinetic modelling. *Trends Food Sci. Technol.* **2000**, *11*, 364–373. [CrossRef]

foods

MDPI

Article

Dietary Supplementation with Green Alga (*Chlorella pyrenoidosa*) Enhances the Shelf Life of Refrigerated Nile Tilapia (*Oreochromis niloticus*) Fillets

Leticia Franchin Rodrigues [1], Mayumi Fernanda Aracati [1], Susana Luporini de Oliveira [1], Camila Carlino-Costa [1], Romário Alves Rodrigues [1], Mateus Roberto Pereira [1], Hirasilva Borba [1], Cleber Fernando Menegasso Mansano [2], Gabriel Augusto Marques Rossi [3], Jorge Galindo-Villegas [4], Gabriel Conde [5], Luiz Arthur Malta Pereira [6], Hélio José Montassier [1] and Marco Antonio de Andrade Belo [6,7,*]

1 Department of One Health, Sao Paulo State University (UNESP), Jaboticabal 14884-900, SP, Brazil; leticia.franchin@unesp.br (L.F.R.); mayumi.aracati@unesp.br (M.F.A.); susana.luporini@unesp.br (S.L.d.O.); camila.carlino@unesp.br (C.C.-C.); romario.a.rodrigues@unesp.br (R.A.R.); mateus.roberto@unesp.br (M.R.P.); hirasilva.borba@unesp.br (H.B.); helio.montassier@unesp.br (H.J.M.)

2 Animal Nutrition Department, Brazil University, Fernadópolis 15600-000, SP, Brazil; cleber.mansano@ub.edu.br

3 Microbiology Department, University Vila Velha (UVV), Vila Velha 29102-770, ES, Brazil; gabriel.rossi@uvv.br

4 Department of Genomics, Faculty of Biosciences and Aquaculture, 8020 Bodø, Norway; jorge.galindo-villegas@nord.no

5 College of Veterinary Medicine, University of Araraquara (UNIARA), Araraquara 14801-340, SP, Brazil; ga_conde@hotmail.com

6 Animal Production Program, Brazil University (UB), Descalvado 13690-000, SP, Brazil; luiz.pereira@ub.edu.br

7 Laboratory of Animal Pharmacology and Toxicology, Brazil University, 950 Hilário da Silva Passos Avenue, Jardim Universitário, Descalvado 13690-000, SP, Brazil

* Correspondence: marco.belo@ub.edu.br or maabelo@hotmail.com; Tel.: +55-01931996333

Abstract: Given the critical need for strategies to enhance food safety and quality, particularly due to the rapid spoilage of fish, this study investigated the effect of dietary supplementation with green alga (*Chlorella pyrenoidosa*) on the shelf life of Nile tilapia (*Oreochromis niloticus*) fillets. In addition, the microbiological, physicochemical, and sensory qualities of the fillets were evaluated. Eighty-four healthy tilapias (100 ± 2 g) were randomly allocated in three groups (n = 28 per group): a control group (without *C. pyrenoidosa* supplementation) and two treatment groups supplemented with either 5 or 10 g of *C. pyrenoidosa* per kg of feed. After 30 days of feeding, the fish were slaughtered and their fillets stored at 4 °C for evaluation at 0, 7, 15, and 30 days post-slaughter (dps). Fillets from fish supplemented with *C. pyrenoidosa* showed significant lower counts of coliforms, Enterobacteriaceae, and mesophilic and psychrotrophic microorganisms compared to the control. Supplementation also reduced thiobarbituric acid reactive substance (TBARS) levels and pH over the 30-day storage period. Sensory and color analyses indicated improved brightness, appearance, and firmness in fillets from treated groups, while control fillets exhibited poorer odor quality. Our findings demonstrate a dose-dependent effect of *C. pyrenoidosa* supplementation in improving the microbiological, physicochemical, and sensory quality of Nile tilapia fillets during refrigerated storage for up to 30 days.

Keywords: chlorophyta; microalga; antioxidant; food quality; fish flesh

1. Introduction

The shelf life of food products is influenced by multiple factors, including storage duration, environmental conditions such as temperature, and the intrinsic susceptibility of the product to quality degradation. Food spoilage occurs because of physical, chemical, and biological changes that can compromise the nutritional value, safety, and sensory attributes of the product [1]. In fish, spoilage is characterized by the development of off-odors and flavors, mucus accumulation, gas production, abnormal coloration, and texture deterioration. These changes emerge as a product of autolysis, oxidation, bacterial activity, or a combination of these processes, directly impacting the shelf life of fish [2].

To extend shelf life, strategies that prevent or reduce oxidative damage are essential, with antioxidants playing a key role in mitigating both oxidative and microbial spoilage in freshwater seafood products [3,4]. *Chlorella pyrenoidosa* is a green microalga widely used as a nutritional supplement for both humans and animals [5]. Its characteristic green color is attributed to the high concentrations of chlorophyll-a and chlorophyll-b, as well as other pigments such as carotenoids (α-, β-, and γ-carotene) and xanthophylls [6]. Dietary supplementation with carotenoids is an increasingly common practice in aquaculture, as it enhances carcass quality and antioxidant capacity in fish [7]. Additionally, carotenoids have been shown to improve the microbiological, physicochemical, and sensory properties of seafood products [8].

In aquaculture, dietary inclusion of *Chlorella* has been shown to enhance fish health and improve the quality of flesh by modulating antioxidant capacity, immune function, and pigmentation. Several studies have reported that microalgae supplementation can reduce lipid oxidation, inhibit microbial spoilage, and improve the texture, color, and sensory characteristics of fish fillets during storage. For example, Takyar et al. [4] demonstrated that *Chlorella vulgaris* and *Spirulina platensis* extracts extended the shelf life of rainbow trout by improving oxidative stability. Similarly, Rosas et al. [7] showed improved antioxidant performance and fillet coloration in mullets fed with carotenoid-rich microalgae. These findings highlight the potential of *Chlorella pyrenoidosa* as a natural feed additive for improving the microbiological, physicochemical, and sensory quality of fish flesh, making it a promising candidate for sustainable strategies in aquaculture and food preservation. Given the nutritional importance of fish in ensuring food security and the need for effective preservation strategies, this study aimed to evaluate the effect of dietary *C. pyrenoidosa* supplementation, administered orally through feed, on the microbiological, physicochemical, and sensory quality of Nile tilapia (*Oreochromis niloticus*) fillets during 30 days of refrigerated storage.

2. Material and Methods

2.1. Animals

A set of 84 healthy Nile tilapia (*O. niloticus*), each weighing approximately 100 g, were purchased from the same spawning and placed in 12 aquariums ($n = 7$), with each aquarium having a capacity of 100 L of chlorine-free running water from an artesian well, flowing at a rate of 1 L/min. Prior to the start of the experiment, the fish were monitored for physiological condition and general behavior to ensure the inclusion of only clinically healthy individuals. Animals were considered suitable if they exhibited species-specific body coloration, absence of external lesions, intact scales, clear eyes, balanced swimming, responsiveness to stimuli, and no signs of stress such as lethargy, erratic swimming, or prolonged bottom-dwelling behavior. This screening ensured the homogeneity of the experimental groups and the reliability of subsequent results. The fish were acclimated for 15 days to allow their plasma cortisol concentrations and osmolarity to return to baseline levels. During the initial three days of acclimation, the fish were treated with a NaCl solution at a concentration of 6.0 g/L [9]. They received commercial pelleted feed (32% crude protein), constituting the basal diet. The animals were fed three times a day, corresponding to 2% of the biomass of the aquariums. Water quality was monitored at feeding times throughout the experimental period using a YSI-63 pH meter (YSI incorporated®, Yellow Springs, OH, USA) and a YSI-55 oximeter (YSI incorporated®, Yellow Springs, OH, USA) and its values remained within the appropriate range for the welfare of tropical fish [10] (dissolved oxygen = 4.07 ± 0.89 mg/L; temperature = 27.64 ± 2.05 °C; pH = 7.64 ± 0.54; and electrical conductivity = 208.29 ± 97.57 µS/cm).

2.2. Experimental Design

Tilapia were randomly distributed among 12 aquariums (100 L of water each, $n = 7$) with the following treatments: T0 (no treatment with *C. pyrenoidosa*), T1, and T2 (treated with 5 g and 10 g of *C. pyrenoidosa* per kg of feed, respectively). After one month, the fish were slaughtered, and fillet samples were collected for analysis of microbiological, physicochemical, and sensory characteristics. These analyses corresponded to the following post-slaughter periods: 0, 7, 15, and 30 days post-slaughter (dps) ($n = 7$ per period, totaling 28 tilapia per treatment).

2.3. Experimental Diets

The basal diet composition consisted of commercial pelleted feed (LAGUNA®, ADM Company, Sao Paulo, Brazil), containing 32% crude protein, 7% ether extract, 5% crude fiber, and 12% mineral matter. The animals were fed three times a day, at 8:00 am, 1:00 pm, and 6:00 pm, corresponding to 2% of the aquarium biomass. To standardize the experimental diet containing *C. pyrenoidosa* (Calêndula Vet Pharmacy, Sao Carlos, SP, Brazil), 5 and 10 g/kg of feed were added to the commercial feed, following the study by Abdulrahman et al. [11]. For diet preparation, the commercial feed was weighed based on the average biomass per tank, and 2% soybean oil was incorporated along with the respective amounts of *C. pyrenoidosa* (5 or 10 g/kg). The control diet received 2% soybean oil without microalgae to ensure energetic balance and consistency in pellet coating across treatments. The diets were stored in dark plastic bags and kept at 9 °C for use during the experimental period of one month prior to the slaughter of the animals. After the 30-day feeding period, no significant performance variation was detected across dietary groups: control fish gained 26.1 g, whereas those supplemented with 5 g and 10 g kg^{-1} of *C. pyrenoidosa* presented 34.1 and 31.7 g, respectively.

2.4. Slaughter of Fish and Removal of Fillets

The stunning method employed was hypothermia, using a mixture of hyperchlorinated water (5 ppm) and ice in a 2:1 ratio. This mixture was placed in a 15 L Styrofoam box lined with sterilized plastic bags, into which the fish were immersed for 3 min to induce stunning. The plastic bags and knives were changed for each group to prevent cross-contamination. After the fish were rendered unconscious, decerebration of the brainstem was performed. The filleting process was carried out by a single individual, who divided each fish into two fillets (right and left, weighing about 50 g each), ensuring the dorsal musculature was preserved along the entire length of the spine and ribs. The left side of each fish was allocated for microbiological analysis, while the right side was designated for physicochemical and sensory analysis. This standardization was maintained throughout the 30-day experimental period. The fillets were stored in sterilized plastic bags and transported to the laboratory where they were stored in a B.O.D. (Biological Oxygen Demand) incubator at a controlled temperature of 4 ± 1 °C.

2.5. Microbiological Analysis

2.5.1. Preparation of Dilutions

Under aseptic conditions, fragments representing the entire surface and depth of each fillet sample were randomly taken until reaching a total of 10 g. After weighing, 90 mL of 0.1% peptone water (HiMedia® Laboratories, Kennett Square, PA, USA) was added, and the mixture was homogenized for approximately 60 s using a Stomacher-type device, resulting in the initial 10^{-1} dilution. Subsequently, serial dilutions of 10^{-2}, 10^{-3}, and 10^{-4} were made by taking a 1 mL aliquot from the initial dilution and transferring it to tubes containing 9 mL of 0.1% peptone water, continuing this process until reaching the 10^{-4} dilution.

2.5.2. Most Probable Number of Total and Thermotolerant Coliforms

The Most Probable Number (MPN) method was used to estimate the quantity of total coliforms and thermotolerant coliforms in the samples according to APHA 9:2015 [12]. In the presumptive test, a 1 mL aliquot from the 10^{-1}, 10^{-2}, and 10^{-3} dilutions was inoculated into a series of three sterile tubes, each containing Lauryl Sulfate Tryptose (HiMedia® Laboratories, USA) broth with inverted Durham tubes. The tubes were incubated for 48 h at 36 ± 1 °C. Tubes were considered positive if they showed gas formation in at least 1/10 of the Durham tube or effervescence when gently shaken and were selected for the confirmatory test. Confirmation of total and thermotolerant coliforms was performed by transferring a platinum loop of the positive tubes from LST to Brilliant Green Bile Lactose broth (HiMedia® Laboratories, USA), incubated at 36 ± 1 °C for 48 h, and to EC broth (HiMedia® Laboratories, USA), incubated at 45 ± 1 °C for 48 h in a water bath, respectively [12].

2.5.3. Counts of Aerobic Mesophilic and Psychrotrophic Microorganisms

For the counting of mesophilic microorganisms, 1 mL from each dilution was inoculated into sterile Petri dishes, and 15–20 mL of Plate Count Agar (Kasvi®, Castelfidardo, Italy), previously melted and maintained at 46–48 °C, was added. The plates were mixed and incubated at 35 ± 1 °C for 48 h, according to the APHA 08:2015 method. After incubation, the plates were read, the number of colonies was multiplied by the corresponding dilution factor, and the result was expressed as colony-forming units per gram (CFU/g) [13]. The same procedure was used for counting psychrotrophic microorganisms, with the plates incubated at 7 ± 1 °C for 10 days, according to the APHA 13.61:2015 method [13].

2.5.4. Counts of Coagulase-Positive Staphylococci

The method ISO 6888-1:1999/Amd 1:2003 was used for the enumeration of coagulase-positive staphylococci. Thus, 0.1 mL of each dilution was inoculated onto the surface of Baird-Parker agar (Kasvi®, Italy) supplemented with egg yolk solution and potassium tellurite (Dinâmica®, Itacoatiara, Brazil), using a Drigalski loop, and evenly spread until complete absorption. The plates were inverted and incubated at 35 ± 1 °C for 30 to 48 h. After bacterial growth, five typical colonies (shiny black with an opaque ring, surrounded by a clear, transparent halo distinct from the medium's opacity) and atypical colonies (grayish or shiny black without a halo) were selected and transferred to Brain Heart Infusion (BHI) broth (Kasvi®, Italy) for biochemical tests (catalase, Gram, and coagulase). Subsequently, the Coagulase Test was performed using BHI culture tubes and lyophilized rabbit plasma (NewProv®, Pinhais, Brazil). These tubes were incubated in a water bath at 36 ± 1 °C for 24 h [14]. Readings were taken at 1, 2, 3, 4, and 24 h.

2.5.5. Enterobacteriaceae Count

According to Mossel [15], MacConkey broth is used as a selective medium for the presumptive detection of bacteria from the Enterobacteriaceae family. Therefore, 1 mL from each dilution was inoculated into sterilized Petri dishes, and approximately 20 mL of MacConkey agar (Kasvi®, Italy), previously melted and maintained at 46–48 °C, was added. The inoculum was then homogenized with the medium. After the medium had completely solidified, the plates were incubated in an inverted position at 36 ± 1 °C for 18 to 24 h. Three to five typical colonies were selected and transferred to tubes containing BHI for Gram staining (as described above) and for the oxidase test [15].

2.6. *Physicochemical Analysis*

2.6.1. Determination of pH

The pH was determined in duplicate in the epaxial muscle using a digital pH meter (Testo® model 205, West Chester, PA, USA) equipped with a penetration electrode for direct insertion.

2.6.2. Colorimetry

Color was determined using the CR-400 Chrome Meter (Konica Minolta Sensing Americas®, Inc., NJ, USA), which utilizes the CIELAB system (L*, a*, and b*) proposed by the Commission Internationale de l'Eclairage [16]. Parameters such as lightness (L*), redness intensity (a*), and yellowness intensity (b*) of the epaxial muscles were evaluated. The assessment was conducted at three different points on each muscle part to obtain an average value. Color intensity is expressed by chroma (*Cab*), while hue (*H°ab*) corresponds to the color name found in its pure state on the spectrum. These values were calculated using the formulas: $Cab = \sqrt{(a^2 + b^2)}$ and $H°ab = \arctan(b^*/a^*)$, according to Hernández et al. [17].

2.6.3. Lipid Oxidation (TBARS)

For each sample, 5 g of fillet was weighed and placed in a Falcon tube for lipid oxidation analysis following the methodology described by Pikul et al. [18]. To the sample, 25 mL of 7.5% trichloroacetic acid (Synth®, Sao Paulo, Brazil) was added (one tube at a time, as the acid dries out the sample). The mixture was blended using a Turrax for 1 min and then filtered through another Falcon tube. All samples were processed in triplicate, and 5 mL of 0.01 M (thiobarbituric acid, Synth®, Brazil) was added to 5 mL of the extract. The mixture was then placed in a water bath at 100 °C for 40 min, with the test tube covered with marbles. The determination of TBARS (thiobarbituric acid reactive substances, expressed as mg of malondialdehyde/kg of sample) was performed using spectrophotometry at 532 nm.

2.6.4. Sensory Analysis

Sensory quality assessments of the fillets were performed at each sampling interval by a panel of five trained evaluators, who were blinded to the treatment conditions. Panelists were trained over three 1-h sessions using fillets with known sensory deviations (e.g., loss of brightness, textural softening, off-odors), following adapted protocols based on Meilgaard et al. [19]. During training, panelists were calibrated using reference samples to ensure consistency in scoring brightness, firmness, appearance, and odor. Fish samples (50 g) from the dorsal region were presented individually to the panelists in individual booths under controlled conditions of light, temperature, and humidity. The panelists were asked to rate the characteristics of brightness, firmness, appearance, and odor on a scale from 1 to 10, with 10 being the best condition and 1 being the worst [8].

2.7. Statistical Analysis

Differences between treatments (10 g, 5 g, and control) and time (0, 7, 15, and 30 days) were determined using the non-parametric Kruskal-Wallis [20] test with GraphPad Prism version 9.0 software. Differences were considered significant when $p < 0.05$. Principal Component Analysis (PCA) was performed using PAST software (version 4.03) to explore patterns among microbiological (total and thermotolerant coliforms, mesophilic and psychrotrophic counts, *Staphylococcus* spp.), physicochemical (pH, TBARS, color parameters), and sensory variables over time and treatments. All variables were auto-scaled (mean-centered and standardized to unit variance) prior to analysis. Components with eigenvalues greater than 1.0 were retained based on the Kaiser criterion. The first two principal components (PC1 and PC2) were used to construct biplots and interpret sample distribution and variable correlations. Factor loadings were examined to identify the most influential variables in each component, allowing the identification of quality degradation patterns and treatment-specific effects during refrigerated storage. The PCA was used as a parameter to perform Spearman correlations (SAS 9.3)

3. Results

3.1. Microbiological Analysis

The results observed in the microbiological analysis of the fish samples are presented in Figure 1. The MPN counts for total coliforms showed significant decrease ($p < 0.05$) in fillets from tilapia treated with 10 g of *C. pyrenoidosa* after 15 days of storage (Figure 1A). The thermotolerant coliforms counts did not show significant differences ($p \geq 0.05$) among treatments during 30 days of storage (Figure 1B).

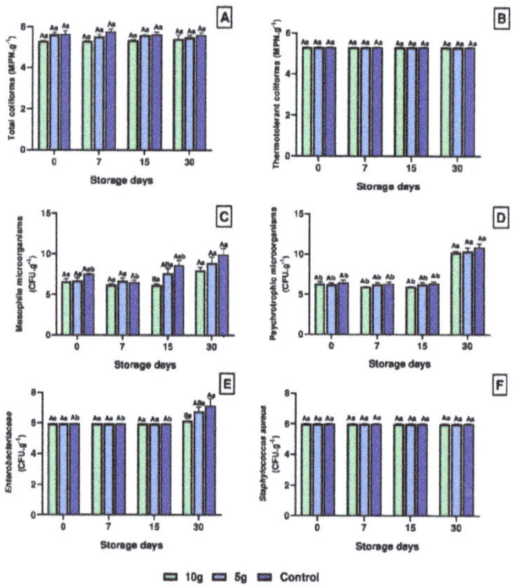

Figure 1. Mean values transformed into log(x + 5) (±SEM) for microbiological analysis. (**A**) Total coliforms; (**B**) Thermotolerant coliforms; (**C**) Mesophilic microorganisms; (**D**) Psychrotrophic microorganisms; (**E**) Enterobacteriaceae; (**F**) coagulase-positive staphylococci. Means (7 fish) followed by the same letter do not differ by the Kruskal-Wallis test ($p < 0.05$). Uppercase letters compare treatments within each experimental period, while lowercase letters evaluate the progression of each treatment across experimental periods. Sampling periods: 0, 7, 15, and 30 days post-slaughter (dps); Treatments: 10 g—treated with 10 g of *Chlorella*; 5 g—treated with 5 g of *Chlorella*; Control—untreated with *Chlorella*.

Tilapia treated with C. *pyrenoidosa* exhibited a reduction in mesophilic microorganism counts throughout the shelf-life period (Figure 1C). Fillets from fish treated with 10 g showed a significant decrease in counts 15 days post-slaughter (dps) compared to the control group (Figure 1C). In control fish fillets, a significant increase in mesophilic microorganism counts was observed at 30 dps compared to the initial storage phase at 7 dps (Figure 1C). Psychrotrophic microorganism counts in tilapia fillets did not differ significantly ($p \geq 0.05$) among treatments (Figure 1D). Microbiological analysis over time revealed a significant increase ($p < 0.05$) in psychrotrophic microorganism counts 30 dps in all treatments (Figure 1D).

The results for enterobacteria colony count showed a significant increase in the fillets from control tilapia 30 dps compared to fish treated with 10 g of C. *pyrenoidosa*, while the counts remained stable in the algae-treated groups (Figure 1E). The genus *Staphylococcus* was not detected in the tilapia fillets stored for 30 days at 4 °C (Figure 1F).

3.2. Physicochemical Analysis

TBARS analysis (thiobarbituric acid reactive substances) over time revealed a significant increase ($p < 0.05$) in all treatments (Figure 2A). Fillets from tilapia treated with 10 g of C. *pyrenoidosa* showed lower values of TBARS throughout the shelf-life period, and these results were significantly lower when compared to those observed in the control group fillets at 15 dps (Figure 2A). During the fillet storage period, the pH presented the lowest values 7 dps for all treatments (Figure 2B). However, the fillets from the control group showed increased pH values at 15 and 30 dps (Figure 2B).

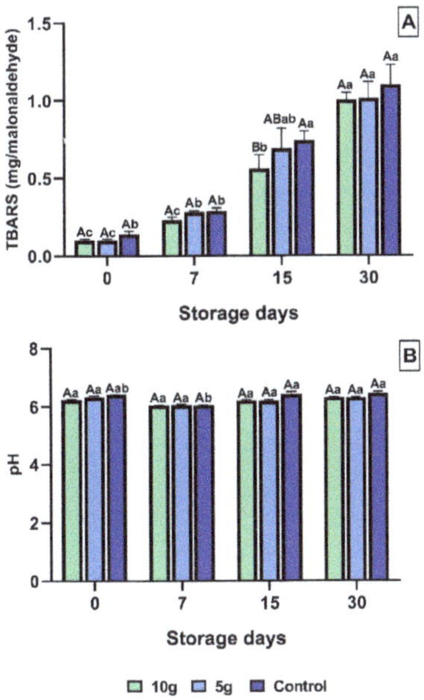

Figure 2. Analysis of TBARS and pH. (**A**) TBARS; (**B**) pH; Means (9 fish) followed by the same letter do not differ by the Kruskal-Wallis test ($p < 0.05$). Uppercase letters compare treatments within each experimental period, while lowercase letters evaluate the progression of each treatment across experimental periods. Sampling periods: 0, 7, 15, and 30 days post-slaughter (dps); Treatments: 10 g—treated with 10 g of *Chlorella*; 5 g—treated with 5 g of *Chlorella*; Control—untreated with *Chlorella*.

Correlation analysis between microorganism counts and pH or TBARS values revealed that control animals showed a significant ($p = 0.0198$ and $p = 0.0197$) positive correlation of 55.83% and 47.26% between the increase in the counts of mesophilic microorganisms and raising pH and TBARS values, respectively (Table 1).

Table 1. Correlation analysis between microorganism counts and pH or TBARS values observed in the tilapia fillets treated or not with *C. pyrenoidosa*.

Correlated Parameters	Experimental Sampling [1]	Correlation Analysis	
		ρ [2]	Prob. > ∣ρ∣ [2]
pH X Mesophilic microorganisms	Control	0.5583	0.0198
	5	−0.1182	0.6626
	10	0.4849	0.0793
TBARS X Mesophilic microorganisms	Control	0.4726	0.0197
	5	0.3094	0.1508
	10	0.1261	0.5662

[1] Correlation between control animals (n = 28); 5—animals treated with 5 g *C. pyrenoidosa*/kg of feed (n = 28); 10—animals treated with 10 g *C. pyrenoidosa*/kg of feed (n = 28). [2] ρ = Coefficient of Spearman Correlation; Prob. > ∣ρ∣—Significance Probability of ρ value.

Colorimetry

In the colorimetry study of tilapia fillets, no significant changes ($p \geq 0.05$) were observed in the analysis of luminosity (L*) and Delta E (ΔE) in the comparison among different treatments (Figure 3A,E). It is worth highlighting in this analysis that there was a significant increase ($p < 0.05$) in both parameters during the refrigerated storage period in all treatments (Figure 3A,E). The analysis of red to green intensity (a*) showed that the fillets initially showed a tendency towards reddish coloration and turned green in all treatments, with these effects being less pronounced in fillets of fish treated with 10 g of green alga, which were statistically significant ($p < 0.05$) at 7 dps when compared to the other treatments (Control and treated with 5 g) (Figure 3B). Fillets from fish treated with 10 g exhibited an initial decrease ($p < 0.05$) in red color intensity (Figure 3B).

The analysis of the blue to yellow color intensity (b*) initially observed that the fillets of all treatments showed a tendency towards a bluish coloration that became yellowish with refrigerated storage, except for the fish fed with 10 g of *C. pyrenoidosa*, which presented a significantly different bluish color at 30 dps (Figure 3C). Chroma index was significantly reduced ($p < 0.05$) in tilapia fillets throughout the shelf-life period, except for fish fed 10 g of *C. pyrenoidosa* (Figure 3D). Fish treated with 5 g demonstrated a significant reduction ($p < 0.05$) in this index after 30 days compared to both the control group and fish fed 10 g (Figure 3D).

The analysis of the Hue angle (H^0_{ab}) over time did not show significant changes between fillets from different treatments throughout the storage period. However, control fish showed a significant decrease in this parameter at 30 dps when compared to fish treated with *C. pyrenoidosa* (Figure 3F). It is worth noting that fish treated with 5 g showed a significant increase in Hue at 7 dps when compared to control fish and fish treated with 10 g (Figure 3F).

Correlation analysis between microorganism counts and colorimetry (Table 2) revealed that fillets from control animals showed a significant ($p = 0.0106$ and $p = 0.0305$) positive correlation with the increase counts of psychrotrophic microorganisms and increase in b* (60.20%), as well as Delta E (ΔE) (55.84%), respectively. Similar results were observed in fish treated with 5 g, which showed 54.68% positive correlation ($p = 0.0430$) between these same parameters (Table 2).

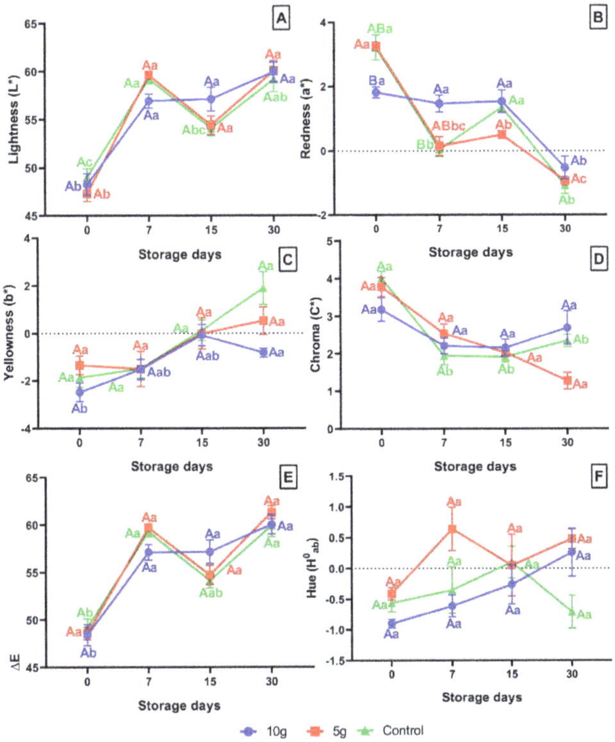

Figure 3. (**A**) Lightness (L*); (**B**) Redness (a*); (**C**) Yellowness (b*); (**D**) Chroma (C*); (**E**) Delta E (ΔE); (**F**) Hue ($H^0{}_{ab}$). Means (9 fish) followed by the same letter do not differ by the Kruskal-Wallis test ($p < 0.05$). Uppercase letters compare treatments within each experimental period, while lowercase letters evaluate the progression of each treatment across experimental periods. Sampling periods: 0, 7, 15, and 30 days post-slaughter (dps); Treatments: 10 g—treated with 10 g of *Chlorella*; 5 g—treated with 5 g of *Chlorella*; Control—untreated with *Chlorella*.

Table 2. Correlation analysis between microorganism counts and colorimetry observed in the tilapia fillets treated or not with *C. pyrenoidosa*.

Correlated Parameters	Experimental Sampling [1]	Correlation Analysis	
		ρ [2]	Prob. > ∣ρ∣ [2]
Psychrotrophic microorganisms X b*	Control	0.6020	0.0106
	5	0.3410	0.1204
	10	0.3450	0.1255
Psychrotrophic microorganisms X ΔE	Control	0.5584	0.0305
	5	0.5468	0.0430
	10	0.1286	0.5785

[1] Correlation between control animals (n = 28); 5—animals treated with 5 g *C. pyrenoidosa*/kg of feed (n = 28); 10—animals treated with 10 g *C. pyrenoidosa*/kg of feed (n = 28). [2] ρ = Coefficient of Spearman Correlation; Prob. > ∣ρ∣—Significance Probability of ρ value.

Correlation analysis between TBARS values and colorimetry revealed that fillets from control animals showed a significant (p = 0.0127) positive correlation with Delta E (ΔE) (50.06%) and a significant (p = 0.0044) negative correlation with redness (a*) (−54.95%) (Table 3). Fillets from tilapia treated with 10 g of *C. pyrenoidosa* showed a positive correlation (p = 0.0255) of 45.50% between the increase in TBARS values to Hue pitch angle during storage (Table 3).

Table 3. Correlation analysis between TBAR's values and colorimetry in the fillet of tilapia treated or not with *Chlorella pyrenoidosa*.

Correlated Parameters	Experimental Sampling [1]	Correlation Analysis	
		ρ [2]	Prob. > ∣ρ∣ [2]
TBARS X $H^0{}_{ab}$	Control	0.0523	0.8036
	5	0.3374	0.1577
	10	0.4550	0.0255
TBARS X ΔE	Control	0.5006	0.0127
	5	0.3405	0.1811
	10	0.3693	0.0757
TBARS X a*	Control	−0.5495	0.0044
	5	−0.5539	0.0092
	10	−0.2962	0.1598

[1] Correlation between control animals (n = 28); 5—animals treated with 5 g *C. pyrenoidosa*/kg of feed (n = 28); 10—animals treated with 10 g *C. pyrenoidosa*/kg of feed (n = 28). [2] ρ = Coefficient of Spearman Correlation; Prob. > ∣ρ∣—Significance Probability of ρ value.

3.3. Sensory Analysis

The results observed in the sensory analysis of the fish highlighted the following characteristics: brightness, firmness, appearance, and odor, as expressed in Figure 4. Over time, a significant worsening ($p < 0.05$) was observed in sensory parameters analyzed for all treatments, with these findings being more pronounced in control fish (Figure 4A–D). At 7 dps, a significant decrease in brightness, firmness, appearance and odor was observed in fillets of control fish when compared to fillets of fish treated with *C. pyrenoidosa* (Figure 4A–D).

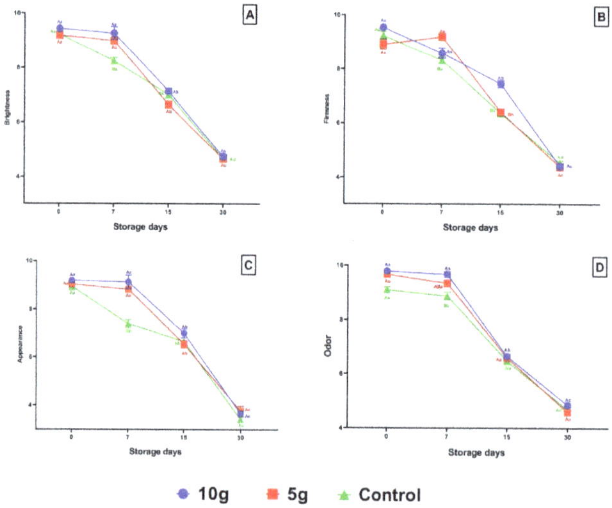

Figure 4. Sensory analysis. (**A**) Brightness; (**B**) Firmness; (**C**) Appearance; (**D**) Odor. Means (9 fish) followed by the same letter do not differ by the Kruskal-Wallis test ($p < 0.05$). Uppercase letters compare treatments within each experimental period, while lowercase letters evaluate the progression of each treatment across experimental periods. Sampling periods: 0, 7, 15, and 30 days post-slaughter (dps); Treatments: 10 g—treated with 10 g of *Chlorella*; 5 g—treated with 5 g of *Chlorella*; Control—untreated with *Chlorella*.

3.4. Principal Component Analysis (PCA)

Principal component analysis (PCA) helps to understand the interactions (correlation) between the variables studied. Figure 5A shows the dispersion of the groups studied and the loads that each vector (variables) presents within the analysis. These loads are illustrated in the heat map (Figure 5B); the more intense the red color, the greater the load attributed to this variable within the analysis and the greater the correlation between these factors. The sum of components (PC) 1 and 2 explains 41.79% of the relationships between the variables studied.

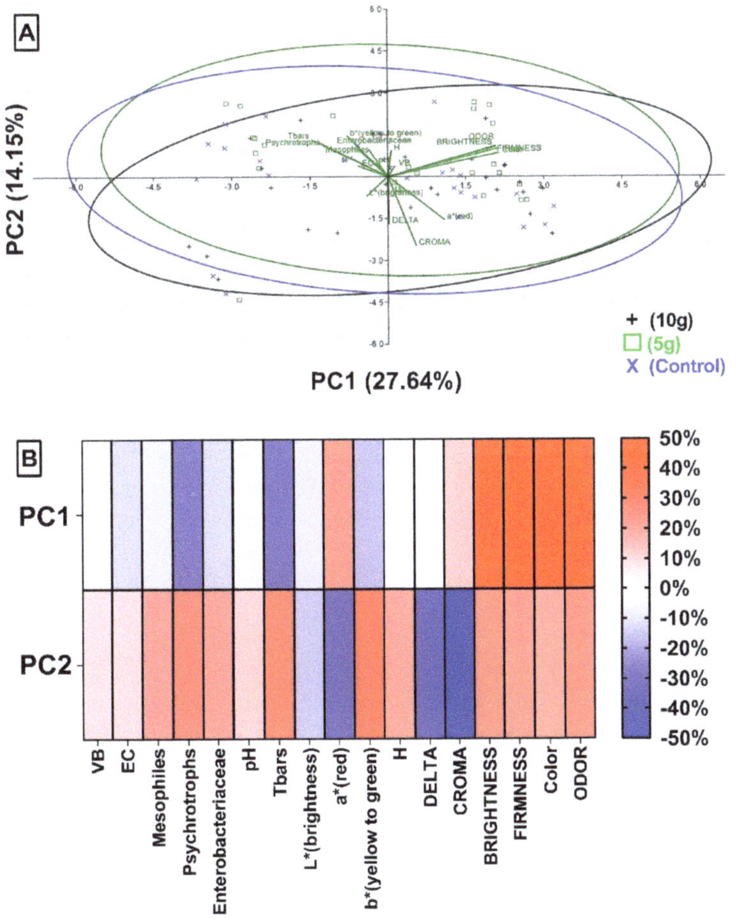

Figure 5. Principal component analysis (PCA) on the distribution of variables obtained after treatment of Nile tilapia with 10 g or 5 g of green alga (*Chlorella pyrenoidosa*) and control (untreated). (**A**) Relationship between variables studied by means of principal components. The directions of the vectors determine high scores in the respective variables. (**B**) Heat Map with the scores obtained in the three-dimensional PCA PC1 and PC2 of the variables used in the Spearman correlation test.

These data show that there may be a correlation between the variables PC1 and PC2, as well as between themselves in the same component. Table 1 shows a positive correlation between pH and mesophiles (p2 = 55.83%) represented by low-intensity red colors. Similarly, the correlation between pH and psychrotrophics (p2 = 72.19%) can be observed, in which the intensity of red shown in Figure 5B is greater in psychrotrophics. It is interesting to note that the Delta E variable in Tables 2 and 3 presents positive correlations with psychotrophics (p2 = 55.84%; 54.68%) and TBARS (p2 = 50.06%) respectively, even though their loadings tend to 0% (PC1) and −34% (PC2).

In accordance with the results obtained in the loadings (the name given in the program) of the 17 variables studied in the first PCA (Figure 5), those that presented 20% or more (in PC1 or PC2) were selected. From this, a new PCA analysis was performed (Figure 6). This second analysis explains 63.65% (45.79% in PC1 and 17.86% in PC2) of the relationships between the data collected in the study. The heat map (Figure 6B) shows intense red colors in brightness, firmness, color and odor in the first principal component, PC1. In PC2, these tones appear in the bacteria evaluated during the study. The correlation between TBARS and mesophiles (p2 = 47.26%) presented in Table 1 can be observed with less intense staining (Figure 5B) and with greater intensity mainly in TBARS in Figure 6B.

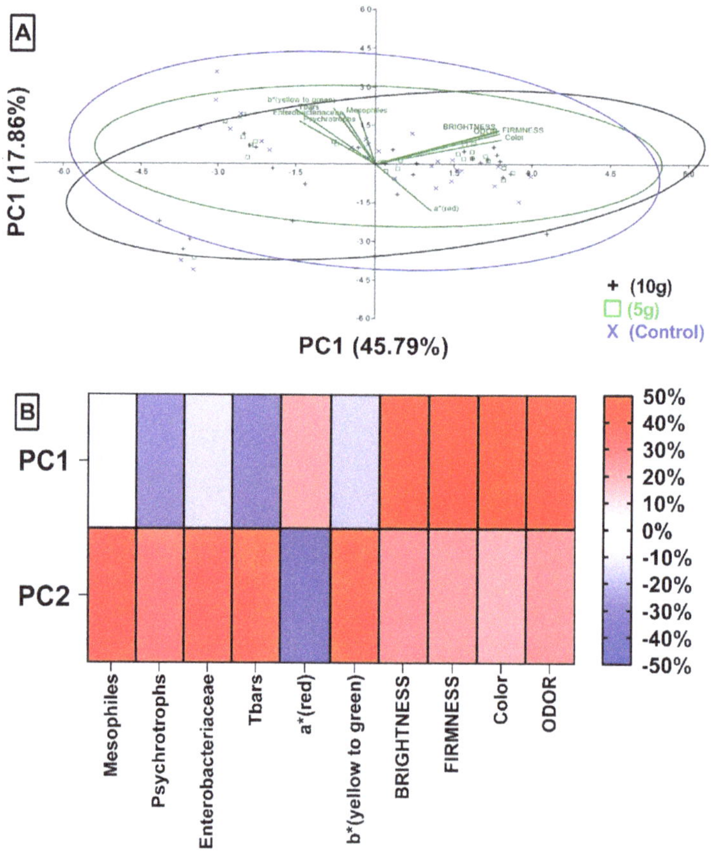

Figure 6. Principal component analysis (PCA) on the distribution of variables selected from the first PCA with loadings of 20% or more in PC1 and PC2 in Nile tilapia with 10 g or 5 g of green alga (*Chlorella pyrenoidosa*) and control (untreated). (**A**) Relationship between variables selected by principal components. The directions of the vectors determine high scores in the respective variables. (**B**) Heat Map with the scores obtained in the three-dimensional PCA PC1 and PC2 of the variables used in the Spearman correlation test.

It is interesting to note that positive correlations with a high probability of significance can be observed when the variables present a higher load in the same component, such as psychotrophic vs. b* (Table 2). Similarly, observing the clusters (Figure 7), we can identify these correlations by the proximity (distance) in which the variables appear. Therefore, the greater the distance between them, the lower the probability of significance or even negative correlation coefficients (p^2).

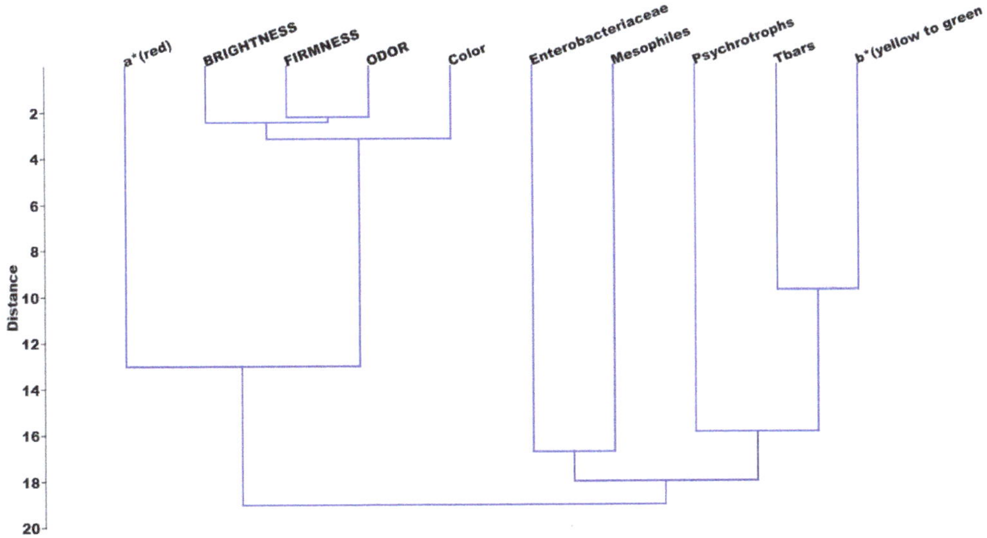

Figure 7. Clusters of loading variables with 20% or more in PC1 and PC2 in Nile tilapia with 10 g or 5 g of green alga (*Chlorella pyrenoidosa*) and control (untreated).

Table 3 shows a negative correlation between TBARS vs. a*. When observing the heat maps (Figures 5B and 6B), the variable a* is colored pink (PC1) and blue (PC2). Therefore, when compared under the same main component or between the components, there is a large difference between them. This difference is evidenced when observing the distance between the variables in Figure 7.

4. Discussion

Nile tilapia supplemented with *C. pyrenoidosa* exhibited reduced microbial counts in fillets during a 30-day monitoring period, with an evident dose-response effect. Total coliforms and mesophilic bacteria count significantly increased in control fish after 15 days of refrigeration, in contrast to the lower counts observed in *C. pyrenoidosa*-treated fish. Total coliforms and thermotolerant coliforms are indicators used to assess the hygienic and sanitary conditions of food, and they can predict the presence of pathogenic microorganisms [3]. Junior et al. [21] emphasized the relevance of monitoring total and thermotolerant coliforms in Nile tilapia, noting a significantly higher prevalence and concentration of these indicators in fillet samples compared to whole fish. According to Ferreira et al. [22], the filleting process in tilapia production is predominantly manual, with limited mechanization. This extensive human handling increases the risk of cross-contamination, particularly by thermotolerant coliforms, due to direct contact with operators and surfaces. Although recent studies have demonstrated the ability of microalgae, including *C. pyrenoidosa*, to remove pathogens from wastewater [23], little is known about their antimicrobial activity. However, to date, no studies have been identified that assess the effects of supplementation with this alga in animal models and its influence on coliforms counts during the shelf life of other foods of animal's origin.

Mesophilic and psychtrophic microorganisms are among the most widely used microbiological indicators of food quality, as they reflect the effectiveness of temperature and hygiene control during processing, transportation, and storage [24]. Stejskal et al. [25] reported that refrigerated hake fillets packaged with active films incorporating *Spirulina platensis* protein concentrate exhibited reduced counts of mesophilic, psychrotrophic, and Enterobacteriaceae microorganisms compared to control samples. These findings align with the results of the present study, in which tilapia supplemented with the green alga *C. pyrenoidosa* demonstrated similarly decreased levels of these microbial groups during refrigerated storage. In a study on the shelf life of tilapia fillets treated with astaxanthin, a carotenoid produced naturally in another freshwater microalga, *Haematococcus pluvialis*, Aracati et al. [8] reported significant reductions in mesophilic and psychrotrophic bacterial counts, attributed to the compound's antioxidant activity. Similar outcomes were reported by Santos et al. [26], who evaluated smoked tilapia fillets coated with chitosan. Furthermore, *C. pyrenoidosa* has demonstrated in vitro antimicrobial and antifungal activity [27], supporting the results of the present study, which showed a reduction in mesophilic and Enterobacteriaceae counts in fillet samples supplemented with *C. pyrenoidosa* at 5 g and 10 g doses, with the most pronounced effect observed at 10 g.

No *Staphylococcus* spp. counts were detected in the tilapia fillets stored for 30 days at 4 °C in the present study. The absence of these microorganisms suggests that the fillets were clean and safe, and that slaughter was performed under good hygienic and sanitary practices, corroborating the study by Atitallah et al. [28]. Coagulase-positive *Staphylococcus*, such as *Staphylococcus aureus*, are responsible for producing a heat-stable toxin and are one of the most prevalent foodborne pathogens [21]. Humans are common carriers of this microorganism, and the authors isolated multiple strains of *Staphylococcus aureus* from the nasal mucosa of food handlers [29], reinforcing the findings of Junior et al. [21], who demonstrated that the presence of *S. aureus* in Nile tilapia may be attributed to improper handling, inadequate hygiene practices, storage deficiencies, and cross-contamination.

Lipid oxidation is one of the main factors contributing to the deterioration of food quality, especially in fish flesh, directly influencing color changes, undesirable flavor, nutritional value, and shelf life [30]. Tilapia fillets from fish supplemented with *C. pyrenoidosa* exhibited significantly lower TBARS levels in comparison to fillets from the control group. Algae contain antioxidant compounds that are efficient oxygen radical scavengers and can diminish the oxidation of fatty acids [4]. The application of an algal-based coating containing *Spirulina platensis* and *Chlorella vulgaris* to refrigerated veal fillets was investigated by Shafiei and Mostaghim [31], who observed a marked decrease in TBARS values, reflecting enhanced oxidative stability in the treated samples. The findings of Takyar et al. [4], who emphasized the antioxidant potential of these algae in delaying lipid oxidation, are consistent with those of the present study, which likewise demonstrated a pronounced reduction in lipid oxidation throughout the storage period.

A positive correlation between the increase in mesophilic bacterial counts and both TBARS levels and pH values was observed exclusively in fillets from the control tilapia group. Elevated pH during storage is known to compromise food quality by facilitating microbial growth, as bacteria metabolize low-molecular-weight compounds present in fish muscle tissue, contributing to pH elevation [8]. The decrease in pH observed during the initial days of storage may be attributed not only to post-mortem glycolysis and residual glycogen degradation but also to the accumulation of low molecular-weight compounds produced by microbial and enzymatic activity. These include organic acids (e.g., lactic and acetic acids), free amino acids, and other small metabolites, which contribute to acidification of the muscle tissue. Over time, as proteolytic degradation progresses, the formation of basic compounds such as ammonia and trimethylamine may lead to pH stabilization or a slight increase, depending on the balance of microbial populations and metabolic pathways [32,33]. These alterations were not detected in fillets from fish supplemented with *C. pyrenoidosa*, suggesting a potential antioxidant effect of the dietary supplementation in enhancing fillet stability under refrigerated conditions. Additionally, our findings, which showed a decline in pH levels across all experimental groups by day 7 of refrigerated storage compared to day 0 (slaughter day), are consistent with previous reports indicating that live fish typically exhibit a natural pH slightly above 7.0, which decreases significantly post-mortem due to rigor mortis and the accumulation of lactic acid resulting from glycogen metabolism [34].

Color plays a crucial role in the acceptance of food products, acting as a key indicator of quality, freshness, preservation status, flavor perception, and commercial appeal. In the present study, both luminosity (L) and total color difference (ΔE) values increased over the storage period across all treatment groups. This trend may be associated with the progressive darkening of the fillets during refrigeration, as the initially fresh fish with high luminosity underwent deterioration, resulting in a gradual loss of brightness [8]. In fish muscle, particularly in Nile tilapia (*Oreochromis niloticus*), the colorimetric parameters lightness (L^*), redness (a^*), and yellowness (b^*) are influenced by the deposition of both endogenous and dietary pigments. The redness (a)* value is strongly associated with the presence of carotenoids such as astaxanthin, β-carotene, and canthaxanthin, which can be absorbed from the diet and deposited in the muscle tissue, enhancing reddish hues. These pigments are lipophilic and accumulate primarily in the subcutaneous and intramuscular fat, influencing a^* values through their antioxidant and color-enhancing effects [7,35]. The yellowness (b)* component is also affected by carotenoids, particularly lutein and zeaxanthin, which impart yellow-orange tones to the flesh. Meanwhile, lightness (L)* is less directly affected by specific pigments and more influenced by muscle structure, water and fat content, and the degree of lipid oxidation, all of which alter the light scattering properties of the muscle. However, high concentrations of deposited pigments may contribute secondarily to light absorption and reflectance. In this study, supplementation with *Chlorella pyrenoidosa*, a microalga rich in chlorophylls and carotenoids, likely contributed to the observed variations in a^* and b^* values in the fillets, especially in the 10 g/kg group, indicating a dose-dependent enhancement of pigment-related color stability.

In the present study, fillets from tilapia treated with 5 g of *C. pyrenoidosa* and those from the control group exhibited a more intense green coloration ($-a^*$). In contrast, fillets from fish supplemented with 10 g of *C. pyrenoidosa* showed a less pronounced green hue ($-a$) only after 30 days of storage. The predominance of green coloration ($-a^*$) observed at the end of the storage period was associated with loss of fillet quality. A negative correlation between increased TBARS levels and decreased redness intensity (a^* values) was observed in both the control group and the 5 g supplementation group. This effect was less evident in fillets from tilapia fed 10 g of *C. pyrenoidosa,* suggesting a dose-dependent protective role of the alga against lipid oxidation-related color loss. These findings are consistent with those reported by Aracati et al. [8] in tilapia fed astaxanthin, and by Majdinasab et al. [36] in rainbow trout supplemented with alginate.

An increase in the yellow component (*+b*) was observed throughout the shelf life in the group supplemented with 10 g of *C. pyrenoidosa*. According to Alfaia et al. [37], the inclusion of *C. vulgaris* in broiler diets led to a yellowish coloration in breast and thigh muscles, attributed to the high carotenoid content of this alga. However, An et al. [38] reported that low inclusion levels (0.05%, 0.15%, and 0.5%) of *Chlorella* in broiler diets did not significantly affect breast meat coloration. These findings help explain the yellowing observed in fillets from fish supplemented with 10 g of *C. pyrenoidosa*, whereas this effect was not evident in the 5 g supplementation group over the 30-day storage period. An increase in the yellow component (+b) and total color difference (ΔE) was positively correlated with the rise in psychrotrophic microbial counts in control fish fillets. These results suggest that the observed colorimetric changes also were associated with the proliferation of psychrotrophic microorganisms capable of growing under refrigeration, thereby compromising fillet quality.

Chroma (C) values decreased during refrigerated storage, indicating a loss in color intensity. This finding is consistent with previous studies by Aracati et al. [8] and Hernández et al. [17], who investigated the shelf life of *Oreochromis niloticus* and *Argyrosomus regius* fillets, respectively, and also reported this alteration during the shelf life. In contrast, the hue angle tended to shift toward a more reddish tone, which may be attributed to the presence of carotenoids, as *C. pyrenoidosa* is known to be a rich source of these pigments. Fradique et al. [39] reported that *C. vulgaris* accumulates substantial amounts of carotenoids in muscle tissue. Regarding total color difference (ΔE), both groups showed an increase over time, with more pronounced changes observed in the control group. Additionally, tilapia fed with *C. pyrenoidosa* exhibited increased hue angle values ($H^{\circ}ab$), indicating a shift toward a more reddish coloration. This observation is consistent with previous findings in *O. niloticus* and *Pagrus pagrus* fed with astaxanthin [8,40]. However, in fish supplemented with 10 g of *C. pyrenoidosa,* the increases in $H^{\circ}ab$ values were positively correlated with elevated TBARS levels, suggesting a relationship between lipid oxidation and color alteration. Mattje et al. [40] also reported that storage duration affects hue, with food products tending to become more yellowish by the end of refrigerated storage, primarily due to oxidative processes. Fish flesh is more perishable compared to red meat and poultry due to high levels of free amino acids, volatile basic nitrogen, and polyunsaturated fatty acids (PUFAs) [4]. Oxidation of PUFAs leads to the deterioration of food quality and consequently reduces shelf life, causing unpleasant and rancid flavors due to the formation of aldehydes and ketones, and diminishing nutritional value through the destruction of essential fatty acids and fat-soluble vitamins. Additionally, it can have adverse health effects due to the formation of free radicals [3]. The addition of algae extracts to fish has antioxidant activity capable of extending shelf life while preserving quality. Thus, *C. pyrenoidosa* can delay lipid oxidation and maintain the sensory properties of refrigerated fish [4]. Sensory analysis results (brightness, firmness, color, and odor) showed that overall acceptability ratings were significantly higher in algae-treated tilapias compared to control animals. Furthermore, a gradual reduction in sensory parameters was observed over time in all groups, with more pronounced effects in the control group, indicating a reflection of fillet quality loss. Lópes-Cánova et al. [41] described similar sensory changes in tilapia during refrigerated storage, including a gradual increase in pH values, corroborating the findings of this study.

The results from microbiological, physicochemical, and sensory evaluations highlight the beneficial role of *C. pyrenoidosa* in extending the quality of refrigerated tilapia fillets. Both the 5 g and 10 g doses improved microbial stability, sensory attributes, and reduced lipid oxidation over 30 days, with the higher dose producing more pronounced effects, suggesting a dose–response correlation. While this study demonstrated the beneficial effects of *Chlorella pyrenoidosa* supplementation on the quality of tilapia fillets, several limitations should be addressed in future research. Detailed chemical profiling of the microalga (e.g., chlorophylls, carotenoids, antioxidants) is needed to clarify the active compounds responsible for the observed effects. Inclusion of a reference antioxidant (e.g., vitamin E) and molecular microbiological tools (e.g., 16S rRNA sequencing) would strengthen causal inference. Moreover, assessing the economic feasibility, nutritional composition of fillets, and storage conditions that reflect commercial realities (e.g., MAP, temperature variation) would enhance practical relevance. Although *C. pyrenoidosa* is generally recognized as safe (GRAS), monitoring for potential contaminants remains essential to ensure food safety.

5. Conclusions

The dietary use of *Chlorella pyrenoidosa* proved effective in enhancing the microbiological stability, oxidative resistance, and sensory quality of Nile tilapia fillets stored under refrigeration. Incorporating this alga into the fish diet offers a practical strategy to naturally extend fillet shelf life and improve product quality, with potential applications in sustainable aquaculture and value-added fish processing.

Author Contributions: L.F.R., M.F.A., S.L.d.O., C.C.-C., C.F.M.M., L.A.M.P. and M.A.d.A.B.: methodology, writing, data curation; R.A.R., M.R.P. and H.B.: methodology and data curation; G.A.M.R.: methodology, review & editing; J.G.-V., M.A.d.A.B. and H.J.M.: supervision and review and editing; C.F.M.M.: review; G.C.: data curation; L.F.R. and M.A.d.A.B.: funding acquisition. All authors have read and agreed to the published version of the manuscript.

Funding: This work was supported by a CNPq grants (Process numbers: 132889/2020-0 and 406600/2023-6). Postgraduate Program in Veterinary Medicine, Unesp FCAV—Jaboticabal.

Institutional Review Board Statement: The sensory analysis conducted in this study was based on the evaluation of organoleptic attributes (brightness, firmness, appearance, and odor) of processed fish samples, with the participation of evaluators who were properly trained for this purpose. The analysis conducted was solely sensory, focusing on the parameters of the raw fillets, without involving the ingestion of the food and consequently with no risks for participants. The animal study protocol was approved by the Ethics Committee on the Use of Animals (CEUA) of the Faculty of Agricultural and Veterinary Sciences (protocol code no. 3423/20 and date of approval 1 December 2020).

Informed Consent Statement: Informed consent was obtained from all subjects involved in the study.

Data Availability Statement: The original contributions presented in the study are included in the article. Further questions can be directed to the corresponding author.

Conflicts of Interest: The authors declare no conflict of interest.

References

1. Giménez, A.; Ares, F.; Ares, G. Sensory shelf-life estimation: A review of current methodological approaches. *Food Res. Int.* **2012**, *49*, 311–325. [CrossRef]
2. Amaral, R.A.; Pinto, C.A.; Lima, V.; Tavares, J.; Martins, A.P.; Fidalgo, L.G.; Silva, A.M.; Gil, M.M.; Teixeira, P.; Barbosa, J.; et al. Chemical-Based Methodologies to Extend the Shelf Life of Fresh Fish—A Review. *Foods* **2021**, *10*, 2300. [CrossRef] [PubMed]
3. Oliveira, S.L.; Costa, C.C.; Aracati, M.F.; Rodrigues, L.F.; Charlies-Silva, I.; Belo, M.A.A. Antioxidantes Naturais como Alternativa Para Aumentar Vida de Prateleira. In *Desafios e Estratégias para Segurança Alimentar Mundial*; Amppla Editora: Florianópolis, Brazil, 2022. [CrossRef]
4. Takyar, M.B.T.; Khajavi, S.H.; Safari, R. Evaluation of antioxidant properties of *Chlorella vulgaris* and *Spirulina platensis* and their application in order to extend the shelf life of rainbow trout (*Oncorhynchus mykiss*) fillets during refrigerated storage. *LWT* **2019**, *100*, 244–249. [CrossRef]
5. Hongyang, S.; Yalei, Z.; Chunmin, Z.; Xuefei, Z.; Jinpeng, L. Cultivation of *Chlorella pyrenoidosa* in soybean processing wastewater. *Bioresour. Technol.* **2011**, *102*, 9884–9890. [CrossRef]
6. Sousa, L.M.D. Effects of Different Food Sources on the Development of *Moina minuta* Hansen (1899) (Crustacea: Branchiopoda). Bachelor's Thesis, Federal Rural University of Amazonia, Capanema Campus, Pará, Brazil, 2021.
7. Rosas, V.T.; Monserrat, J.M.; Bessonart, M.; Magnone, L.; Romano, L.A.; Tesser, M.B. Comparison of β-carotene and Spirulina (*Arthrospira platensis*) in mullet (*Mugil liza*) diets and effects on antioxidant performance and fillet colouration. *J. Appl. Phycol.* **2019**, *31*, 2391–2399. [CrossRef]

8. Aracati, M.F.; Rodrigues, L.F.; de Oliveira, S.L.; Rodrigues, R.A.; Conde, G.; Cavalcanti, E.N.F.; de Andrade Belo, M.A. Astaxanthin improves the shelf-life of tilapia fillets stored under refrigeration. *J. Sci. Food Agric.* **2022**, *102*, 4287–4295. [CrossRef]
9. Carneiro, P.C.F.; Urbinati, E.C. Salt as a stress response mitigator of matrinxã, *Brycon cephalus* (Günther), during transport. *Aquac. Res.* **2001**, *32*, 297–304. [CrossRef]
10. Boyd, C.E. *Water Quality in Ponds for Aquaculture*; Alabama Agricultural Experiment Station/Auburn University: Auburn, AL, USA, 1990.
11. Abdulrahman, N.M. Evaluation of *Spirulina* spp. as food supplement and its effect on growth performance of common carp fingerlings. *Int. J. Fish. Aquat. Stud.* **2014**, *2*, 89–92.
12. Kornacki, J.L.; Gurtler, J.B.; Stawick, B.A. Enterobacteriaceae, coliforms, and *Escherichia coli* as quality and safety indicators. In *Compendium of Methods for the Microbiological Examination of Foods*, 5th ed.; Salfinger, Y., Tortorello, M.L., Eds.; American Public Health Association: Washington, DC, USA, 2015; Chapter 9; pp. 103–120.
13. Vasavada, P.C.; Critzer, F.J. Psychrotrophic microrganisms. In *Compendium of Methods for the Microbiological Examination of Foods*, 5th ed.; Salfinger, Y., Tortorello, M.L., Eds.; American Public Health Association: Washington, DC, USA, 2015; Chapter 13; pp. 175–189.
14. *ISO 6888-1*; Microbiology of Food and Animal Feeding Stuffs—Horizontal Method for the Enumeration of Coagulase-Positive Staphylococci (Staphylococcus aureus and Other Species)—Part 1: Technique Using Baird-Parker Agar Medium. International Organization for Standardization: Geneva, Switzerland, 1999.
15. Mossel, D.A.A.; Mengerink, W.H.J.; Scholts, H.H. Use of a modified MacConkey agar medium for the selective growth and enumeration of Enterobacteriaceae. *J. Bacteriol.* **1962**, *84*, 381. [CrossRef]
16. Commission Internationale de l'Eclairage. *Colorimetry*, 2nd ed.; Publication CIE No. 15.2; Central Bureau of the CIE: Vienna, Austria, 1986.
17. Hernández, M.D.; López, M.B.; Álvarez, A.; Ferrandini, E.; García, B.G.; Garrido, M.D. Sensory, physical, chemical and microbiological changes in aquacultured meagre (*Argyrosomus regius*) fillets during ice storage. *Food Chem.* **2009**, *114*, 237–245. [CrossRef]
18. Pikul, J.; Leszczynski, D.E.; Kummerow, F.A. Evaluation of three modified TBA methods for measuring lipid oxidation in chicken meat. *J. Agric. Food Chem.* **1989**, *37*, 1309–1313. [CrossRef]
19. Meilgaard, M.; Civille, G.V.; Carr, B.T. *Sensory Evaluation Techniques*, 5th ed.; CRC Press: Boca Raton, FL, USA, 2015.
20. Kruskal, W.H.; Wallis, W.A. Use of ranks in one-criterion variance analysis. *J. Am. Stat. Assoc.* **1952**, *47*, 583–621. [CrossRef]
21. Junior, P.G.; Assunção, A.W.; Baldin, J.C.; Amaral, L.A. Microbiological quality of whole and filleted shelf-tilapia. *Aquaculture* **2014**, *433*, 196–200. [CrossRef]
22. Ferreira, A.C.A.; da Silva Monteiro, E.; de Oliveira Sousa, D.; de Souza Silva, C.M.; da Silva, I.C.R.; Orsi, D.C. Microbiological quality assessment of fresh tilapia marketed in the Federal District and of the ice used for its conservation. *Sci. Plena* **2021**, 17. [CrossRef]
23. Bhatt, A.; Arora, P.; Prajapati, S.K. Unveiling mechanistic intricacies of *Chlorella pyrenoidosa*-mediated pathogen removal from sewage. *Npj Clean Water* **2024**, *7*, 116. [CrossRef]
24. Soares, V.M.; Pereira, J.G.; Izidoro, T.B.; Martins, O.A.; Pinto, J.P.D.A.N.; Biondi, G.F. Qualidade microbiológica de filés de peixe congelados distribuídos na cidade de Botucatu-SP. *UNOPAR Cien.Ciências Biol. Saúde* **2011**, *13*, 85–88.
25. Stejskal, N.; Miranda, J.M.; Martucci, J.F.; Ruseckaite, R.A.; Barros-Velázquez, J.; Aubourg, S.P. Quality enhancement of refrigerated hake muscle by active packaging with a protein concentrate from *Spirulina platensis*. *Food Bioprocess Technol.* **2020**, *13*, 1110–1118. [CrossRef]
26. Santos, F.M.S.; da Silva, A.I.M.; Vieira, C.B.; de Araújo, M.H.; da Silva, A.L.C.; Carneiro-da-Cunha, M.D.G.; Bezerra, R.S. Use of chitosan coating in increasing the shelf life of liquid smoked Nile tilapia (*Oreochromis niloticus*) fillet. *J. Food Sci. Technol.* **2017**, *54*, 1304–1311. [CrossRef]
27. Ali, I.H.; Doumandji, A. Comparative phytochemical analysis and in vitro antimicrobial activities of the cyanobacterium *Spirulina platensis* and the green alga *Chlorella pyrenoidosa*: Potential application of bioactive components as an alternative to infectious diseases. *Bull. De Linstitut Sci. Rabat Sect. Sci. Vie* **2017**, *39*, 41–49.
28. Atitallah, A.B.; Barkallah, M.; Hentati, F.; Dammak, M.; Hlima, H.B.; Fendri, I.; Abdelkafi, S. Physicochemical, textural, antioxidant and sensory characteristics of microalgae-fortified canned fish burgers prepared from minced flesh of common barbel (*Barbus barbus*). *Food Biosci.* **2019**, *30*, 100417. [CrossRef]
29. Acco, M.; Ferreira, F.S.; Henriques, J.A.P.; Tondo, E.C. Identification of multiple strains of *Staphylococcus aureus* colonizing nasal mucosa of food handlers. *Food Microbiol.* **2003**, *20*, 489–493. [CrossRef]
30. Hematyar, N.; Rustad, T.; Sampels, S.; Kastrup Dalsgaard, T.K. Relationship between lipid and protein oxidation in fish. *Aquac. Res.* **2019**, *50*, 1393–1403. [CrossRef]

31. Shafiei, R.; Mostaghim, T. Improving shelf life of calf fillet in refrigerated storage using edible coating based on chitosan/natamycin containing *Spirulina platensis* and *Chlorella vulgaris* microalgae. *J. Food Meas. Charact.* **2022**, *16*, 145–161. [CrossRef]
32. Gill, T.A. Objective analysis of seafood quality. *Food Rev. Int.* **1990**, *6*, 681–714. [CrossRef]
33. Özogul, F.; Polat, A.; Özogul, Y. The effects of modified atmosphere packaging and vacuum packaging on chemical, sensory and microbiological changes of sardines (*Sardina pilchardus*). *Food Chem.* **2004**, *85*, 49–57. [CrossRef]
34. Khalafalla, F.A.; Ali, F.H.; Hassan, A.R.H. Quality improvement and shelf-life extension of refrigerated Nile tilapia (*Oreochromis niloticus*) fillets using natural herbs. *Beni-Suef Univ. J. Basic Appl. Sci.* **2015**, *4*, 33–40. [CrossRef]
35. Kalinowski, C.T.; Robaina, L.E.; Fernandez-Palacios, H.; Schuchardt, D.; Izquierdo, M.S. Effect of different carotenoid sources and their dietary levels on red porgy (*Pagrus pagrus*) growth and skin colour. *Aquaculture* **2005**, *244*, 223–231. [CrossRef]
36. Majdinasab, M.; Hosseini, S.M.H.; Sepidname, M.; Negahdarifar, M.; Li, P. Development of a novel colorimetric sensor based on alginate beads for monitoring rainbow trout spoilage. *J. Food Sci. Technol.* **2018**, *55*, 1695–1704. [CrossRef]
37. Alfaia, C.M.; Pestana, J.M.; Rodrigues, M.; Coelho, D.; Aires, M.J.; Ribeiro, D.M.; Prates, J.A.M. Influence of dietary *Chlorella vulgaris* and carbohydrate-active enzymes on growth performance, meat quality and lipid composition of broiler chickens. *Poult. Sci.* **2021**, *100*, 926–937. [CrossRef]
38. An, B.K.; Kim, K.E.; Jeon, J.Y.; Lee, K.W. Effect of dried *Chlorella vulgaris* and *Chlorella growth* factor on growth performance, meat qualities and humoral immune responses in broiler chickens. *Springerplus* **2016**, *5*, 718. [CrossRef]
39. Fradique, M.; Batista, A.P.; Nunes, M.C.; Gouveia, L.; Bandarra, N.M.; Raymundo, A. Incorporation of *Chlorella vulgaris* and *Spirulina maxima* biomass in pasta products. Part 1: Preparation and evaluation. *J. Sci. Food Agric.* **2010**, *90*, 1656–1664. [CrossRef] [PubMed]
40. Mattje, L.G.B.; Tormen, L.; Bombardelli, M.C.M.; Corazza, M.L.; Bainy, E.M. Ginger essential oil and supercritical extract as natural antioxidants in tilapia fish burger. *J. Food Process. Preserv.* **2019**, *43*, 13942. [CrossRef]
41. López-Cánovas, A.E.; Cabas, I.; Chaves-Pozo, E.; Ros-Chumillas, M.; Navarro-Segura, L.; López-Gómez, A.; Fernandes, J.M.O.; Galindo-Villegas, J.; García-Ayala, A. Nanoencapsulated clove oil applied as an anesthetic at slaughtering decreases stress, extends the freshness, and lengthens shelf life of cultured fish. *Foods* **2020**, *9*, 1750. [CrossRef] [PubMed]

foods

MDPI

Article

Preservative Potential of *Anethum graveolens* Essential Oil on Fish Fillet Quality and Shelf Life During Refrigerated Storage

Aya Tayel, Faten S. Hassanin, Shimaa N. Edris, Ahmed Hamad and Islam I. Sabeq *

Department of Food Hygiene and Control, Faculty of Veterinary Medicine, Benha University, Benha 13511, Egypt; aya.magdy@fvtm.bu.edu.eg (A.T.); faten.mohamed@fvtm.bu.edu.eg (F.S.H.); shimaa.edrees@fvtm.bu.edu.eg (S.N.E.); ahmed.alhussaini@fvtm.bu.edu.eg (A.H.)

* Correspondence: islam.sabek@fvtm.bu.edu.eg

Abstract: This study estimated the preservative potential of Dill essential oil (DEO, *Anethum graveolens*) in terms of the quality and shelf life of *Pangasius bocourti* (basa fish) fillets during cold storage. GC-MS analysis of DEO's chemical composition identified monoterpenes, including α-phellandrene (21.81%), d-limonene (18.54%), carvone (17.42%), and Dill ether (14.82%). DEO showed concentration-dependent antioxidant properties in the DPPH assay, with an IC50 of 48.3 ± 0.9 μg/mL (mean ± SE). Its antibacterial efficacy against various foodborne pathogens was evaluated using the resazurin turbidimetric microdilution method. Fish fillets were treated with DEO at 200, 2000, and 4000 ppm, and compared to the untreated control and 200 ppm butylhydroxytoluene (BHT)-treated groups. Physicochemical parameters, microbial growth, and sensory characteristics were assessed over a 15-day period at 2.5 °C ± 0.5 °C. Higher concentrations of DEO effectively preserved the pH, water-holding capacity, and color stability of the fillets. Microbial analysis showed that DEO, particularly at 4000 ppm, significantly inhibited the growth of aerobic bacteria, lactic acid bacteria, coliforms, and staphylococci compared with the control. Sensory evaluation revealed that DEO treatment, especially at 4000 ppm, maintained the odor, color, texture, and overall acceptability of fish fillets throughout storage. These results suggest that *Anethum graveolens* L. essential oil can serve as an effective natural preservative to enhance the quality and prolong the shelf life of refrigerated fish fillets.

Keywords: fish fillet; dill essential oil; food safety; food quality; natural preservative; quality preservation

1. Introduction

Recognized for their health benefits, fish and seafood products are crucial components of human nutrition, supplying at least 20% of the protein consumption for one-third of the global population. The reliance on these food sources is particularly pronounced in developing nations [1]. Fish has been demonstrated to play a crucial role in human nutrition, offering significant advantages for both food security and addressing malnutrition and micronutrient deficiencies in less-developed nations. The value of incorporating fish into diets has been well established, particularly in its ability to combat nutritional challenges faced by developing countries [2]. Fish also has beneficial characteristics, including nutrient-dense protein content and unique flavor. Fish has a clear advantage over other animal and plant products because of the vast diversity of species and wide range of prices. This variety makes fish accessible to individuals across all economic levels, including the high-, middle-, and low-income groups. Consequently, fish has become one of the most widely consumed sources of animal protein [1]. Seafood, especially fish, is highly susceptible to spoilage owing to enzymatic and microbiological degradation. This rapid deterioration can lead to waste with fish product losses potentially reaching up to 40% [3]. Fish and other seafood are particularly prone to decay caused by enzymes and microorganisms. This quick decomposition can result in significant waste, with potential losses of fish products as high as 40% [4]. Fish fillets rapidly deteriorate in nutritional quality, even when stored in refrigeration, due to various processes, including enzymatic browning, non-enzymatic lipid oxidation, and protein breakdown. This deterioration is one of the adverse effects of spoilage and can lead to both health-related and economic issues [5]. Chilling and freezing are the conventional methods for preserving fish fillets. While chilled fish can retain a high sensory quality that consumers find acceptable, it faces potential microbial safety issues due to the temperature range in which it is stored. This is because psychotropic pathogens can proliferate without noticeably affecting the sensory characteristics of the fish [6]. Although freezing can extend the shelf life of fish and fish products, it may lead to changes in texture. This occurs because of the formation of ice crystals and denaturation of proteins, resulting in increased dryness and toughness. These effects are more pronounced in lean fish species compared to fatty or semi-fatty varieties [7].

The majority of preservatives used in meat products, especially in seafood, rely on chemical additives to prolong shelf life and improve sensory appeal [8]. Among the chemical preservatives used in seafood, butylated hydroxytoluene (BHT) stands out as the most frequently utilized option [9]. The utilization of BHT is attributed to its potent antioxidant properties and antimicrobial effects [8]. Nevertheless, BHT has been shown to pose risks to public health including its potential to cause toxic and cancer-inducing effects [10]. Due to these detrimental effects on both the sensory qualities of fresh seafood and consumer health, food safety authorities and certain nations have banned the use of these substances in meat production [10]. Furthermore, both consumers and professionals in the field have recognized the urgent importance of investigating and implementing "bio-preservatives" [11]. EOs and plant extracts have been described in [12].

Recent research has shown that plant essential oils contain antimicrobial properties that can reduce food spoilage bacteria both in vitro and in vivo [13]. Furthermore, the broad antimicrobial properties of essential oils may contribute to preserving seafood freshness, an aspect highly valued by consumers [14]. Moreover, the wide-ranging antimicrobial effects of essential oil could help maintain the freshness of seafood, a quality highly prized by customers [15].

The antioxidant activity of essential oils is mainly associated with the presence of oxygenated monoterpenes and other volatile bioactive compounds that can act as radical scavengers or metal chelators. Additionally, the presence of unpaired electrons allows phenolic compounds to form complexes with metal ions and oxygen, which contributes to the prevention of lipid oxidation [16]. The use of essential oils as antimicrobial agents offers two key advantages: their natural origin makes them generally safer for consumers compared to synthetic chemicals, and they have a reduced likelihood of promoting resistance to dangerous microorganisms. These benefits make essential oils an attractive alternative to conventional antimicrobial substances [17].

One of the annual plants is Dill (*Anethum graveolens* L.), which contains essential oil and belongs to the *Umbelliferae* family. Dill contains compounds such as essential oils, fatty acids, mineral elements (Mn, Ca, Na, Cu, Mg, Fe, K, and phosphorus), vitamins, carbohydrates, proteins, fiber, flavonoids, carotenoids, and phenolic compounds [18]. The primary constituents of DEO, including d-carvone, d-limonene, and α-phellandrene, are responsible for its antibacterial and antioxidant properties [19]. These compounds can protect lipids from oxidation by scavenging reactive oxygen species and free radicals, contributing to fish spoilage [19].

The antimicrobial and antioxidant Dill components enhance the shelf life, sensory, and physicochemical properties of Atlantic bonito fish using Dill leaves [20], fresh fish (*Cyprinus carpio*) meat quality [12], minced meat using Dill EO [21], beef burger using Dill EO [8], refrigerated storage beef using Dill EO [22], and cherry tomatoes using Dill EO [23].

These potent antibacterial qualities make Dill (*Anethum graveolens* L.) essential oil (DEO) a good candidate for prolonging the shelf life of fish products and could prevent the proliferation of foodborne pathogens. To date, few studies have assessed the preservation characteristics of *A. graveolens* EO in fish fillets. Therefore, the current study was conducted to assess the preservative effects of *Anethum graveolens* EO on fillets of Basa fish (*Pangasius bocourti*).

2. Material and Methods

2.1. Ethical Approval

The research protocol received approval from the Research Ethics Care and Use Committee at Banha University's Faculty of Veterinary Medicine, with the reference number BUFVTM 18-08-24.

2.2. Dill Essential Oil Preparation

At the beginning of the vegetative period, fresh Dill was acquired from a local market. The hydrodistillation process involved placing 50 g of powdered Dill and 750 mL of distilled water into a glass Clevenger apparatus. Essential oil extraction proceeded for 3 h at approximately 100 °C at a distillation rate of 1 mL/min. The resulting essential oil was then collected in pre-weighed vials, using a balance with 0.0001 precision, and stored at a temperature of 4 °C [24].

2.3. Determination of the Chemical Compounds of Dill Essential Oil

Gas Chromatography–Mass Spectrometry (GC-MS) Analysis

A Trace GC-TSQ mass spectrometer (Thermo Scientific, Austin, TX, USA) equipped with a direct capillary column TG–5MS (30 m × 0.25 mm × 0.25 μm film thickness) was utilized to analyze the chemical composition of the samples. The column oven temperature was initially set at 50 °C, then raised by 5 °C/min to 250 °C and maintained for 2 min. It was further increased to 300 °C at a rate of 30 °C/min and held for 2 min. The injector and MS transfer line temperatures were maintained at 270 °C and 260 °C, respectively. Helium served as the carrier gas, flowing at a constant rate of 1 mL/min. With a solvent delay of 4 min, diluted samples of 1 μL were automatically injected using an Autosampler AS1300 coupled with GC in split mode (Thermo Scientific, Austin, TX, USA). EI mass spectra were gathered in full scan mode at 70 eV ionization voltages across the m/z range of 50–650. The ion source temperature was configured at 200 °C. Component identification was achieved by comparing their mass spectra with those found in WILEY 09 and NIST 14 mass spectral databases [25].

2.4. Dill Oil Antioxidant Activity

The extract's antioxidant properties were evaluated at Al-Azhar University's Regional Center for Mycology and Biotechnology (RCMB) using the 2,2-diphenyle-1-picrylhydrazyl (DPPH) free radical scavenging assay. The test was conducted three times, and the mean values were used. The method for assessing DPPH radical scavenging activity followed the protocol outlined by Ang et al. (2015) [26] with some modifications. In summary, a fresh methanol solution (0.004% *w*/*v*) of 2,2-diphenyl-1-picrylhydrazyl (DPPH) radical was created and kept in dark conditions at 10 °C. The test compound was also dissolved in methanol. A 40 uL portion of this solution was introduced to 3 ml of the DPPH solution. A UV–visible spectrophotometer (Milton Roy, Spectronic 1201, Houston, TX, USA) was used to immediately record absorbance measurements. The decline in absorbance at 515 nm was continuously monitored, with data recorded every minute until stabilization occurred (16 min). The absorbance of DPPH radical without antioxidant (control) and the reference compound ascorbic acid were also measured. Each measurement was performed in triplicate, and the results are expressed as mean ± standard error (SE). The percentage inhibition (PI) of the DPPH radical was calculated according to the following formula:

$$\mathrm{PI} = [((AC - AT)/AC) \times 100]$$

where AC = absorbance of the control at t = 0 min and AT = absorbance of the sample +DPPH at t = 16 min.

2.5. In Vitro Antibacterial Activity

The minimum inhibitory concentration (MIC) of DEO was assessed using a resazurin-based turbidimetric microdilution method [27]. A stock solution containing 100 mg/mL of Dill essential oil was prepared using sterile distilled water. The initial column of a 96-well plate received 100 μL of this solution, which was then serially diluted twice across Columns 2 to 10. Following this, 50 μL of the stock bacterial suspension (6 log CFU/mL) was added to each well in Columns 1–10, yielding a final concentration of 5–5.7 log CFU/mL. The tested Gram-negative strains included *Escherichia coli* ATCC 25,922 and four field isolates of Salmonella enterica—BS26, BS29, N7, and N9—that were previously isolated and identified by Sabeq et al. (2022) [28] and Gamil et al. (2024) [29]. The Gram-positive bacterial panel consisted of *Listeria monocytogenes* ATCC 19,115 and two strains of *Staphylococcus aureus*, ST62 and NC15. Column 11 contained 100 μL of the diluted standardized inoculum, while Column 12 held 100 μL of medium broth as a sterility control. After incubation at 37 °C for 24 h, 30 μL of resazurin (0.015%) was introduced to each well. The wells were then incubated for an additional 2 to 4 h to monitor color changes. Columns that retained the blue resazurin color and showed no change at the end of the incubation period were considered to have surpassed the MIC value. To determine the minimum biocidal concentration (MBC), the contents of wells with concentrations exceeding the MIC value were immediately plated.

2.6. Fish Sample Collection and Preparation

The research utilized fresh Basa fish fillets (*Pangasius bocourti*) acquired from a seafood supplier in Banha City, Egypt, during October and November 2024. A total of 7.5 kg of fillets were promptly transported to the laboratory in ice at 0 °C. The fillets, weighing 300 g each, were randomly allocated into five groups. The control samples underwent immersion in sterile distilled water, while BHT-treated fillets were submerged in a 200 ppm (*w*/*v*) BHT solution. The DEO dipping solution was created by dissolving DEO in a 0.8% (*w*/*v*) tween 80 solution using a homogenizer mixer (Hielscher, Teltow, Germany) at 1500 rpm for 5 min. The remaining three groups were treated with Dill essential oil (EO) at varying concentrations: 200 ppm (DEO-1), 2000 ppm (DEO-2), and 4000 ppm (DEO-3). All samples were immersed for a duration of 30 min (Figure 1).

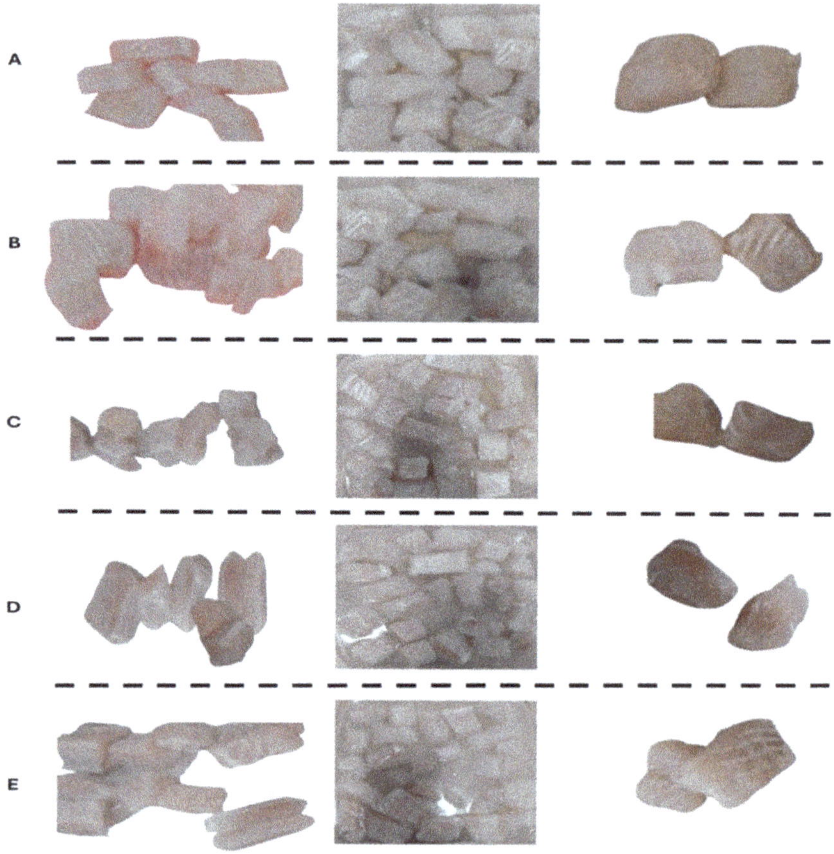

Figure 1. Fish sample preparation and treatment using BHT and three different concentrations of Dill essential oil. (**A**) Control group; (**B**) BH-treated group; (**C**) Dill essential oil-treated group (200 ppm); (**D**) Dill essential oil-treated group (2000 ppm); and (**E**) Dill essential oil-treated group (4000 ppm).

After extraction from the dipping solutions, the specimens were allowed to drain on a sanitized sieve for 15 min. They were then placed in dual zipper food-grade low-density polyethylene (LDPE) bags and distributed across 6 monitoring locations. The samples were stored in a programmable cooling incubator at 2.5 ± 0.5 °C (BINDER GmbH, Tuttlingen, Germany). At designated intervals (0, 3, 6, 9, 12, and 15 days post-treatment), the samples underwent analysis for physicochemical properties, microbial load, and sensory attributes throughout the 15-day period. The experiment was conducted in duplicate.

2.7. Physicochemical Analysis of Fish Fillets

2.7.1. pH Measurement

In summary, the pH of each fish fillet sample was directly assessed using electrodes from a pH meter (Jenway 3510 pH meter, Cole-Parmer, Staffordshire, UK). The device was calibrated at ambient temperature using three distinct pH levels (10, 4, and 7) in combination with a metal temperature probe [30].

2.7.2. WHC Estimation

The filter paper press method (FPPM) was employed to assess the water-holding capacity (WHC) of fish fillets. This technique involved applying a 5 kg weight for 30 s to compress a 0.2–0.5 g sample on Whatman No. 1 filter paper. The WHC of the meat was determined by calculating the percentage of water retained in the sample after deducting the forced loose moisture from its initial weight [31].

2.7.3. Purge Loss (PL) and Cooking Loss (CL) Estimation

For the assessment of purge loss, two cuboidal samples weighing 50 ± 5 g each were designated. The evaluation of purge loss during various storage intervals was conducted by calculating the percentage decrease in fish fillet weight from the initial measurement taken on the first day of refrigeration (at 0, 3, 6-, 9-, 12-, and 15-days following treatment) [32]. Following the estimation of purge loss at each storage interval, the same two samples were utilized to assess cooking loss (CL). Fish fillets, previously weighed and shaped into cuboids (50 ± 5 g), were individually sealed in thin-walled, heat-resistant plastic bags. These were then immersed in a water bath at 80 °C for 10 min. After heating, the samples were cooled to room temperature using tap water, further chilled to 5 °C in an ice bath, dried, and weighed again. The CL was calculated as the percentage difference between the initial raw weight and the final cooked weight [32].

2.7.4. Warner–Bratzler Shear Force (WBSF)

The Warner–Bratzler Shear Force (WBSF) measurement was conducted on cooked fish fillets utilizing the 3343 Universal Test Device Mono column (Instron, Norwood, MA, USA). The samples were cut perpendicular to the muscle fibers' longitudinal orientation. The WBSF value, expressed in kilogram-force (KGF), was calculated as the mean of six core measurements taken from each fish fillet sample [33].

2.7.5. Instrumental Color Analysis

The chromometer CR-410 (Konica Minolta Sensing INC., Osaka, Japan) was utilized to measure three color parameters: L^*, a^*, and b^* in the samples. The device was set to the L^*, a^*, b^* color space with illuminant D65, using a 2° observer angle and an 8.0 mm aperture size with a closed cone. Prior to taking measurements, the instrument was calibrated using a standardized white tile. Readings were taken across the cut surface of peeled shrimp after allowing it to bloom for 30 min. The obtained color values were then employed to calculate color saturation (Hue angle (h°) = arctg b^*/a^*) and color intensity (C = $(a^{*2} + b^{*2})^{0.5}$). For each group, an average of six measurements was taken. Increased Chroma levels signify higher saturation of the sample's primary Hue, while greater Hue angle (or color intensity) values indicate a reduced amount of meat [34], and the total color difference (ΔE), which indicates the amount of color difference between shrimp before and after storage, was calculated as follows [35]:

$$(\Delta E) = [(L^* - L^*{}_0)2 + (a^* - a^*{}_0)\,2 + (b^* - b^*{}_0)2]^{1/2}$$

Whiteness indices (WIs) are mathematical formulas that combine measurements of lightness, yellow, and blue into a single value to assess the level of whiteness [36]. The WI was determined according to the following formula reported in [37]:

$$\mathrm{WI} = 100 - \sqrt{(100 - L)^2 + a^2 + b^2}$$

And the yellowness index (YI) measures how much a sample's surface deviates from pure white in terms of its yellow coloration [20]:

$$\text{YI} = 142.86 * \frac{b}{L}$$

Browning index (BI) $= \frac{100}{0.17}\left(\frac{a^*+1.75L^*}{5.645L^*+a^*-0.012b^*} - 0.31\right)$ [38]

2.8. *Fish Fillet Microbial Analysis*

The aerobic plate count (APC), coliform count, lactic acid bacteria count, and staphylococcal count of Basa fillets generated from compared groups were assessed over 12 days (0, 3, 6, 9, and 12 days) in a binder incubator (BINDER GmbH, Tuttlingen, Germany) at 2.5 ± 0.5 °C.

2.8.1. Determination of Aerobic Plate Count

The aerobic plate count (APC) in Basa fillet samples was assessed in the same manner as for ground beef products [39]. To prepare each sample, a 10% homogenate was created by combining 10 g of the sample with 90 mL of sterile distilled water using a Stomacher 400R (Seward, West Sussex, UK) under aseptic conditions. The resulting homogenates underwent serial tenfold dilutions in sterile distilled water. For each dilution, 1 mL was spread on the surface of two separate sterile plate count agar plates (Condalab, Madrid Spain). Following inoculation and solidification, the plates were incubated at 37 °C for 24 h [40]. Colonies were counted and reported as log colony-forming units per gram of food (CFU/g).

2.8.2. Lactic Acid Bacteria Count (LAB)

To identify lactic acid bacteria (LAB), researchers utilized Man, Rogosa, and Sharpe medium agar (MRS, HiMedia, Kennett Square, PA, USA) with previously prepared tenfold serial dilutions. The samples were applied using the spread plating technique and then placed in a Gas Pak Jar for anaerobic incubation at 30 °C for a duration of 72 h [39]. The number of colonies was measured and expressed as log CFU/g of food.

2.8.3. Determination of Coliform Count

For coliform enumeration, 1 mL of previously prepared tenfold serial dilutions from Basa fillet homogenates was introduced into two separate sterile Petri dishes containing Violet red bile agar (Himedia Laboratories, Maharashtra, India). The dishes were then incubated at 37 °C [41]. The number of colonies was measured and expressed as log CFU/g of food.

2.8.4. Determination of *Staphylococcus* Count

Staphylococcus counts in Basa fillets were determined using the surface-plating method on a Baird Parker agar plate (Oxoid, Hampshire, UK), as previously published for milk [42]. A sterile disposable spreader was employed to evenly distribute one milliliter of each previously prepared serial dilution across a Baird Parker agar plate. The inoculated plates were then placed upright in an incubator for a period ranging from approximately 10 min to one hour, allowing sufficient time for the agar to absorb the inoculum. The injections were inverted and incubated at 37 °C for 48 h [43]. The number of colonies was measured and expressed as log colony-forming units (CFUs) per gram of food.

2.9. Sensory Evaluation

Sensory evaluation of raw fish fillets was conducted by ten-member panels that underwent necessary training and testing. The training was considered complete when the panelists were comfortable with the evaluation process and individual scores were within one unit of the mean score. The laboratory assessment involved serving representative fish fillet samples on porcelain plates in an open area, without disclosing the treatment type, with three replicates per sample. With the provision of proper natural light, panelists employed a nine-point hedonic scale to evaluate freshness, assigning scores from 1 to 9 for each attribute based on sensory quality criteria. The evaluation encompassed color, odor, appearance, and texture. Overall sensory quality scores were used to categorize fish fillets as very good, good, acceptable, unacceptable, or bad, corresponding to the scale: 9 (like extremely), 8 (like very much), 7 (like moderately), 6 (like slightly), 5 (neither like nor dislike), 4 (dislike slightly), 3 (dislike moderately), 2 (dislike very much), and 1 (dislike extremely). The samples were evaluated in their natural state without additional preparation [44].

2.10. Statistical Analysis

SPSS Version 22 (SPSS Inc. Chicago, IL, USA) was used for data analysis. The effects of antimicrobial dipping, storage periods (1, 3, 6, 9, 12, and 15 d), and their interaction on the fish fillet's physicochemical, microbiological, and antioxidant attributes were examined using general linear models (GLMs), with fillets being treated as a random variable and antimicrobial dipping and chilling time as fixed effects. The means and standard errors of the results were determined. The statistical model employed Tukey's b multiple comparison test to assess the impact of antimicrobial dipping compared to the control and compare different monitoring point averages within the same group. Significant differences were defined as $p < 0.05$. All experiments were performed in triplicate, and data are presented as mean $\pm$ standard error (SE). For each treatment group and storage day, three fish fillet samples ($n = 3$) were used for microbial and physicochemical analyses, including pH, water-holding capacity, cooking and purge losses, shear force, and color measurements. Antioxidant assays (e.g., DPPH) and MIC determinations were conducted using three independent replicates ($n = 3$) for each concentration. For the sensory evaluation, a trained panel of ten assessors ($n = 10$) evaluated the samples at designated intervals.

Principal component analysis (PCA) was employed to assess the relationships among various physicochemical and microbiological parameters of fish meat samples treated with different antioxidants, including DEO at three concentrations, a control (CON), and BHT (a synthetic antioxidant). The analyzed parameters included color change, pH, water-holding capacity (WHC), CL, PL and microbial assessment. Prior to the PCA, all data were standardized to ensure comparability across variables with different measurement scales. The PCA was conducted using R (FactoMineR package, Version 4.5.0), where the covariance matrix was computed, and eigenvalues and eigenvectors were extracted to determine the principal components (PCs). The selection of significant PCs was based on eigenvalues greater than 1 (Kaiser's criterion) and scree plot analysis. The contribution of each variable to the total variance was assessed through loading scores, and a biplot was generated to visualize the clustering of treatments and the influence of different parameters. Bartlett's test of sphericity and the Kaiser–Meyer–Olkin (KMO) measure of sampling adequacy were performed to validate the appropriateness of the PCA, with a KMO value greater than 0.7 indicating suitability. The results of the PCA provided insights into the major factors influencing the quality attributes of fish meat under different antioxidant treatments.

3. Results

3.1. Dill Essential Oil Chemical Composition

Gas chromatography–mass spectrometry (GC/MS) analysis of DEO provided a comprehensive chemical profile of its constituents, offering insights into its potential functional and therapeutic applications (Table 1). GC/MS analysis revealed the presence of multiple bioactive compounds in DEO, each characterized by distinct retention times and mass spectral patterns. The chemical profile indicated that monoterpenes dominated the volatile fraction and accounted for most of the identified compounds.

Table 1. Chemical composition of Dill essential Oil.

No.	Compound	Chemical Family	RT	RI	Area, %
1	2-thujene	Monoterpene	4.71	915	0.32
2	β-ocimene	Monoterpene	4.84	944	1.84
3	Sabinene	Monoterpene	5.60	820	0.17
4	α-pinene	Monoterpene	6.04	886	0.36
5	β-pinene	Monoterpene	6.04	878	0.36
6	α-phellandrene	Monoterpene	6.33	938	21.81
7	p-cymene	Monoterpene	6.71	936	8.89
8	d-limonene	Monoterpene	6.93	941	18.54
9	Dill ether	Monoterpene	10.83	947	14.82
10	Camphor	Monoterpene	11.10	911	0.93
11	Camphene	Monoterpene	11.33	883	0.61
12	Carvone	Monoterpene	12.21	933	17.42
13	Terpineol	Monoterpene	12.48	868	1.21
14	Carvyl acetate	Monoterpene	14.69	779	1.73
15	Isodillapiole	phenylpropanoid	16.38	700	0.58
16	α-sinensal	Sesquiterpenoid	26.77	810	0.23
17	Apiol	Phenylpropene	21.72	879	10.30

3.2. Dill Oil Antioxidant Activity

The antioxidant activity of DEO was evaluated using the DPPH (2,2-diphenyl-1-picrylhydrazyl) radical scavenging assay (Figure 2). The DEO demonstrated dose-dependent antioxidant activity, with DPPH scavenging percentages ranging from 17.6% at 1.95 μg/mL to 80.4% at 1000 μg/mL. The IC50 value, which represents the concentration required to inhibit 50% of DPPH radicals, was determined to be 48.3 ± 0.9 μg/mL.

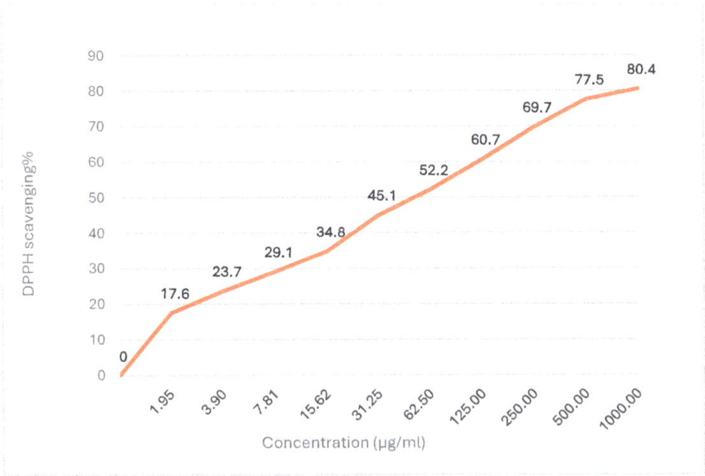

Figure 2. DDPH scavenging % of Dill essential oil.

3.3. *In Vitro Antimicrobial Assay*

Dill (*Anethum graveolens* L.) essential oil exhibited moderate antibacterial activity, particularly against Gram-positive bacteria such as *Listeria monocytogenes* and *Staphylococcus aureus*, with MIC values as low as 0.195 mg/mL (Figure 3). The antibacterial activities of DEO on Gram-negative bacteria, including *E. coli* and Salmonella isolates, were less pronounced, requiring higher concentrations (up to 6.25 mg/mL) for growth inhibition. The MIC results revealed varied antimicrobial activity of DEO across the tested bacterial strains. For *Escherichia coli* ATCC 25922, growth inhibition was observed at concentrations $\geq$ 3.9 mg/mL. For the *Salmonella enterica* isolates BS26 and BS29, the MIC was 6.25 mg/mL, while the MIC of both the N7 and N9 isolates was 3.12 mg/mL. In contrast, the *Listeria monocytogenes* ATCC 19,115 and *Staphylococcus aureus* field isolates ST62 and NC15 were more sensitive, with MICs as low as 0.195 mg/mL. These concentrations were calculated based on a stock solution of DEO prepared at 100 mg/mL (Figure 3).

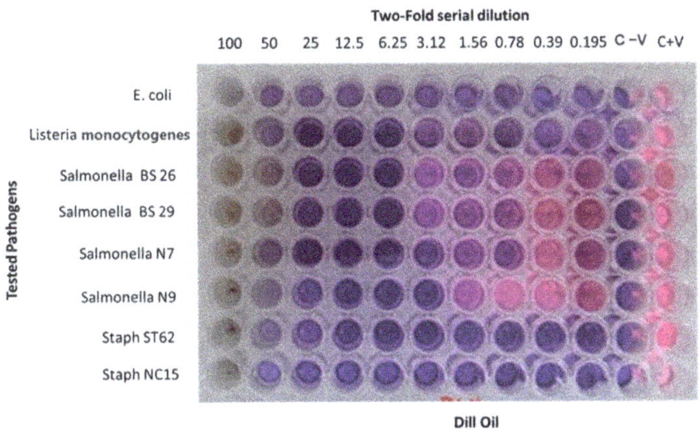

Figure 3. Microdilution assay for determining the minimum inhibitory concentration (MIC) of Dill essential oil (DEO) against bacterial strains.

3.4. Physicochemical Analysis of Fish Fillets

Statistical results indicated a significant interaction between treatment and storage time, affecting all estimated physicochemical attributes and color ($p < 0.05$). DEO treatment influenced all physicochemical attributes except CL and DL. Additionally, all physicochemical attributes were significantly affected by the storage interval ($p < 0.05$). The pH values of the fish samples varied significantly across the different treatment groups and storage times ($p < 0.001$) (Figure 4). On Day Zero, the highest pH was observed in the DEO-3 group (9.24), followed by DEO-2 (8.94) and BHT (8.87), whereas the lowest pH was recorded in the DEO-1 group (8.44). By the third day, the pH values remained significantly different, with BHT-treated samples exhibiting the highest pH (9.02) and DEO-1 samples displaying a lower pH (8.74) ($p < 0.05$). On the sixth day post-treatment, All DEO-supplied groups, particularly DEO-3, had lower pH values than both the control and BHT ($p < 0.05$). By the ninth day, the control and DEO-1 groups had the highest pH values (9.31 and 9.33, respectively), while BHT-treated samples exhibited a significantly lower pH (9.10) ($p < 0.05$), indicating better preservation. On the twelfth day, all groups exhibited comparable pH values, ranging from 8.44 to 8.77, demonstrating a general stabilization of pH across treatments. However, on day 15, the control group exhibited the highest pH (9.67), indicative of spoilage progression, whereas the BHT-treated samples maintained a significantly lower pH (9.00), reinforcing their effectiveness in preserving fish quality. The overall trends suggest that the BHT and DEO treatments influenced pH fluctuations over time, with BHT being more effective in maintaining stability, whereas DEO treatments showed concentration-dependent variations. Moreover, WHC values varied on the first day, while all other groups were not different from the control ($p > 0.05$). On the third day post-treatment, DEO-2-supplied fish exhibited the highest WHC value compared to all other groups ($p < 0.05$). On all of the days following post-treatment, the BHT- and DEO-treated groups had similar WHC values compared to the control ($p > 0.05$). The longer storage period decreased the WHC values in the control fish, which was obvious in the last storage period compared to the first day ($p < 0.05$). Compared to the first day, all other fish treated with BHT or different DEO levels, particularly DEO-1, exhibited similar WHC values across the entire storage period ($p > 0.05$). Purge loss was not affected by BHT and/or DEO treatment ($p > 0.05$), except on the third and ninth days post-treatment where DEO-3- and BHT-supplied fish exhibited higher purge loss than the control ($p < 0.05$), respectively. All DEO-treated fish exhibited similar purge loss (PL) values across the entire storage period compared to the first day ($p > 0.05$). However, the control fish showed an ascending curve that reached the peak at six days post-treatment ($p < 0.05$) and then dropped to similar loss values compared to the first day ($p > 0.05$). BHT-supplied fish, compared to the first day, exhibited a declining PL curve and reached the lowest losses at 12 and 15 days post-treatment ($p < 0.05$). Cooking losses (CLs) were not affected by BHT and/or DEO treatment ($p > 0.05$), except on the 12th day post-treatment where BHT-supplied fish exhibited higher loss than all other groups ($p < 0.05$). Compared to the first day, all groups did not show significant differences in CLs across the entire storage period ($p > 0.05$), except DEO-1- and DEO-2-supplied fish ($p < 0.05$). The DEO-1-supplied fish showed an ascending CL curve and reached the peak a day six post-treatment, followed by a sharp declining trend compared to the first day ($p < 0.05$). While DEO-2-supplied fish had similar CL values until the ninth day post-treatment ($p > 0.05$), followed by a sharp lower CL trend on the last two storage periods compared to the first day ($p < 0.05$).

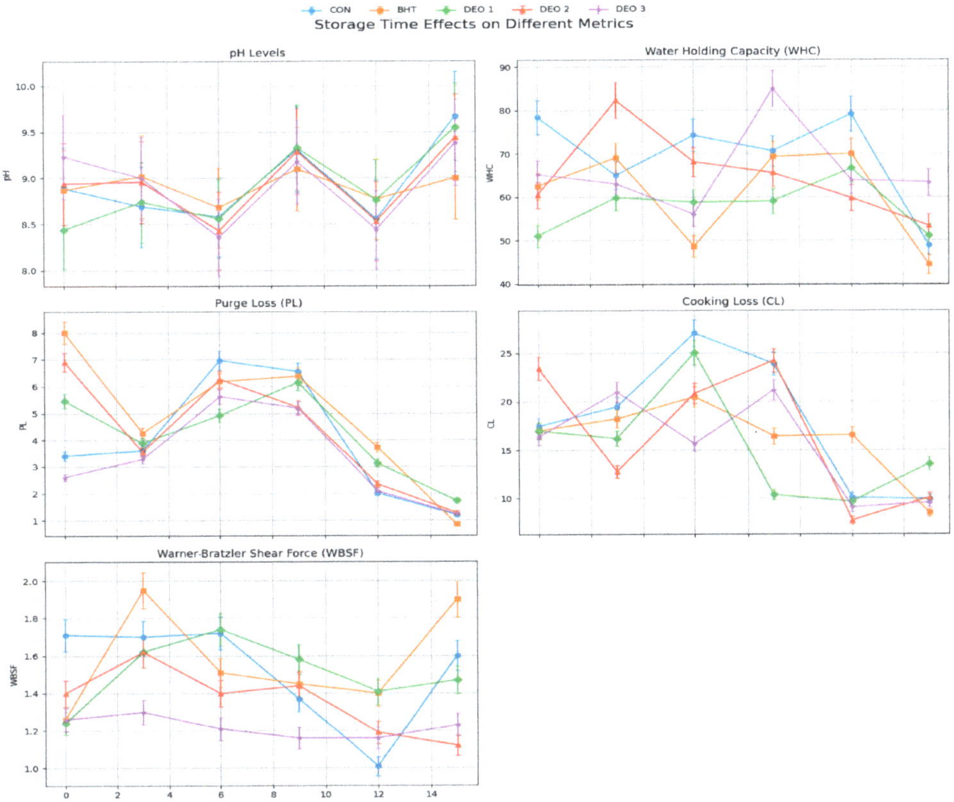

Figure 4. Changes in physicochemical parameters over time in fish subjected to different Dill essential oil treatments.

The Warner–Bratzler Shear Force (WBSF) values were assessed at multiple time points to evaluate the impact of different antioxidant treatments (control, BHT, DEO-1, DEO-2, and DEO-3) on texture retention (Figure 4). In the initial measurement (D-Zero), the control group exhibited the highest WBSF value (1.71), indicating a firmer texture. In contrast, fish treated with BHT (1.26), DEO-1 (1.24), and DEO-3 (1.26) demonstrated significantly lower WBSF values, suggesting an initial softening effect compared with the control ($p < 0.05$). Over time, the control group exhibited a declining WBSF trend, reaching its lowest value at D-12 (1.01) before slightly increasing at D-15 (1.60). This pattern indicates loss of firmness, likely due to protein degradation and structural changes during storage. The BHT-treated group showed fluctuations in WBSF, with a notable peak at D-3 (1.95) and another increase at D-15 (1.90). This suggests that BHT contributed to the preservation of structural integrity, delaying the softening process. DEO-1-treated samples exhibited moderate fluctuations, with a peak at D-9 (1.58) and a subsequent increase at D-15 (1.47), indicating partial resistance to texture deterioration. Similarly, DEO-2-treated fish maintained a relatively stable pattern, peaking at D-12 (1.41) and decreasing at D-15 (1.12), suggesting that DEO-2 delayed texture degradation until later storage periods. The most distinct pattern was observed in the DEO-3-treated group, which demonstrated an ascending WBSF trend, reaching its peak at D-6 (1.74) before declining at D-9 (1.16). This suggests that DEO-3 initially enhanced muscle firmness, but this effect diminished after prolonged storage.

Statistical analysis confirmed that treatment (G), time (T), and their interaction (G*T) significantly influenced the WBSF values ($p < 0.001$), highlighting the role of antioxidant supplementation in texture modification.

Concerning meat color, it was noted that DEO-3-supplied fish exhibited the lowest lightness (L^*) ($p < 0.05$) on the first day; at the same time, both the BHT group and other DEO-supplied groups had higher L^* compared to the control ($p < 0.05$) (Figure 5). On the third day post-treatment, the BHT group and all DEO-supplied groups had higher L^* compared to the control, particularly DEO-2 ($p < 0.05$). On the ninth day post-treatment, it was noted that DEO-2-supplied fish exhibited the lowest L^* ($p < 0.05$), while all other groups were not different from the control ($p > 0.05$). On the twelfth day post-treatment, all groups had higher L^* ($p < 0.05$) compared to the control. On the 15th day post-treatment, it was noted that L^* sharply decreased in BHT-supplied fish ($p < 0.05$), while DEO-supplied fish displayed higher values compared to the control, clearly in DEO-2 ($p < 0.05$). Compared to the first day, all fish groups, except DEO-2-supplied fish, changed with a longer storage period ($p < 0.05$). Also, all fish groups displayed lower L^* values on the third day post-treatment ($p < 0.05$) compared to the first day, and then started to ascend followed by declining curves that reached their peaks either on the sixth day post-treatment in DEO-3 or on the ninth day post-treatment in the control, BHT, and DEO-1-supplied fish ($p < 0.05$). Moreover, the redness index (a^*) was not affected by any of the treatments on the first day compared to the control ($p > 0.05$) (the redness index was higher in DEO-2 than the BHT-supplied fish). On the following days, changes started to appear where the BHT-, and DEO-1-treated fish had lower and higher a^* values compared to the control, respectively ($p < 0.05$). On the sixth day post-treatment, the BHT-treated fish still had a lower a^* value, while the DEO-3-supplied fish exhibited the highest a* value compared to the control ($p < 0.05$). On the ninth day post-treatment, all treated fish groups, particularly DEO-3, exhibited higher a^* values compared to the control group ($p < 0.05$) with a longer storage period; the a^* value increase in the control fish was obvious on the last two storage periods compared to the first day ($p < 0.05$). Both the BHT group and all of the DEO-supplied groups generated an ascending curve that reached the highest values at either day 12 or 15 post-treatment, particularly BHT and DEO-1 compared to the first day ($p < 0.05$) (Figure 5). Both the BHT- and DEO-supplied fish, except DEO-3, showed higher yellowness (b^*) values compared to the control on the first day ($p < 0.05$). On the third day post-treatment, both DEO-1 and DEO-3 exhibited the highest b^* values compared to the control and other fish-treated groups ($p < 0.05$). On day six post-treatment, both DEO-2 and DEO-3 exhibited the highest b^* values compared to the control and other fish-treated groups ($p < 0.05$). On the ninth day post-treatment, while both the BHT- and DEO-1-supplied groups showed higher b^* values compared to the control, DEO-3 exhibited the lowest b^* value ($p < 0.05$). On the twelfth day post-treatment, all fish-treated groups exhibited a sharp decrease in b^* value compared to the control, except the DEO-1-supplied fish, which showed the highest b^* value compared to the control ($p < 0.05$). On the 15th day post-treatment, it was noted that the b^* value increased in all fish-treated groups compared to the control ($p < 0.05$). With a longer storage period, the control fish showed an ascending curve that reached the peak at the middle storage periods, six and nine days post-treatment ($p < 0.05$), and then dropped to similar b^* values of the first day ($p > 0.05$). Across the entire storage period, except the ninth day, the BHT-supplied fish exhibited declining b^* values compared to the first day ($p < 0.05$). Compared to the first day, the DEO-1-supplied group showed stable yellowness values that dropped at the last storage period, 15 days post-treatment ($p < 0.05$). Across the storage period, DEO-2 exhibited comparable b^* values to the first day, but higher and lower values were recorded on the sixth and twelfth days post-treatment ($p < 0.05$). With longer storage, the DEO-3-supplied fish displayed an increasing b^* trend compared to the first day, which was apparent on the third, sixth, and fifteenth days post-treatment ($p < 0.05$) (Figure 5).

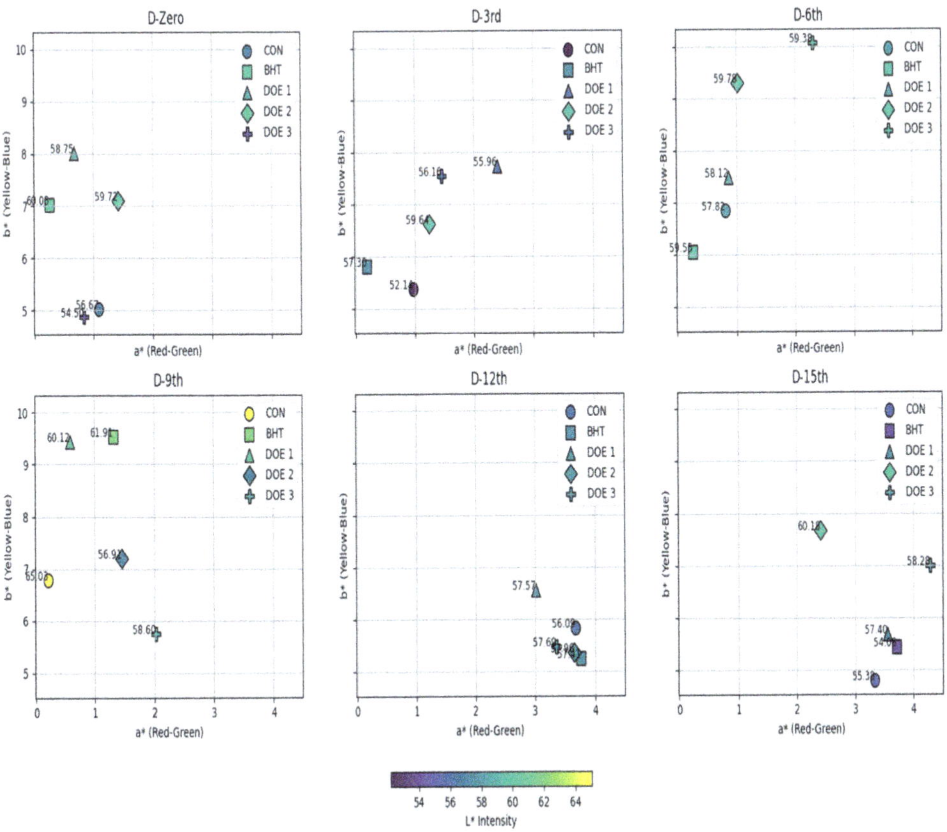

Figure 5. Changes in color parameters of fish meat samples subjected to different treatments over a 15-day storage period. a^* (Red–Green Axis): measures the degree of redness ($+a^*$) or greenness ($-a^*$), which is primarily influenced by myoglobin oxidation; b^* (Yellow–Blue Axis): represents yellowness ($+b^*$) or blueness ($-b^*$), which is often associated with lipid oxidation and pigment degradation; L^* (Lightness Intensity): indicated by the color gradient, with darker shades representing lower L^* values and lighter shades indicating higher L^* values.

Hue (color saturation): It was noted that the BHT-treated fish exhibited a higher Hue than the control and DEO-treated fish from the first day up to the sixth day post-treatment ($p < 0.05$). On the third and sixth days post-treatment, it was noted that the DEO-1- and DEO-3-supplied fish had the lowest Hue than all the other groups, respectively ($p < 0.05$). On the ninth day post-treatment, it was noted that all fish-treated groups had a lower Hue value than the control, particularly the DEO-3-supplied fish ($p < 0.05$). On the twelfth and fifteenth days post-treatment, the DEO-1-supplied and all DEO-treated fish showed a higher Hue, respectively ($p < 0.05$), compared to the control and BHT-treated fish groups. With a longer storage period, the control fish showed an ascending curve that reached the peak on the ninth day post-treatment ($p < 0.05$), and then the curve sharply decreased until the completion of the storage period ($p < 0.05$) compared to the first day. Both the BHT- and DEO-3-supplied fish, compared to the first day, exhibited a declining Hue curve and reached the lowest losses at 12 and 15 days post-treatment ($p < 0.05$). Compared to the first day, the DEO-2-supplied fish had a comparable Hue during the entire storage period ($p > 0.05$), except the lowest value was observed at 12 days post-treatment ($p < 0.05$) (Figure 6).

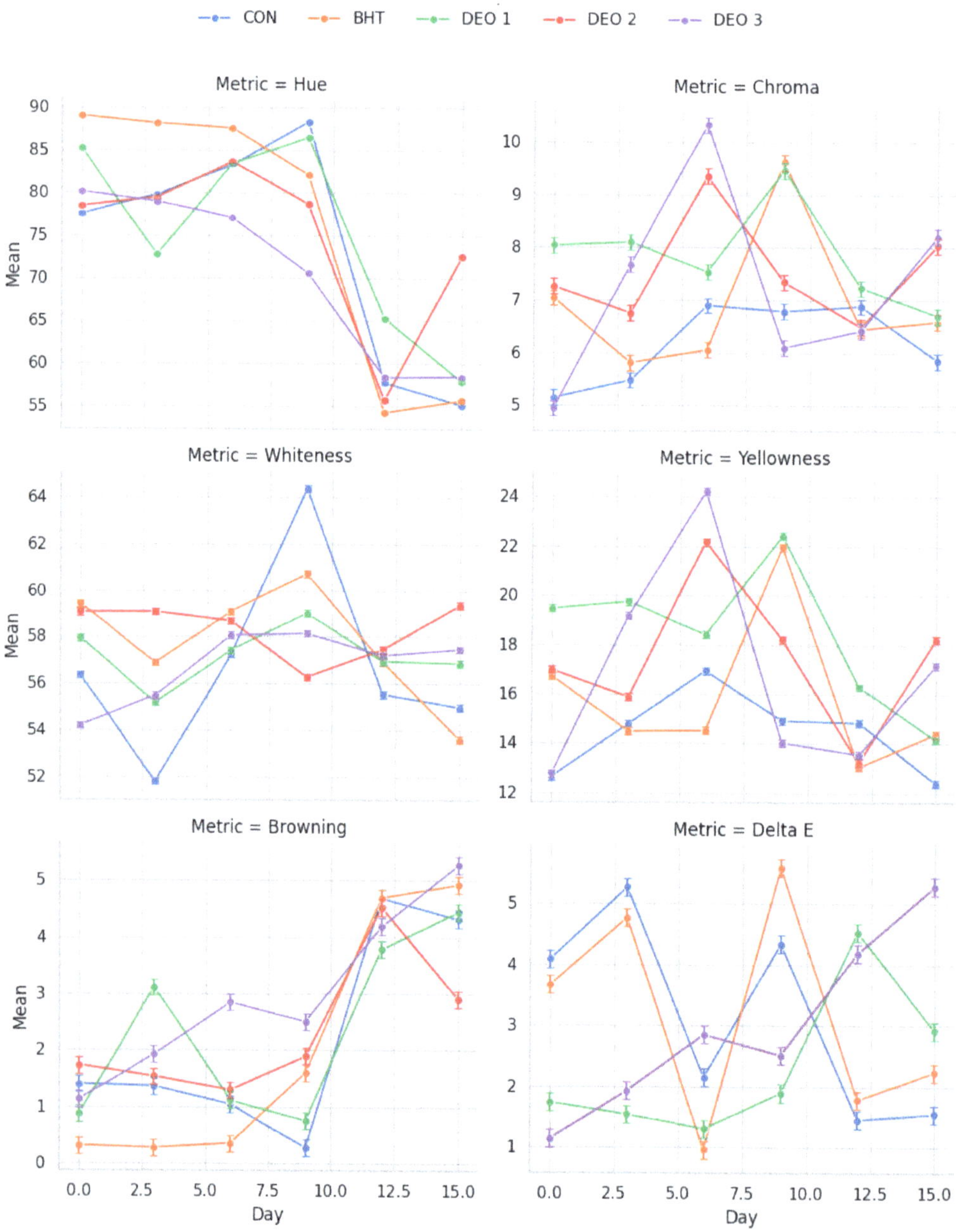

Figure 6. Changes in color stability parameters of fish meat during storage under different Dill essential oil treatments.

Chroma: Compared to the control group on the first day, both the BHT- and DEO-supplied fish groups, except DEO-3, exhibited a higher Chroma ($p < 0.05$). This higher Chroma trend was also noticed in the DEO-1- and DEO-3-supplied groups on the third day post-treatment compared to the control, particularly DEO-1. Six days post-treatment, the DEO-2- and DEO-3-supplied groups exhibited a higher Chroma value ($p < 0.05$) than the other groups. On the ninth day post-treatment, both the BHT- and DEO-1-supplied groups exhibited a higher Chroma value ($p < 0.05$) than all other groups. On the twelfth day, all groups differed, where DEO-1-supplied fish had higher values while other treated groups had lower Chroma values compared to the control ($p < 0.05$). Similarly, on the 15th day post-treatment, the BHT- and DEO-1-supplied groups exhibited a lower Chroma value, but the DEO-2- and DEO-3-supplied groups exhibited a higher Chroma value than the control ($p < 0.05$). With a longer storage period, it was noted that both the control and DEO-1-supplied groups did not show differences in Chroma across the entire storage period ($p > 0.05$), while the BHT group exhibited a sharp declining Chroma curve on the third and sixth days post-treatment ($p < 0.05$). Compared to the first day, DEO-2 exhibited a constant curve along the storage experiment ($p > 0.05$) but had a peak on the sixth day post-treatment ($p < 0.05$). While DEO-3-supplied fish showed an ascending curve until six days post-treatment ($p < 0.05$), and then a declining curve compared to the first day ($p < 0.05$) (Figure 6).

Regarding the whiteness index, on the first day, the BHT- and DEO-2 treated groups had significantly higher WI values than the other groups ($p < 0.05$). The DEO-2-fortified group exhibited substantially greater WI values than the other groups on the third and fifteenth days of storage ($p < 0.05$). However, the WI values of the DEO-2-treated group decreased significantly on the ninth day of storage in comparison to the control group ($p < 0.05$). Furthermore, all treated groups, regardless of BHT or DEO, exhibited significantly elevated WI values on the 12th day of cold storage in comparison to the control group ($p < 0.05$). It was observed that the WI values across groups, except the DEO-2-fortified group, displayed a comparable trend. On the first day, there was a downward trend in WI values and then an upward trend, reaching the highest values on days 6 and 9 in the DEO-3-fortified groups and day 9 only in the remaining groups. Then, the WI value decreases until the end of the storage period (Figure 6). For the yellowness index, the yellowness index (YI) in both the BHT- and DEO-1-fortified groups demonstrated a descending trend from the third day, reaching its maximum values on the ninth day and subsequently decreasing until the end of the storage period. Simultaneously, the DEO-2- and DEO-3-fortified groups exhibited a comparable trend as previously observed and achieved their peak YI values on the sixth and fifteenth days of storage. In comparison to other groups, on the sixth day of storage, both the DEO-2- and DEO-3-fortified groups showed the highest YI values ($p < 0.05$). Moreover, the BHT- and DEO-1-fortified groups had higher YI values than the other groups on the 9th day of storage ($p < 0.05$), but DEO-1 displayed a substantially greater YI value than the other groups on the 12th day of cold storage. Furthermore, for the browning index, on the first day of cold storage, the browning index (BI) for the DEO-2-fortified group exhibited a statistically significant increase ($p < 0.05$) in comparison to the BHT-fortified group. The DEO-1-fortified group exhibited the highest BI value, whereas the BHT-fortified group displayed the lowest BI value in comparison to the other groups ($p < 0.05$) on the third day of cold storage. Furthermore, the DEO-3-fortified group exhibited the highest BI value relative to the other groups ($p < 0.05$) on the 6th, 9th, and 15th days of cold storage; however, on day 12, both the control and BHT-treated groups demonstrated the highest BI values compared to the other groups ($p < 0.05$). The browning index values within the groups exhibited a similar pattern, with a decrease during cold storage and an upward trend until they reached their maximum level on day 12 for the control, BHT-fortified group, and DEO-2-fortified group, and on day 15 for both the DEO-1- and DEO-3-fortified groups (Figure 6). The Delta E (ΔE) level of the DEO-3-treated group was lower than that of the BHT-treated group on the initial day of cold storage ($p < 0.05$). Nevertheless, the ΔE level of both the BHT- and DEO-1-treated groups increased in comparison to the other treated groups on the third day of cold storage ($p < 0.05$) (Figure 6). Furthermore, the DEO-3-treated groups exhibited the maximum ΔE level in comparison to the other treated groups ($p < 0.05$) on days 6 and 15 of cold storage. On the ninth day of cold storage, the ΔE levels of both the BHT- and DEO-1-treated groups were significantly higher than those of the other treated groups ($p < 0.05$). On the twelfth day of cold storage, DEO-2 exhibited higher ΔE levels than those of the other groups ($p < 0.05$). The ΔE levels within the group exhibited a similar pattern to that of WI, YI, and BI, as they initially decreased and then increased in an ascending manner until they reached their maximum levels on day 9 for the BHT- and DEO-1-treated groups, day 12 for the DEO-2-treated groups, and day 15 for the DEO-3-treated groups, respectively.

3.5. Fish Fillet Microbial Analysis

The application of BHT and/or DEO generally did not yield significant effects on APC ($p > 0.05$) from the first day to day 15 post-treatment, with one notable exception. On day nine after treatment, higher DEO concentrations were found to reduce APC below 5 log CFU/g ($p < 0.05$) compared with the control and other treated fish groups. Although not statistically significant, the DEO-treated groups, particularly DEO-3, exhibited numerically lower APC by approximately one log CFU/g compared to the control group ($p > 0.05$) until the ninth day post-treatment ($p < 0.05$). As the storage duration increased, all groups, including the positive control and BHT-treated samples, showed an increasing trend, reaching over 6 log CFU/g at 12 and 15 d post-treatment ($p < 0.05$) (Figure 7A). Likewise, the BHT and/or DEO treatments did not significantly affect LAB ($p > 0.05$) during the initial 15 d post-treatment. However, with increased storage time, all groups, including the positive and negative controls, displayed an upward trend, exceeding 6 log CFU/g at 12- and 15- days post-treatment ($p < 0.05$). Although the DEO-treated groups, especially DEO-3, showed numerically lower lactic acid bacteria counts (~1.5 log CFU/g) on day 9 compared to the control, these reductions were not statistically significant ($p > 0.05$). Therefore, no definitive antimicrobial effect can be concluded at this storage point. Nevertheless, by the twelfth day post-treatment, LAB counts in all groups, including DEO-treated fish, surpassed six logs CFU/g ($p < 0.05$) (Figure 7B). The BHT and/or DEO treatments generally did not affect coliform levels ($p > 0.05$), with two exceptions: on the third and ninth days after treatment, fish treated with DEO-2 and BHT exhibited higher coliform counts than the control and other treated groups ($p < 0.05$). As the storage time increased, an upward trend in coliform counts was observed across all fish groups, particularly in the negative control ($p < 0.05$) (Figure 7C). Regarding the impact on staphylococcal counts, all groups showed similar counts, a pattern that was repeated in the final two post-treatment storage periods. However, variations emerged during the intermediate periods. On the third day post-treatment, the BHT-treated fish displayed higher staphylococcal counts than those of the control. By the sixth day, DEO-1 showed differences, and on the ninth day, all DEO-treated groups exhibited lower counts than the control and other groups ($p < 0.05$). As storage time increased, most groups maintained stable staphylococcal counts relative to day one, except for the control and BHT-treated fish. These two groups showed significant increases ($p < 0.05$) on the 9th and 15th days post-treatment. Notably, the DEO-treated fish remained unaffected by extended storage ($p > 0.05$) (Figure 7D).

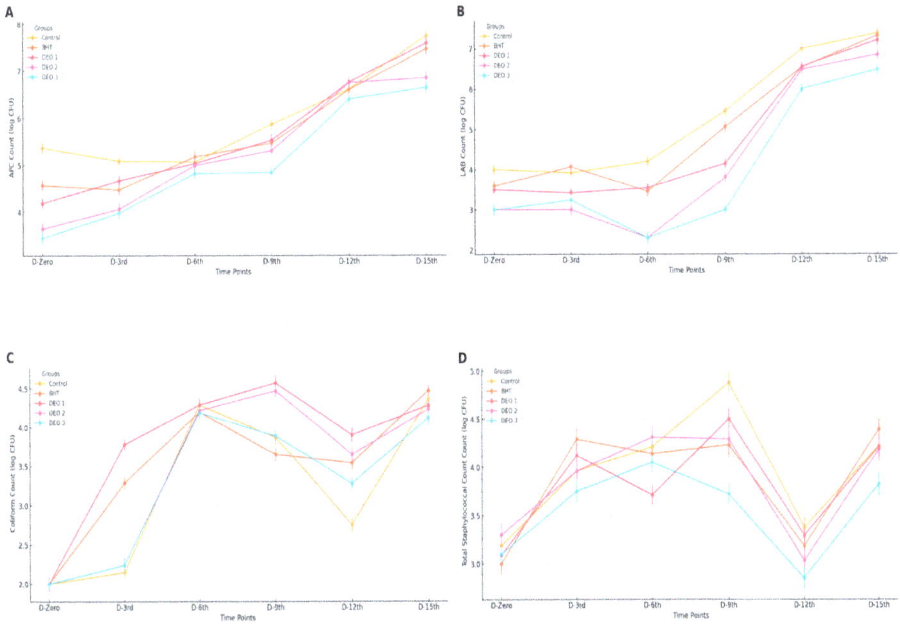

Figure 7. Microbial analysis of fish fillets stored at refrigeration (4 °C) for 15 days after treatment with Dill essential oil (DEO) at three concentrations (DEO-1, DEO-2, and DEO-3) compared with the control and BHT groups. (**A**) Aerobic plate count (APC); (**B**) lactic acid bacteria (LAB) count; (**C**) coliform count; and (**D**) total staphylococcal count. Error Bars represent the standard error (SE) of the mean microbial counts (log CFU/g) at each storage period.

3.6. Sensory Evaluation

Figure 8 illustrates a detailed comparison of the sensory evaluation results for fish meat subjected to various treatments. These treatments included three different concentrations of DEO, butylated hydroxytoluene (BHT), and a control. The assessment, conducted over a storage period, examined four key sensory attributes: odor, color, texture, and overall acceptability. Evaluations were performed at the start of the experiment (Day Zero) and subsequently on days 3, 6, 9, 12, and 15. As anticipated, all treatment groups exhibited a gradual decrease in sensory scores throughout the storage period. This decline was attributed to the natural degradation of fish meat quality, primarily due to processes such as lipid oxidation and microbial growth. In the odor assessment, all the treatment groups initially showed comparable scores. However, the control group experienced the quickest deterioration in odor quality, especially after day 6. While the BHT-treated samples maintained better odor scores than the control, they still showed a consistent decline by day 15. The DEO treatments demonstrated a concentration-dependent effect, with DEO-3 preserving the odor quality for the longest period, followed by DEO-2 and DEO-1. The coloration changed over time. By the ninth day, both the control and BHT groups exhibited a significant decrease in color scores. In contrast, the samples treated with Dill, particularly DEO-3, demonstrated superior color retention. Texture analysis revealed a gradual decrease in firmness and overall textural quality across all treatments over time. The samples in the control group softened most rapidly, whereas those treated with DEO-3 maintained a firmer consistency for an extended period. A comprehensive sensory evaluation, which integrates individual sensory characteristics into a single assessment, was initially favorable for all samples. However, by day 9, the acceptability ratings of the control group had decreased significantly. Although the BHT-treated samples outperformed the control, they showed a gradual decline. In contrast, the Dill-treated groups demonstrated superior preservation of the overall sensory quality, with DEO-3 consistently receiving the highest scores throughout the experiment.

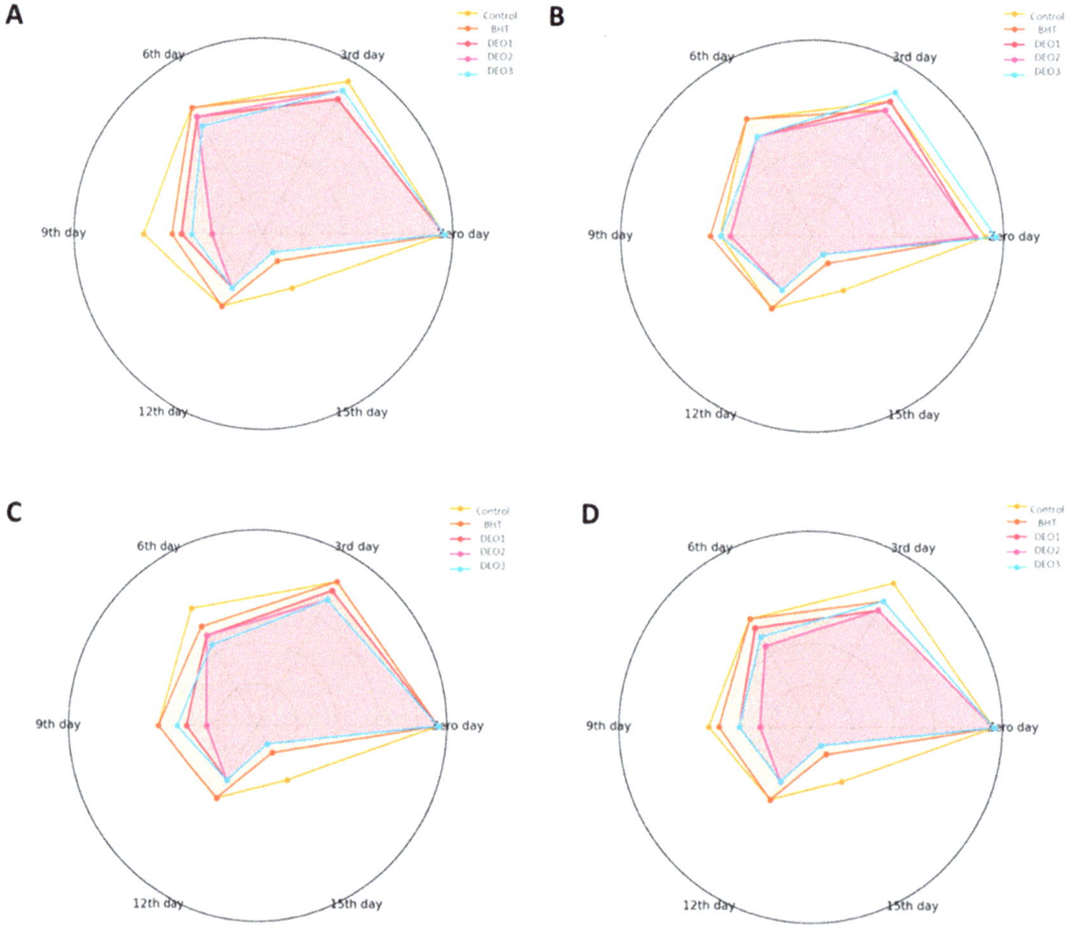

Figure 8. Sensory evaluation of fish meat treated with different concentrations of Dill essential oil compared to BHT and the control group over a 15-day storage period. The sensory attributes assessed included odor (**A**), color (**B**), texture (**C**), and overall acceptability (**D**) at different time intervals.

3.7. Correlations Between Physicochemical Parameters

The scatter plot matrix reveals several key relationships among the physicochemical parameters analyzed (Figure 9). The correlation between pH and WHC appears weak or inconsistent. Additionally, a moderate negative correlation was observed between WHC and CL, whereby samples with higher WHC tend to exhibit lower CLs. Furthermore, a positive correlation between PL and CL underscores the interconnected nature of protein integrity and CL.

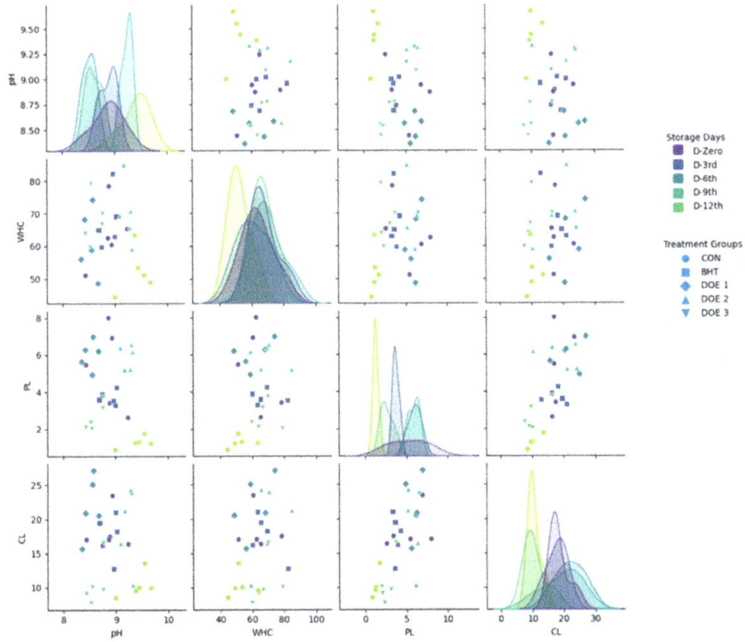

Figure 9. Scatter plot matrix of physicochemical parameters (pH, WHC, PL, and CL) in stored fish meat under different treatment groups and storage durations. Each subplot represents the relationship between two variables, with diagonal plots displaying kernel density estimations (KDEs) for distribution analysis. Storage days are color-coded (D-Zero, D-3rd, D-6th, D-9th, and D-12th), while treatment groups are represented by different marker shapes (• CON, ■ BHT, ♦ DEO-1, ▲ DEO-2, and ▼ DEO-3).

3.8. PCA Analysis

The principal component analysis (PCA) presented in Figure 10 illustrates the distribution of different treatment groups—control (CON), BHT (a synthetic antioxidant), and three concentrations of Dill essential oil (DEO-1, DEO-2, and DEO-3)—based on key physicochemical and microbiological parameters, including color change, pH, water-holding capacity (WHC), CL, PL, and microbial load. The first principal component (PC1) accounts for 42.3% of the total variance, while the second principal component (PC2) explains 27.8%, collectively capturing a substantial portion of the dataset's variability. The traditional treatments, CON and BHT, are positioned in distinct quadrants. The CON group, located in the lower-left quadrant, suggests minimal alterations in physicochemical properties, likely reflecting untreated fish meat's natural deterioration over time. In contrast, BHT appears in the upper-left quadrant, indicating a different trajectory in response to antioxidant treatment, possibly due to its effects on oxidative stability and microbiological control. The experimental treatments (DEO-1, DEO-2, and DEO-3) cluster within the right half of the PCA plot, signifying distinct physicochemical and microbiological characteristics compared to traditional treatments. DEO-1, located in the upper-right quadrant, demonstrates a unique response, potentially reflecting stronger antioxidant or antimicrobial effects at this concentration. DEO-2 and DEO-3, closely positioned in the lower-right quadrant, indicate similar influences on fish meat properties, suggesting a trend in treatment response as the DEO concentration increases.

Figure 10. PCA analysis of treatment effects.

4. Discussion

The current study aimed to assess the preservation effects of *Anethum graveolens* essential oil (DEO) on the quality and shelf life of Basa fish fillets in refrigerated storage. The main components of Dill essential oil are monoterpenes, primarily α-phellandrene, d-limonene, carvone, and Dill ether. These findings align with those of [45], who found that carvone (42.47%), limonene (29.04%), and α-phellandrene (13.12%) are essential components of Dill essential oil. Similarly, ref. [46] identified significant concentrations of carvone (34.33%), α-phellandrene (22.03%), Dill ether (18.84%), limonene (6.93%), and Dill apiol (5.01%) as primary constituents. In addition, refs. [8,47] also found that the most common monoterpenes in DEO are carvone and limonene. Nevertheless, ref. [48] disclosed that the majority of the components of DEO are dillapiole (44.01%), d-limonene (19.47%), and carvotanacetone (14.03%). The variations in the concentration and composition of various components can be ascribed to varied weather conditions and geographical regions, as well as metabolism, maturity, and the portion of the plant from which the essential oils (EOs) are extracted [49]. Furthermore, harvesting time, storage conditions, and extraction techniques all have an impact on essential oil yield and composition [50,51]. Although DEO lacks classical phenolic compounds, its moderate DPPH scavenging activity may be attributed to oxygenated monoterpenes such as carvone and limonene, which have demonstrated radical scavenging activity in other studies despite not possessing aromatic hydroxyl groups [52,53]. So, DEO acts as a promising source of antioxidants due to its high level of phenolic compounds. The DPPH antioxidant assay results indicated a concentration-dependent variation in the suppression of DPPH radicals by the DEO. The antioxidant activity may be ascribed to the presence of monoterpenes in DEO, which may act as radical scavenging agents. The majority of earlier experiments concluded that the essential oils contained monoterpene hydrocarbons, oxygenated monoterpenes, and/or sesquiterpenes, which have strong antioxidative properties [53]. Limonene from the class of monoterpene phenols has been reported to possess antioxidant activity [54]. Thus, the current high content of limonene component might contribute to estimated high antioxidant activity. Limonene is listed in the Code of Federal Regulations as generally recognized as a safe (GRAS) antioxidant in human foodstuff [55].

The DEO demonstrated a lower antimicrobial potential on Gram-negative bacteria, particularly *Salmonella* spp., in the resazurin-based turbidimetric assay. Nevertheless, it demonstrated a slightly higher antimicrobial potential on *E. coli* at 0.098 mg/mL than the value reported by Mujović et al. (2024) [8], who demonstrated that the MIC of DEO for *E. coli* was 28.41 μL/mL. Conversely, Gram-positive bacteria demonstrated a higher degree of susceptibility to DEO than Gram-negative bacteria, which is likely attributable to variations in cell wall structure. DEO flavonoids can form a complex with bacteria's outer membrane and soluble proteins that are related to it [56]. Nevertheless, the synergistic effect of beneficial essential oil components has previously been shown to be effective in terms of antibacterial activity [57]. Gram-negative bacteria have both an outer membrane and a periplasmic space, whereas gram-positive bacteria lack either of these components. This membrane stops hydrophiles from entering the bacteria. The periplasmic region includes large enzymes capable of decomposing foreign substances [56,58]. These findings align with previous studies highlighting dill oil's moderate antimicrobial efficacy against foodborne pathogens [8,59]. Biological activities of EOs are related to their chemical compositions [60]. The current principal constituents of DEO are monoterpenes, notably limonene and carvone, which have potent antibacterial effects [8,61,62]. Monoterpenes lipophilic characteristics can permeate cell membranes, enhance fluidity, and inhibit embedded enzymes [63]. Current DEO promising antimicrobial activities against multidrug resistant pathogens support the potential application as a natural antimicrobial agent in food preservation and safety strategies.

Fish muscle is unlike mammalian muscle, which experiences a significant post-mortem pH decline due to glycogen breakdown into lactic acid and often exhibits a more moderate pH drop due to its lower glycogen reserve [64]. The higher initial pH of DEO-3-treated fish may be attributed to the buffering capacity of certain bioactive compounds in DEO, such as terpenes and phenolic compounds, which can neutralize the acidic metabolites produced during post-mortem changes [8]. Another justification for higher pH might be attributed to the antimicrobial activity of DEO. By inhibiting the growth of native microflora, including current estimated *LAB*, which is capable of acidic byproduct production, DEO treatment could result in a relative increase in pH compared with untreated samples. This DEO antimicrobial effect has also been documented in previous studies on fungal pathogens, where DEO disrupted cellular processes and inhibited acid production [65]. BHT-treated samples showed the highest pH, especially from the 3rd to 6th days compared to the control and DEO-treated groups. This may be due to the antioxidant properties of synthetic BHT, which likely prevents the formation of acidic byproducts from lipid oxidation and proteolysis, maintaining a higher pH [66]. DEO contains bioactive compounds with similar antioxidant properties, but these natural compounds may degrade over time, leading to a decrease in antioxidative and antimicrobial efficacy. The degradation of natural compounds might trigger spoilage involving proteolytic activity and lipid oxidation, resulting in acidic metabolites with a decreased pH. Previous research indicated that while DEO exhibits initial antimicrobial and antioxidant effects, its efficacy diminished with extended storage, leading to quality degradation of fish products [8]. In the untreated control samples and those treated with low concentrations of DEO (DEO-1), the higher pH observed during later storage stages (days 9–15) is likely due to microbial spoilage. Microbial activity leads to the production of basic compounds such as ammonia and biogenic amines, resulting in increased pH. Studies have shown that inadequate antioxidant protection allows microbial proliferation and spoilage, contributing to pH elevation in stored fish products [67]. Higher concentrations of DEO (DEO-3) appeared to be more effective in inhibiting oxidative and proteolytic processes, maintaining muscle quality, and preventing pH fluctuations.

The enhanced preservation effect at higher DEO concentrations is supported by studies demonstrating that increased levels of natural extracts can improve the shelf life and quality of fish products during storage [68]. The current study indicates that DEO may enhance the water-holding capacity of fish muscle. The impact of DEO and its primary components, such as carvone, on the water-holding capacity (WHC) of fish muscle has been explored in various studies [8,69]. These earlier investigations highlighted the potential of DEO to enhance WHC through its antioxidant properties and interactions with muscle proteins. On the third day post-treatment, DEO-2-treated fish exhibited the highest WHC value compared to all other groups, including the control. DEO-2 likely contains an optimal concentration of bioactive compounds, such as carvone and limonene, which stabilize muscle proteins and enhance their ability to retain moisture. These compounds may interact with myofibrillar proteins like actin and myosin, reducing proteolytic degradation and maintaining the integrity of the muscle structure. Additionally, DEO-2's antioxidative properties may prevent lipid oxidation, which can lead to the formation of hydrophobic products that reduce WHC [69]. From day 6 onward, all treated groups (BHT, DEO-1, DEO-2, and DEO-3) exhibited similar WHC values compared to the control group. The control group showed a significant decrease in WHC by the end of the storage period (day 15), indicating progressive dehydration due to proteolytic activity and oxidative damage [70]. Endogenous enzymes like calpains and cathepsins degrade myofibrillar proteins, leading to increased water release and reduced WHC [71]. Regarding purge loss (PL), the current findings revealed significant differences in PL trends among the groups, particularly at specific time points. The interplay between WHC and PL is driven by protein oxidation, proteolysis, and structural modifications in muscle fibers. As storage progresses, oxidative stress leads to carbonylation and aggregation of myofibrillar proteins, reducing their ability to retain water and causing increased PL [72]. Essential oil components, such as carvone and limonene, have been reported to mitigate these effects by scavenging free radicals and inhibiting pro-oxidant metal ions, thereby stabilizing protein structure and maintaining hydration capacity [73]. Cooking loss (CL) is a critical parameter that reflects the ability of muscle proteins to retain water during heat processing. It is influenced by factors such as protein denaturation, oxidative modifications, and muscle structure integrity. The results indicate that while the BHT and DEO treatments did not significantly affect CL throughout most of the storage period ($p > 0.05$), certain differences emerged at specific time points. On day 12, fish treated with BHT exhibited significantly higher CL compared to all other groups ($p < 0.05$). This increase in CL suggests that lipid and protein oxidation may have compromised the muscle's water-holding properties. Previous studies have reported that synthetic antioxidants like BHT primarily target lipid oxidation but may have limited effects on protein oxidation, which is a key factor influencing WHC and, consequently, CL [72]. In contrast, DEO-1 and DEO-2 exhibited distinct trends, with DEO-2 showing superior late-stage stability. These results underscore the potential of DEO as a natural alternative for preserving water retention properties in stored fish meat, particularly in mitigating protein oxidation-related cooking losses. Oxidative modifications can lead to protein cross-linking, aggregation, and reduced WHC, resulting in higher CL [74].

The scatter plot matrix provides insights into the effects of storage time and treatment formulations on meat physicochemical properties. The correlations suggest that acidity alone does not significantly influence the water retention capacity of stored meat. Instead, other physicochemical and structural factors, such as protein oxidation, myofibrillar interactions, and lipid stability, likely play a more dominant role in determining WHC. This observation aligns with previous findings, indicating that protein denaturation and oxidative modifications can impact the ability of muscle fibers to retain water, rather than pH being the sole determinant. The results indicate that DEO-2 and DEO-3 exhibit potential benefits in maintaining WHC and reducing CL, which could contribute to improved meat quality during extended storage. The observed trends in pH, WHC, and PL suggest that oxidative and protein degradation processes are key determinants of quality deterioration over time. For food scientists and industry professionals, these findings underscore the importance of incorporating targeted antioxidant and protein-stabilizing interventions in meat preservation strategies. Future research should include controlled oxidative stability tests, rheological assessments, and sensory evaluations to further validate the efficacy of these treatments. Additionally, future research should focus on elucidating the molecular mechanisms underlying protein–water interactions and their implications for texture, juiciness, and overall consumer acceptability of stored and processed meat products.

The observed variations in the Warner–Bratzler Shear Force (WBSF) values among the different treatment groups can be attributed to the interplay between post-mortem proteolytic activities and the antioxidative properties of the applied treatments. In the control group, the progressive decline in WBSF values over time was indicative of muscle softening, primarily due to endogenous proteolytic enzymes such as calpains and cathepsins degrading myofibrillar proteins. This degradation leads to the weakening of the muscle structure, resulting in decreased shear force measurements [75,76]. By contrast, the BHT-treated group maintained higher WBSF values throughout the storage period. BHT, a synthetic antioxidant, inhibits lipid oxidation and the subsequent oxidative damage to muscle proteins. BHT preserves the integrity of myofibrillar structures by preventing the formation of ROS, thereby mitigating the extent of proteolysis and maintaining muscle firmness [77,78]. The DEO-treated group exhibited an initial increase in WBSF values followed by a decline. This pattern suggests that DEO may initially enhance muscle firmness, possibly through the stabilization of muscle proteins by its antioxidative components. The initial increase in WBSF in the DEO-treated group could be attributed to several mechanisms, including the presence of bioactive compounds, such as carvone, limonene, and other terpenoids, which are known for their antioxidant properties. These compounds can scavenge reactive oxygen species (ROS) and prevent oxidative damage to myofibrillar proteins such as actin and myosin, which are critical for maintaining muscle texture [8,79]. Certain terpenes in DEO have been shown to form hydrogen bonds or hydrophobic interactions with myofibrillar proteins, potentially stabilizing their conformations [8,80,81]. However, direct evidence linking DEO components to the promotion of cross-linking in myofibrillar proteins is limited, and studies on similar natural compounds, particularly polyphenols, provide relevant insights [82,83]. These interactions and cross-linking interactions between DEO components and muscle proteins can increase the resistance of muscle fibers to shear forces, resulting in higher WBSF values, which are indicative of tougher meat texture in the initial storage period. However, over time, the protective effect diminishes due to the degradation of DEO bioactive compounds during storage, leading to increased proteolytic activity and subsequent muscle softening [8,84,85].

The color parameters—L^*, a^*, b^*—and color stability parameters such as Hue, Chroma, and Delta E (ΔE) are crucial indicators of fish quality and shelf-life stability. The lightness parameter (L^*) reflects the brightness or whiteness of the fish muscle, which is an important indicator of freshness and consumer appeal. On Day Zero, fish treated with the highest concentration of Dill essential oil (DEO-3) exhibited the lowest L^* values, indicating reduced lightness. This initial decrease in lightness was also noticed with the application of Dill seed essential oil (DSEO) in meat products and was attributed to interactions of DEO's bioactive compounds (such as carvone and limonene), which present with the highest concentration in DEO-3 than other levels, with muscle pigments or lipids [8]. In contrast, the BHT and other DEO treatments preserved lightness better, likely due to their antioxidative properties that prevent pigment degradation. By day 3, all treated groups exhibited higher L^* values than the control, indicating that these treatments effectively slowed down oxidative processes that generated darker meat. DEO-2 was particularly effective at this stage [86]. On day 9, DEO-2-treated fish showed the lowest L^*, suggesting some degree of browning or discoloration, while other groups were not significantly different from the control. This could be attributed to the transient nature of DEO-2's antioxidative effects or interactions with muscle pigments or the enzymatic breakdown of DEO active components [8]. By day 12, all treated groups maintained higher L^* values than the control, highlighting their ability to preserve lightness during mid-storage. On day 15, the sharp decrease in L^* for BHT-treated fish was perhaps attributable to the onset of spoilage, advanced protein oxidation, and pigment degradation, likely due to the restricted antioxidant activity of BHT to inhibit lipid oxidation. DEO-2-treated fish maintained higher L^* values, suggesting sustained antioxidative effects and better lightness preservation [8]. The redness index (a^*) measures the intensity of red color in fish muscle, primarily influenced by myoglobin stability and oxidation. On Day Zero, no treatment significantly affected redness, indicating that myoglobin stability was not immediately impacted by the treatments [87]. By days 3–6, the BHT-treated fish showed lower a^* values, possibly due to reduced myoglobin stabilization [8]. In contrast, the DEO-1-treated fish exhibited higher a^* values, suggesting enhanced preservation of red pigments. On day 6, the DEO-3-treated fish displayed the highest a^* values, likely due to the highest level of antioxidant components that might generate higher myoglobin protection than other levels [88]. By day 9, all treated groups, particularly DEO-3, exhibited higher a^* values than the control, confirming better preservation of red color. This suggests that DEO treatments may inhibit oxidative reactions that degrade myoglobin. The increased redness in the treated groups indicates that the antioxidative effects of DEO treatments were more pronounced, preventing the oxidation of myoglobin and maintaining the red color. Carvone, the major component of DEO may chelate pro-oxidant metal ions, such as iron and copper [73]. The iron and copper ions are responsible for catalyzing oxidative reactions leading to myoglobin oxidation [89]. Therefore, by binding these metals, carvone can inhibit their oxidative activity, thereby preserving myoglobin in its reduced, oxygen-binding state. On days 12–15, redness increased in all groups, with BHT- and DEO-1-treated fish reaching their highest values. The increased redness in untreated fish could be due to oxidation-induced changes in myoglobin, leading to altered color perception. The treated groups' higher a^* values at later stages suggest that their antioxidative effects became more apparent over time, stabilizing and enhancing redness. The yellowness index (b^*) reflects the intensity of yellow tones in fish muscle, often influenced by lipid oxidation and pigment interactions. On Day Zero, BHT- and DEO-treated fish (except DEO-3) showed higher b^* values, indicating initial yellowing effects. This could be attributable to the DEO contents of yellowish pigments such as carotenoids and their potential isomers [90,91] or the formation of yellowish oxidation products [92]. The higher contents of DEO of yellow pigments such as carotenoids and their potential isomers, which simultaneously had strong antioxidant actions, clearly contributed to sustained higher yellowness values in DEO-2- and DEO-3-treated fish until the ninth storage day compared to other groups. On day 12, the sharp decrease in b^* values for most treated groups suggests that antioxidative processes slowed down or stabilized after an initial peak [67]. By day 15, all treated groups exhibited higher b^* values than the control. This observation may be the result of the deterioration of DEO natural yellow pigment with extended storage, loss of antioxidant effectiveness, and subsequent production of yellowish oxidation products [92].

The native aerobic plate count (APC) microorganisms were suppressed by higher concentrations of DEO, particularly DEO-3, maintaining microbial levels below critical thresholds until the ninth day, while the control samples exceeded these limits earlier. Similar inhibitory effects were observed for *lactic acid bacteria* (*LAB*) and coliforms, with Dill oil treatments, especially DEO-3, showing notable antimicrobial activity by decelerating microbial proliferation. On day 9, the staphylococcal counts were noticeably lower in the DEO-treated fillets than in the control samples. The current noticeable DEO antimicrobial properties are aligned with the findings reported on the microbiological quality of beef burgers [8]. The strong antimicrobial efficacy of DEO can be attributed to the high contents of bioactive components and their specific modes of action. The antimicrobial efficacy of DEO in suppressing spoilage-causing microorganisms is evident. Dill oil contains a high concentration of biologically active substances, primarily terpenes (such as carvone, limonene, and myrcene), phenolic compounds, and flavonoids [8]. These components are recognized for their ability to combat microbes, effectively targeting both Gram-positive and Gram-negative bacterial strains [93]. These compounds disrupt the microbial cell membranes, leading to increased permeability and subsequent cell death [94]. Additionally, bioactive components interfere with bacterial enzyme systems, thereby hindering essential metabolic pathways. Studies have demonstrated the effectiveness of DEO against various microorganisms, including *Aspergillus niger*, *Saccharomyces cerevisiae*, and *Candida albicans* [94]. Furthermore, research has indicated that Dill oil exhibits antibacterial activity against strains such as *Staphylococcus aureus*, *Escherichia coli*, and *Salmonella* Typhimurium. Current study findings support the potential application of Dill oil as a natural preservative in food products, enhancing safety and extending shelf life [95]. By day 9, the cumulative antimicrobial activity of DEO may have peaked, with sustained interactions with microbial cells leading to a noticeable reduction in staphylococcal counts [96]. At this point, spoilage microbes that were initially resistant may have been overwhelmed by the consistent antimicrobial pressure of DEO. The prolonged action of DEO likely caused irreparable damage to bacterial cells, reducing their ability to recover and proliferate, even under favorable conditions, which explains the lower count in the control on days 12 and 15.

The sensory evaluation results highlight the effectiveness of DEO treatments, particularly at higher concentrations (DEO-3), in preserving the quality attributes of fish meat during storage. Several studies support the efficacy of DEO in preserving the sensory qualities of fish meat during storage [21,97]. For instance, research has demonstrated that DEO exhibits significant antimicrobial and antioxidant activities, which contribute to maintaining the chemical and sensory properties of minced meat over an 18-day storage period [98]. These findings suggest that DEO, particularly at optimal concentrations, can serve as a natural alternative to synthetic antioxidants like BHT, offering improved sensory preservation and extended shelf life for fish meat.

The findings of this study have promising practical implications for the meat processing industry. The demonstrated efficacy of DEO in maintaining meat quality—by reducing purge and cooking losses, preserving pH, and improving tenderness during storage—suggests its potential use as a natural alternative to synthetic antioxidants such as BHT. With increasing consumer demand for clean-label and natural food products, incorporating DEO into meat preservation protocols could offer a safe, effective, and sustainable solution. Furthermore, its antimicrobial and antioxidant properties could reduce reliance on artificial preservatives, enhancing both product shelf life and consumer appeal in retail environments.

5. Conclusions

This study investigated the preservative effects of *Anethum graveolens* essential oil (DEO) on the quality and shelf life of Basa fish fillets during refrigerated storage. GC-MS analysis revealed the presence of monoterpenes such as α-phellandrene, d-limonene, carvone, and Dill ether in DEO. DEO exhibited dose-dependent antioxidant activity and antibacterial efficacy against various foodborne pathogens. Fish fillets treated with DEO, particularly at higher concentrations, effectively maintained pH, water-holding capacity, and color stability compared to the control. Microbial analysis showed that DEO significantly reduced the growth of aerobic plate count, lactic acid bacteria, coliforms, and Staphylococci. Sensory evaluation indicated that DEO treatments preserved the odor, color, texture, and overall acceptability of the fish fillets throughout the storage period. These findings demonstrate the potential of *Anethum graveolens* essential oil as a natural preservative to enhance the quality and extend the shelf life of fish fillets during refrigerated storage. Finally, the current study focused on only one type of fish and one constant storage setting; such constraints may be considered in future investigations to clarify the consequences of DEO preservative traits. Future research should also consider defining mechanisms of action and estimating in vivo safety data.

Author Contributions: Conceptualization, I.I.S., S.N.E. and A.H.: methodology, A.T., I.I.S. and S.N.E.; software, A.H. and S.N.E.; validation, F.S.H. and I.I.S.; formal analysis, A.T., I.I.S. and S.N.E.; investigation, A.T., I.I.S. and S.N.E.; resources, A.T., data curation; A.H., writing—original draft preparation; A.T., I.I.S. and S.N.E.; writing—review and editing, F.S.H. and A.H.; visualization, A.H.; supervision, F.S.H., I.I.S. and S.N.E. All authors have read and agreed to the published version of the manuscript.

Funding: This research received no external funding.

Institutional Review Board Statement: Not Applicable.

Informed Consent Statement: Not Applicable.

Data Availability Statement: The original contributions presented in this study are included in the article. Further inquiries can be directed to the corresponding author.

Conflicts of Interest: The authors declare no conflicts of interest.

References

1. Mohanty, B.P.; Mahanty, A.; Ganguly, S.; Mitra, T.; Karunakaran, D.; Anandan, R. Nutritional Composition of Food Fishes and Their Importance in Providing Food and Nutritional Security. *Food Chem.* **2019**, *293*, 561–570. [CrossRef] [PubMed]
2. Kawarazuka, N.; Béné, C. The Potential Role of Small Fish Species in Improving Micronutrient Deficiencies in Developing Countries: Building Evidence. *Public Health Nutr.* **2011**, *14*, 1927–1938. [PubMed]
3. Lorenzo, J.M.; Batlle, R.; Gómez, M. Extension of the Shelf-Life of Foal Meat with Two Antioxidant Active Packaging Systems. *LWT* **2014**, *59*, 181–188. [CrossRef]
4. de Alencar, M.G.; de Quadros, C.P.; Luna, A.L.L.P.; Neto, A.F.; da Costa, M.M.; Queiroz, M.A.Á.; de Carvalho, F.A.L.; da Silva Araújo, D.H.; Gois, G.C.; dos Anjos Santos, V.L.; et al. Grape Skin Flour Obtained from Wine Processing as an Antioxidant in Beef Burgers. *Meat Sci.* **2022**, *194*, 108963. [CrossRef]
5. Chauhan, O.P. *Advances in Food Chemistry: Food Components, Processing and Preservation*; Springer Nature: Berlin/Heidelberg, Germany, 2022; ISBN 9789811947964.
6. Brackett, R.E. Microbiological Safety of Chilled Foods: Current Issues. *Trends Food Sci. Technol.* **1992**, *3*, 81–85.
7. Tavares, J.; Martins, A.; Fidalgo, L.G.; Lima, V.; Amaral, R.A.; Pinto, C.A.; Silva, A.M.; Saraiva, J.A. Fresh Fish Degradation and Advances in Preservation Using Physical Emerging Technologies. *Foods* **2021**, *10*, 780. [CrossRef]
8. Mujović, M.; Šojić, B.; Peulić, T.; Kocić-Tanackov, S.; Ikonić, P.; Božović, D.; Teslić, N.; Županjac, M.; Novaković, S.; Jokanović, M.; et al. Effects of Dill (*Anethum graveolens*) Essential Oil and Lipid Extracts as Novel Antioxidants and Antimicrobial Agents on the Quality of Beef Burger. *Foods* **2024**, *13*, 896. [CrossRef]

9. Blundell, R.; Shah, M.A.; Azzopardi, J.I.; Iqbal, S.; Rasul, A.; Shah, G.M. Butylated Hydroxytoluene. In *Antioxidants Effects in Health: The Bright and the Dark Side*; Elsevier: Amsterdam, The Netherlands, 2022; pp. 195–200. ISBN 9780128190968.
10. Najaran, Z.T.; Hassanzadeh, M.K.; Nasery, M.; Emami, S.A. Dill (*Anethum graveolens* L.) Oils. In *Essential Oils in Food Preservation, Flavor and Safety*; Elsevier: Amsterdam, The Netherlands, 2015; pp. 405–412. ISBN 9780124166417.
11. Šojić, B.; Milošević, S.; Savanović, D.; Zeković, Z.; Tomović, V.; Pavlić, B. Isolation, Bioactive Potential, and Application of Essential Oils and Terpenoid-Rich Extracts as Effective Antioxidant and Antimicrobial Agents in Meat and Meat Products. *Molecules* **2023**, *28*, 2293. [CrossRef]
12. Eshaghi, R.; Mohsenzadeh, M.; Ayala-Zavala, J.F. Bio-Nanocomposite Active Packaging Films Based on Carboxymethyl Cellulose, Myrrh Gum, TiO_2 Nanoparticles and Dill Essential Oil for Preserving Fresh-Fish (*Cyprinus carpio*) Meat Quality. *Int. J. Biol. Macromol.* **2024**, *263*, 129991. [CrossRef]
13. Amiri, A.; Dugas, R.; Pichot, A.L.; Bompeix, G. In Vitro and in Vitro Activity of Eugenol Oil (*Eugenia caryophylata*) against Four Important Postharvest Apple Pathogens. *Int. J. Food Microbiol.* **2008**, *126*, 13–19. [CrossRef]
14. Carlucci, D.; Nocella, G.; De Devitiis, B.; Viscecchia, R.; Bimbo, F.; Nardone, G. Consumer Purchasing Behaviour towards Fish and Seafood Products. Patterns and Insights from a Sample of International Studies. *Appetite* **2015**, *84*, 212–227. [CrossRef] [PubMed]
15. Baptista, R.C.; Horita, C.N.; Sant'Ana, A.S. Natural Products with Preservative Properties for Enhancing the Microbiological Safety and Extending the Shelf-Life of Seafood: A Review. *Food Res. Int.* **2020**, *127*, 108762. [CrossRef] [PubMed]
16. Gramza, A.; Khokhar, S.; Yoko, S.; Gliszczynska-Swiglo, A.; Hes, M.; Korczak, J. Antioxidant Activity of Tea Extracts in Lipids and Correlation with Polyphenol Content. *Eur. J. Lipid Sci. Technol.* **2006**, *108*, 351–362. [CrossRef]
17. Cardile, V.; Russo, A.; Formisano, C.; Rigano, D.; Senatore, F.; Arnold, N.A.; Piozzi, F. Essential Oils of Salvia Bracteata and Salvia Rubifolia from Lebanon: Chemical Composition, Antimicrobial Activity and Inhibitory Effect on Human Melanoma Cells. *J. Ethnopharmacol.* **2009**, *126*, 265–272. [CrossRef]
18. Peerakam, N.; Wattanathorn, J.; Punjaisee, S.; Buamongkol, S.; Sirisa-Ard, P.; Chansakaow, S. Chemical Profiling of Essential Oil Composition and Biological Evaluation of *Anethum graveolens* L. (Seed) Grown in Thailand. *J. Nat. Sci. Res.* **2014**, *4*, 34–41.
19. Zolfaghari, A.; Bazargani-Gilani, B.; Aghajani, N. Edible Film Based on Corn Zein Containing Dill Extract and Essential Oil/β-Cyclodextrin Inclusion Complex: Shelf Life Enhancement of Common Carp Fillet. *Food Sci. Nutr.* **2023**, *11*, 4275–4288. [CrossRef]
20. Altan, C.O.; Köstekli, B.; Çorapcı, B.; İpar, M.S.; Kocatepe, D.; Turan, H. The Sensory Characteristics, Nutritional Profile and Physical Changes of the Atlantic Bonito (Sarda Sarda Bloch, 1793) Gravlax: Effect of Dill (*Anethum graveolens*) and Garden Cress (*Lepidium sativum*). *Int. J. Gastron. Food Sci.* **2022**, *28*, 100490. [CrossRef]
21. Anvar, N.; Nateghi, L.; Shariatifar, N.; Mousavi, S.A. The Effect of Essential Oil of *Anethum graveolens* L. Seed and Gallic Acid (Free and Nano Forms) on Microbial, Chemical and Sensory Characteristics in Minced Meat during Storage at 4 °C. *Food Chem. X* **2023**, *19*, 100842. [CrossRef]
22. Behbahani, B.A.; Shahidi, F.; Yazdi, F.T.; Mortazavi, S.A.; Mohebbi, M. Use of Plantago Major Seed Mucilage as a Novel Edible Coating Incorporated with *Anethum graveolens* Essential Oil on Shelf Life Extension of Beef in Refrigerated Storage. *Int. J. Biol. Macromol.* **2017**, *94*, 515–526. [CrossRef]
23. Tian, J.; Ban, X.; Zeng, H.; Huang, B.; He, J.; Wang, Y. In Vitro and in Vivo Activity of Essential Oil from Dill (*Anethum graveolens* L.) against Fungal Spoilage of Cherry Tomatoes. *Food Control* **2011**, *22*, 1992–1999. [CrossRef]
24. Zandi-Sohani, N.; Hojjati, M.; Carbonell-Barrachina, Á.A. Insecticidal and Repellent Activities of the Essential Oil of *Callistemon citrinus* (Myrtaceae) Against *Callosobruchus maculatus* (F.) (Coleoptera: Bruchidae). *Neotrop. Entomol.* **2013**, *42*, 89–94. [CrossRef] [PubMed]
25. El-Kareem, M.S.M.A.; Rabbih, M.A.E.F.; Selim, E.T.M.; Elsherbiny, E.A.E.; El-Khateeb, A.Y. Application of GC/EIMS in Combination with Semi-Empirical Calculations for Identification and Investigation of Some Volatile Components in Basil Essential Oil. *Int. J. Anal. Mass Spectrom. Chromatogr.* **2016**, *4*, 14–25. [CrossRef]
26. Ang, L.Z.P.; Hashim, R.; Sulaiman, S.F.; Coulibaly, A.Y.; Sulaiman, O.; Kawamura, F.; Salleh, K.M. In Vitro Antioxidant and Antidiabetic Activites of Gluta Torquata. *Ind. Crops Prod.* **2015**, *76*, 755–760. [CrossRef]
27. Elsheikh, M.; Osman, A.; Edris, S.; Dawam, W.; Sitohy, M.; Sabeq, I. Soybean Glycinin's Antibacterial Properties Provide a Feasible Natural Alternative for Improving the Overall Quality and Shelf-Life of Beef Steaks and Combating Foodborne Pathogens. *Food Bioprocess Technol.* **2024**, *18*, 1777–1792. [CrossRef]
28. Sabeq, I.; Awad, D.; Hamad, A.; Nabil, M.; Aboubakr, M.; Abaza, M.; Fouad, M.; Hussein, A.; Shama, S.; Ramadan, H.; et al. Prevalence and Molecular Characterization of Foodborne and Human-derived Salmonella Strains for Resistance to Critically Important Antibiotics. *Transbound. Emerg. Dis.* **2022**, *69*, e2153–e2163. [CrossRef]
29. Gamil, B.; Salem, A.M.; Arab, W.S.; Sabeq, I.I. The Microbiological Quality, Shelf-Life, and Multidrug-Resistant Salmonella Contamination Rates Assessment in Chicken Giblets Purchased from Live Poultry Shops. *Microbe* **2024**, *3*, 100057. [CrossRef]

30. Dawam, W.; Elsheikh, M.; Edris, S.; Osman, A.; Sitohy, M.; Sabike, I.I. Soybean 11S Globulin Spray Boosts Chilled Shrimp Shelf-Life, Physicochemical, and Sensory Attributes beyond Twelve Days. *Benha Vet. Med. J.* **2024**, *46*, 69–73. [CrossRef]
31. Elsheikh, M.; Dawam, W.; Edris, shimaa; Osman, A.; Sitohy, M.; Sabike, I.I. Cowpea 11S Globulin Spray Quadruples Shrimp Shelf-Life and Maintains Freshness. *Benha Vet. Med. J.* **2024**, *46*, 74–79. [CrossRef]
32. Honikel, K.O. Reference Methods for the Assessment of Physical Characteristics of Meat. *Meat Sci.* **1998**, *49*, 447–457. [CrossRef]
33. Wheeler, T.L.; Shackelford, S.D.; Koohmaraie, M. *Warner-Bratzler Shear Force Protocol*; USDA-ARS US Meat Animal Research Center: Clay Center, NE, USA, 2009; pp. 1–16.
34. Youssuf, H.; Soror, E.I.; Shehab, A.; El-daim, A.M.; Abo-Gamil, Z.H.; Ahmed-Farid, O.; Hamad, A.; Edris, S.; Matter, A.F. Amelioration of Hypoxia and Cold Stress in Nile Tilapia: Comparative Effect of Chlorella Vulgaris and Its Nanoparticle Dietary Supplementation on Performance, Antioxidant, Hepatic Functions, and Meat Quality. *Aquac. Int.* **2025**, *33*, 66. [CrossRef]
35. Saricoban, C.; Yilmaz, M.T. Modelling the Effects of Processing Factors on the Changes in Colour Parameters of Cooked Meatballs Using Response Surface Methodology. *World Appl. Sci. J.* **2010**, *9*, 14–22.
36. Rhim, J.W.; Wu, Y.; Weller, C.L.; Schnepf, M. Physical Characteristics of a Composite Film of Soy Protein Isolate and Propyleneglycol Alginate. *J. Food Sci.* **1999**, *64*, 149–152. [CrossRef]
37. Qian, Y.F.; Xie, J.; Yang, S.P.; Wu, W.H. Study of the Quality Changes and Myofibrillar Proteins of White Shrimp (*Litopenaeus vannamei*) under Modified Atmosphere Packaging with Varying CO_2 Levels. *Eur. Food Res. Technol.* **2013**, *236*, 629–635. [CrossRef]
38. Heires, M. The International Organization for Standardization (ISO). *New Political Econ.* **2008**, *13*, 357–367. [CrossRef]
39. Sabike, I.I.; Fujikawa, H.; Edris, A.M. The Growth Kinetics of Salmonella Enteritidis in Raw Ground Beef. *Biocontrol Sci.* **2015**, *20*, 185–192. [CrossRef]
40. *ISO 4833-1:2013*; Microbiology of the Food Chain- Horizontal Method for the Enumeration of Microorganisms—Part 1: Colony Count at 30 °C by the Pour Plate Technique. International Organization for Standardization: Geneva, Switzerland, 2013; pp. 1–8.
41. *ISO 4832:2006*; Microbiology of Food and Animal Feeding Stuffs. Horizontal Method for the Enumeration of Coliforms. Colony-Count Technique. International Organization for Standardization: Geneva, Switzerland, 2006; p. 30.
42. Sabike, I.I.; Fujikawa, H.; Sakha, M.Z.; Edris, A.M. Production of Staphylococcus Aureus Enterotoxin a in Raw Milk at High Temperatures. *J. Food Prot.* **2014**, *77*, 1612–1616. [CrossRef]
43. Léguillier, V.; Pinamonti, D.; Chang, C.-M.; Gunjan; Mukherjee, R.; Himanshu; Cossetini, A.; Manzano, M.; Anba-Mondoloni, J.; Malet-Villemagne, J.; et al. A Review and Meta-Analysis of Staphylococcus Aureus Prevalence in Foods. *Microbe* **2024**, *4*, 100131. [CrossRef]
44. Fik, M.; Leszczynska-Fik, A. Microbiological and Sensory Changes in Minced Beef Treated with Potassium Lactate and Sodium Diacetate during Refrigerated Storage. *Int. J. Food Prop.* **2007**, *10*, 589–598. [CrossRef]
45. Kostić, I.; Lazarević, J.; Šešlija Jovanović, D.; Kostić, M.; Marković, T.; Milanović, S. Potential of Essential Oils from Anise, Dill and Fennel Seeds for the Gypsy Moth Control. *Plants* **2021**, *10*, 2194. [CrossRef]
46. Benlembarek, K.; Lograda, T.; Ramdani, M.; Figueredo, G.; Chalard, P. Chemical Composition and Biological Activities of *Anethum graveolens* L. Essential Oil from Algeria. *J. Essent. Oil Bear. Plants* **2022**, *25*, 728–740. [CrossRef]
47. Ma, B.; Ban, X.; Huang, B.; He, J.; Tian, J.; Zeng, H.; Chen, Y.; Wang, Y. Interference and Mechanism of Dill Seed Essential Oil and Contribution of Carvone and Limonene in Preventing Sclerotinia Rot of Rapeseed. *PLoS ONE* **2015**, *10*, e0131733. [CrossRef] [PubMed]
48. El-Sayed, K.K.; El-Sheikh, E.-S.A.; Sherif, R.M.; Gouhar, K.A. Chemical Composition and Bio-Efficacy of Essential Oils Isolated from Seeds of *Anethum graveolens* L., Leaves of *Thymus vulgaris* L., and Nuts of *Myristica fragrans* Houtt. Against *Callosobruchus maculatus* (Fab.) (Coleoptera: Bruchidae). *J. Essent. Oil Bear. Plants* **2021**, *24*, 1402–1414. [CrossRef]
49. Bhatia, S.; Al-Harrasi, A.; Jawad, M.; Shah, Y.A.; Al-Azri, M.S.; Ullah, S.; Anwer, M.K.; Aldawsari, M.F.; Koca, E.; Aydemir, L.Y. A Comparative Study of the Properties of Gelatin (Porcine and Bovine)-Based Edible Films Loaded with Spearmint Essential Oil. *Biomimetics* **2023**, *8*, 172. [CrossRef]
50. Jianu, C.; Stoin, D.; Cocan, I.; David, I.; Pop, G.; Lukinich-Gruia, A.T.; Mioc, M.; Mioc, A.; Șoica, C.; Muntean, D.; et al. In Silico and In Vitro Evaluation of the Antimicrobial and Antioxidant Potential of Mentha × Smithiana R. GRAHAM Essential Oil from Western Romania. *Foods* **2021**, *10*, 815. [CrossRef]
51. Choi, K.-C.; Son, Y.-O.; Hwang, J.-M.; Kim, B.-T.; Chae, M.; Lee, J.-C. Antioxidant, Anti-Inflammatory and Anti-Septic Potential of Phenolic Acids and Flavonoid Fractions Isolated from *Lolium multiflorum*. *Pharm. Biol.* **2016**, *55*, 611–619. [CrossRef]
52. Cai, Y.-Z.; Sun, M.; Xing, J.; Luo, Q.; Corke, H. Structure–Radical Scavenging Activity Relationships of Phenolic Compounds from Traditional Chinese Medicinal Plants. *Life Sci.* **2006**, *78*, 2872–2888. [CrossRef]

53. Ben Hsouna, A.; Ben Halima, N.; Smaoui, S.; Hamdi, N. Citrus Lemon Essential Oil: Chemical Composition, Antioxidant and Antimicrobial Activities with Its Preservative Effect against Listeria Monocytogenes Inoculated in Minced Beef Meat. *Lipids Health Dis.* **2017**, *16*, 146. [CrossRef]
54. Yu, L.; Yan, J.; Sun, Z. D-Limonene Exhibits Anti-Inflammatory and Antioxidant Properties in an Ulcerative Colitis Rat Model via Regulation of INOS, COX-2, PGE2 and ERK Signaling Pathways. *Mol. Med. Rep.* **2017**, *15*, 2339–2346. [CrossRef]
55. Sun, J. D-Limonene: Safety and Clinical Applications. *Altern. Med. Rev.* **2007**, *12*, 259.
56. Jannesar, N.; Bassiri, A.; Ghavami, M.; Chenarbon, H.A.; Tarzi, B.G. Investigation of Physicochemical and Antibacterial Properties of Dill (*Anethum graveolens* L.) Microencapsulated Essential Oil Using Fluidized Bed Method. *Food Chem. X* **2024**, *23*, 101708. [CrossRef]
57. Shahidi, F.; Janitha, P.K.; Wanasundara, P.D. Phenolic Antioxidants. *Crit. Rev. Food Sci. Nutr.* **1992**, *32*, 67–103. [CrossRef] [PubMed]
58. Elgayyar, M.; Draughon, F.A.; Golden, D.A.; Mount, J.R. Antimicrobial Activity of Essential Oils from Plants against Selected Pathogenic and Saprophytic Microorganisms. *J. Food Prot.* **2001**, *64*, 1019–1024. [CrossRef] [PubMed]
59. Derakhshan, S.; Navidinia, M.; Ahmadi, A. Antibacterial Activity of Dill (*Anethum graveolens*) Essential Oil and Antibiofilm Activity of Cumin (*Cuminum cyminum*) Alcoholic Extract. *Infect. Epidemiol. Microbiol.* **2017**, *3*, 122–126.
60. Pandey, A.K.; Singh, P.; Tripathi, N.N. Chemistry and Bioactivities of Essential Oils of Some Ocimum Species: An Overview. *Asian Pac. J. Trop. Biomed.* **2014**, *4*, 682–694. [CrossRef]
61. Mutlu-Ingok, A.; Karbancioglu-Guler, F. Cardamom, Cumin, and Dill Weed Essential Oils: Chemical Compositions, Antimicrobial Activities, and Mechanisms of Action against *Campylobacter* spp. *Molecules* **2017**, *22*, 1191. [CrossRef]
62. Huang, L.; Wang, Y.; Li, R.; Wang, Q.; Dong, J.; Wang, J.; Lu, S. Thyme Essential Oil and Sausage Diameter Effects on Biogenic Amine Formation and Microbiological Load in Smoked Horse Meat Sausage. *Food Biosci.* **2021**, *40*, 100885. [CrossRef]
63. Sikkema, J.; de Bont, J.A.; Poolman, B. Mechanisms of Membrane Toxicity of Hydrocarbons. *Microbiol. Rev.* **1995**, *59*, 201–222. [CrossRef]
64. Abbas, K.A.; Mohamed, A.; Jamilah, B.; Ebrahimian, M. A Review on Correlations between Fish Freshness and PH during Cold Storage. *Am. J. Biochem. Biotechnol.* **2008**, *4*, 416–421. [CrossRef]
65. Tian, J.; Ban, X.; Zeng, H.; He, J.; Chen, Y.; Wang, Y. The Mechanism of Antifungal Action of Essential Oil from Dill (*Anethum graveolens* L.) on Aspergillus Flavus. *PLoS ONE* **2012**, *7*, e30147. [CrossRef]
66. Özalp Özen, B.; Soyer, A. Effect of Plant Extracts on Lipid and Protein Oxidation of Mackerel (Scomber Scombrus) Mince during Frozen Storage. *J. Food Sci. Technol.* **2017**, *55*, 120–127. [CrossRef]
67. Kurek, M.; Pišonić, P.; Ščetar, M.; Janči, T.; Čanak, I.; Vidaček Filipec, S.; Benbettaieb, N.; Debeaufort, F.; Galić, K. Edible Coatings for Fish Preservation: Literature Data on Storage Temperature, Product Requirements, Antioxidant Activity, and Coating Performance—A Review. *Antioxidants* **2024**, *13*, 1417. [CrossRef] [PubMed]
68. Caglak, E.; Karsli, B. Use of Dill Extracts as a Natural Preservative on Shelf-life Extension of Rainbow Trout Croquettes during Refrigerator Storage. *Food Sci. Nutr.* **2023**, *11*, 7330–7340. [CrossRef] [PubMed]
69. Santos, H.M.C.; Méndez, L.; Secci, G.; Parisi, G.; Martelli, R.; Medina, I. Pathway-Oriented Action of Dietary Essential Oils to Prevent Muscle Protein Oxidation and Texture Deterioration of Farmed Rainbow Trout. *Animal* **2019**, *13*, 2080–2091. [CrossRef] [PubMed]
70. Singh, A.; Benjakul, S. Proteolysis and Its Control Using Protease Inhibitors in Fish and Fish Products: A Review. *Compr. Rev. Food Sci. Food Saf.* **2018**, *17*, 496–509. [CrossRef]
71. Li, N.; Xie, J.; Chu, Y.M. Degradation and Evaluation of Myofibril Proteins Induced by Endogenous Protease in Aquatic Products during Storage: A Review. *Food Sci. Biotechnol.* **2023**, *32*, 1005–1018. [CrossRef]
72. Estévez, M.; Luna, C. Dietary Protein Oxidation: A Silent Threat to Human Health? *Crit. Rev. Food Sci. Nutr.* **2017**, *57*, 3781–3793. [CrossRef]
73. Wojtunik-Kulesza, K.A.; Rudkowska, M.; Klimek, K.; Mołdoch, J.; Agacka-Mołdoch, M.; Budzyńska, B.; Oniszczuk, A. S-(+)-Carvone, a Monoterpene with Potential Anti-Neurodegenerative Activity—In Vitro, In Vivo and Ex Vivo Studies. *Molecules* **2024**, *29*, 4365. [CrossRef]
74. Lund, M.N.; Heinonen, M.; Baron, C.P.; Estévez, M. Protein Oxidation in Muscle Foods: A Review. *Mol. Nutr. Food Res.* **2010**, *55*, 83–95. [CrossRef]
75. Bragadóttir, M. Endogenous Antioxidants in Fish. Ph.D. Thesis, University of Iceland, Reykjavík, Iceland, 2001.
76. Singh, J.; Singh, B. Inhibition of Post-Mortem Fish Muscle Softening and Degradation Using Legume Seed Proteinase Inhibitors. *J. Food Sci. Technol.* **2019**, *57*, 1–11. [CrossRef]
77. Lee, S.; Kim, M.-G.; Hur, S.-W.; Katya, K.; Kim, K.-W.; Lee, B.-J. Assessment of Safety, Effects, and Muscle-Specific Accumulation of Dietary Butylated Hydroxytoluene (BHT) in *Paralichthys olivaceus*. *Aquac. Nutr.* **2023**, *2023*, 1381923. [CrossRef]

78. Wu, H.; Richards, M.P.; Undeland, I. Lipid Oxidation and Antioxidant Delivery Systems in Muscle Food. *Compr. Rev. Food Sci. Food Saf.* **2022**, *21*, 1275–1299. [CrossRef] [PubMed]
79. Saleh-e-In, M.M.; Sultana, N.; Rahim, M.M.; Ahsan, M.A.; Bhuiyan, M.N.H.; Hossain, M.N.; Rahman, M.M.; Kumar Roy, S.; Islam, M.R. Chemical Composition and Pharmacological Significance of *Anethum sowa* L. Root. *BMC Complement. Altern. Med.* **2017**, *17*, 127. [CrossRef]
80. Qian, R.; Sun, C.; Bai, T.; Yan, J.; Cheng, J.; Zhang, J. Recent Advances and Challenges in the Interaction between Myofibrillar Proteins and Flavor Substances. *Front. Nutr.* **2024**, *11*, 1378884. [CrossRef]
81. Zengin, H.; Baysal, A. Antibacterial and Antioxidant Activity of Essential Oil Terpenes against Pathogenic and Spoilage-Forming Bacteria and Cell Structure-Activity Relationships Evaluated by SEM Microscopy. *Molecules* **2014**, *19*, 17773–17798. [CrossRef]
82. Guo, A.; Jiang, J.; True, A.D.; Xiong, Y.L. Myofibrillar Protein Cross-Linking and Gelling Behavior Modified by Structurally Relevant Phenolic Compounds. *J. Agric. Food Chem.* **2021**, *69*, 1308–1317. [CrossRef]
83. Prodpran, T.; Benjakul, S.; Phatcharat, S. Effect of Phenolic Compounds on Protein Cross-Linking and Properties of Film from Fish Myofibrillar Protein. *Int. J. Biol. Macromol.* **2012**, *51*, 774–782. [CrossRef]
84. Burt, S. Essential Oils: Their Antibacterial Properties and Potential Applications in Foods—A Review. *Int. J. Food Microbiol.* **2004**, *94*, 223–253. [CrossRef]
85. Domínguez, R.; Pateiro, M.; Munekata, P.E.S.; McClements, D.J.; Lorenzo, J.M. Encapsulation of Bioactive Phytochemicals in Plant-Based Matrices and Application as Additives in Meat and Meat Products. *Molecules* **2021**, *26*, 3984. [CrossRef]
86. Torres-Arreola, W.; Soto-Valdez, H.; Peralta, E.; Cárdenas-López, J.L.; Ezquerra-Brauer, J.M. Effect of a Low-Density Polyethylene Film Containing Butylated Hydroxytoluene on Lipid Oxidation and Protein Quality of Sierra Fish (*Scomberomorus sierra*) Muscle during Frozen Storage. *J. Agric. Food Chem.* **2007**, *55*, 6140–6146. [CrossRef]
87. Sohn, J.; Taki, Y.; Ushio, H.; Kohata, T.; Shioya, I.; Ohshima, T. Lipid Oxidations in Ordinary and Dark Muscles of Fish: Influences on Rancid Off-odor Development and Color Darkening of Yellowtail Flesh During Ice Storage. *J. Food Sci.* **2005**, *70*, s490–s496. [CrossRef]
88. Zhu, W.; Han, M.; Bu, Y.; Li, X.; Yi, S.; Xu, Y.; Li, J. Plant Polyphenols Regulating Myoglobin Oxidation and Color Stability in Red Meat and Certain Fish: A Review. *Crit. Rev. Food Sci. Nutr.* **2022**, *64*, 2276–2288. [CrossRef] [PubMed]
89. Faustman, C. Meat Quality Myoglobin Chemistry and Modifications That Influence (Color and) Color Stability. In Proceedings of the American Meat Science Association, 67th Annual Reciprocal Meat Conference, Madison, WI, USA, 15–18 June 2014.
90. Khoo, H.-E.; Prasad, K.N.; Kong, K.-W.; Jiang, Y.; Ismail, A. Carotenoids and Their Isomers: Color Pigments in Fruits and Vegetables. *Molecules* **2011**, *16*, 1710–1738. [CrossRef] [PubMed]
91. Lancaster, J.E.; Lister, C.E.; Reay, P.F.; Triggs, C.M. Influence of Pigment Composition on Skin Color in a Wide Range of Fruit and Vegetables. *J. Am. Soc. Hortic. Sci.* **1997**, *122*, 594–598.
92. Tongnuanchan, P.; Benjakul, S.; Prodpran, T. Effects of Oxygen and Antioxidants on the Lipid Oxidation and Yellow Discolouration of Film from Red Tilapia Mince. *J. Sci. Food Agric.* **2012**, *92*, 2507–2517. [CrossRef]
93. Noumi, E.; Ahmad, I.; Adnan, M.; Merghni, A.; Patel, H.; Haddaji, N.; Bouali, N.; Alabbosh, K.F.; Ghannay, S.; Aouadi, K.; et al. GC/MS Profiling, Antibacterial, Anti-Quorum Sensing, and Antibiofilm Properties of *Anethum graveolens* L. Essential Oil: Molecular Docking Study and In-Silico ADME Profiling. *Plants* **2023**, *12*, 1997. [CrossRef]
94. Jirovetz, L.; Buchbauer, G.; Stoyanova, A.S.; Georgiev, E.V.; Damianova, S.T. Composition, Quality Control, and Antimicrobial Activity of the Essential Oil of Long-Time Stored Dill (*Anethum graveolens* L.) Seeds from Bulgaria. *J. Agric. Food Chem.* **2003**, *51*, 3854–3857. [CrossRef]
95. Ruangamnart, A.; Buranaphalin, S.; Temsiririrkkul, R.; Chuakul, W.; Pratuangdejkul, J. Chemical Compositions and Antibacterial Activity of Essential Oil from Dill Fruits (*Anethum graveolens* L.) Cultivated in Thailand. *Mahidol Univ. J. Pharm. Sci.* **2015**, *42*, 135–143.
96. Popa, M.; Măruțescu, L.; Oprea, E.; Bleotu, C.; Kamerzan, C.; Chifiriuc, M.C.; Grădișteanu Pircalabioru, G. In Vitro Evaluation of the Antimicrobial and Immunomodulatory Activity of Culinary Herb Essential Oils as Potential Perioceutics. *Antibiotics* **2020**, *9*, 428. [CrossRef]
97. Snuossi, M.; Trabelsi, N.; Ben Taleb, S.; Dehmeni, A.; Flamini, G.; De Feo, V. Laurus Nobilis, Zingiber Officinale and *Anethum graveolens* Essential Oils: Composition, Antioxidant and Antibacterial Activities against Bacteria Isolated from Fish and Shellfish. *Molecules* **2016**, *21*, 1414. [CrossRef]
98. Kannaiyan, S.; Gunasekaran, J.; Kannuchamy, N.; Thachil, M.T.; Gudipati, V. Antioxidant and Antibacterial Activities of Dill Extracts and Their Preservative Effect on Mackerel Fillets during Refrigerated Storage. *Indian. J. Nat. Prod. Resour.* **2015**, *6*, 106–113.

foods

MDPI

Article

Integrating Life Cycle Assessment in Innovative Berry Processing with Edible Coating and Osmotic Dehydration

Alexandra Mari *, Tryfon Kekes, Christos Boukouvalas and Magdalini Krokida

School of Chemical Engineering, National Technical University of Athens, Zografou, 15780 Athens, Greece
* Correspondence: alexandramari@mail.ntua.gr; Tel.: +30-2107723149

Abstract: This study presents a Life Cycle Assessment (LCA) of a berry production system using osmotic dehydration and edible coating to extend the shelf life and improve the nutritional value. The goal is to evaluate environmental impacts, identify hotspots, and propose improvements. Osmotic dehydration is the main contributor to environmental impact, particularly due to the energy and resources required by apple juice as the osmotic agent. It contributes up to 0.64 kg CO_2 eq. per kg of blueberries, 1.36 kg CO_2 eq. per kg of raspberries, and 0.66 kg CO_2 eq. per kg of strawberries. The edible coating, however, has minimal environmental impact due to its low energy consumption and biodegradable materials. Packaging has a lower carbon footprint but contributes more to fossil fuel depletion and human toxicity. Raspberries show the highest human health impact (3.5×10^{-6} DALY/kg) and ecosystem impact (9.5×10^{-8} species.yr/kg), followed by strawberries (1.78×10^{-6} DALY/kg, 4.97×10^{-8} species.yr/kg) and blueberries (1.7×10^{-6} DALY/kg, 5.1×10^{-8} species.yr/kg), highlighting the greater environmental and health costs of raspberries. Despite the environmental burden of osmotic dehydration, it offers economic benefits by extending the shelf life, reducing losses, improving supply chain efficiency, and enhancing product quality, which leads to higher prices and profit margins. The study concludes that, while the environmental impacts of osmotic dehydration should be optimized, its economic and logistical benefits make it a promising preservation solution. Further research into eco-friendly practices is recommended to reduce ecological costs while maintaining commercial advantages.

Keywords: Life Cycle Assessment; berries processing; sustainability; osmotic dehydration; edible coating

1. Introduction

Modern lifestyles are intricately linked to dietary habits, with fruits and vegetables being integral components of a healthy diet due to their abundance of bioactive compounds [1]. Berries are a valuable component of a healthy diet, known for their rich nutritional profile, including essential vitamins (C and K), minerals (manganese and potassium), and bioactive compounds such as flavonoids and anthocyanins [2]. These phytochemicals offer potent antioxidant and anti-inflammatory benefits, supporting heart health, brain function, and reducing the risk of chronic diseases like diabetes and cancer [3,4]. Due to their low-calorie content and natural sweetness, berries are increasingly popular among health-conscious consumers [5,6].

The shelf life of berries is a crucial factor in ensuring their availability and minimizing losses throughout the food supply chain. Due to their high perishability and rapid deterioration, berries are particularly susceptible to spoilage, which leads to significant food waste, especially during the transportation and storage stages [7]. The seasonal nature of berry production often results in periods of oversupply, which, if not efficiently managed, can lead to excess product that spoils before reaching consumers [8,9]. The losses incurred not only affect the economic viability of berry producers but also contribute to the broader environmental issue of food waste [10,11]. According to the United Nations Food and Agriculture Organization (FAO), 13.8% of global food production is lost annually before it even reaches consumers, with fruits and vegetables being particularly vulnerable, experiencing a loss rate of 21.6% [12]. This loss has profound economic implications, particularly in the berry industry, where oversupply and inadequate storage systems can drastically reduce profitability [13].

To mitigate these losses, various preservation techniques have been developed to extend the shelf life of berries while maintaining their quality and nutritional value. The ability to extend the freshness of berries provides significant economic benefits, reducing the volume of product waste and improving supply chain efficiency [5,14]. Preservation methods such as osmotic dehydration and edible coatings are gaining attention for their effectiveness in prolonging berry freshness and minimizing spoilage. Osmotic dehydration works by immersing berries in hypertonic solutions, which reduces the moisture content and inhibits microbial growth, thus slowing down spoilage and extending the shelf life [15]. Edible coatings, typically made from natural biopolymers such as polysaccharides, proteins, and lipids, create a protective barrier that limits moisture loss, prevents oxygen ingress, and reduces microbial contamination [16]. These coatings help maintain the quality and texture of berries, extending their freshness during storage and transportation [17,18]. Additionally, the extended shelf life made possible by these methods not only reduces food waste but also improves the overall efficiency of the berry supply chain, allowing for better inventory management, fewer transport-related losses, and greater profitability for producers [19].

Life Cycle Assessment (LCA) plays a crucial role in evaluating the environmental impacts of these preservation methods [20]. By examining the entire lifecycle of a product or process—from raw material extraction to disposal—LCA provides valuable insights into the resource use, energy consumption, and emissions associated with different preservation strategies [21]. In the case of osmotic dehydration and edible coatings, LCA can help determine their environmental sustainability by comparing their ecological footprint and identifying opportunities for improvement [22]. This comprehensive evaluation enables the identification of the most sustainable practices within the food industry, particularly in the context of berry preservation.

The aim of this study is to conduct an LCA of sustainable techniques for preserving the quality and extending the shelf life of berries. Specifically, the research analyzes the environmental impacts of osmotic dehydration and edible coatings as methods to prolong the shelf life of berries, with the primary goal being to assess the feasibility and sustainability of these technologies. This study takes a comprehensive approach to evaluating these preservation methods from an environmental perspective, focusing on their long-term sustainability and potential to reduce the ecological footprint of berry preservation.

2. Materials and Methods

Life Cycle Assessment (LCA) was performed following the guidelines outlined in ISO 14040 and 14044:2006 [23], and it consists of four steps: (i) Goal & Scope Definition, (ii) Inventory Analysis, (iii) Impact Assessment, and (iv) Interpretation.

2.1. Goal and Scope

The Goal of the LCA analysis was to determine the effect of the implementation of mild processing methods, such as osmotic dehydration and edible coatings, in the development of innovative berries with an increased shelf-life and high nutritional value.

A gate-to-gate approach was selected for the evaluation of the environmental footprint of processed berries. Specifically, the system boundaries encompass production processes from berries pickup to processing and packaging.

The data utilized in this study were sourced from the GaBi professional and Ecoinvent databases, which pertain to the geographical area of the European Union 28 (EU-28). All the studies and data collected are relevant to the past five years.

The Scope of the LCA analysis involves defining the goals and boundaries of the study, collecting data on resource inputs and environmental outputs throughout all stages of the life cycle, evaluating potential environmental impacts, and interpreting the results to guide decision-making and foster sustainability.

The Life Cycle Assessment (LCA) was performed following the recommendations proposed by the ISO 14040 recommendations series (14040:2006 and 14044:2006) [23]. ReCiPe 2016 (H, hierarchist) was selected as a method to perform the impact assessment, with its main objective being the transformation of Life Cycle Inventory results into a limited number of environmental impact scores using characterization factors. Finally, GaBi ts software (v10.6.2.9, Sphera Solutions GmbH, Echterdingen, Stuttgart, Germany) was used for the calculation of the impact categories [24].

The ReCiPe 2016 methodology defines impact indicators at two levels: midpoint and endpoint indicators. Midpoint indicators focus on specific environmental impacts, providing detailed insights into areas such as climate change, particulate matter formation, and resource depletion. These indicators cover issues like climate change (with and without biogenic carbon), particulate matter, mineral resource depletion, freshwater consumption and ecotoxicity, eutrophication, human toxicity (carcinogenic and non-carcinogenic), ionizing radiation, land use, marine ecotoxicity, photochemical ozone formation, stratospheric ozone depletion, and terrestrial acidification and ecotoxicity.

Endpoint indicators aggregate the midpoint indicators to simplify result interpretation. However, as aggregation increases, so does the uncertainty of the results. Endpoint indicators offer a broader overview of environmental impacts, assessing damage to human health (measured in Disability-Adjusted Life Years, DALY), ecosystems (measured in species.yr), and resource availability (measured in monetary terms, $).

2.1.1. Product Systems and System's Boundaries

The evaluation was conducted on the innovative berry production line with extended shelf life. The system studied includes the process of receiving, processing, and packaging berries in an innovative production line, as shown in Figure 1. Specifically, compared to conventional berries that are received and directly packaged, processing methods of osmotic dehydration and edible coating have been added to the production phase.

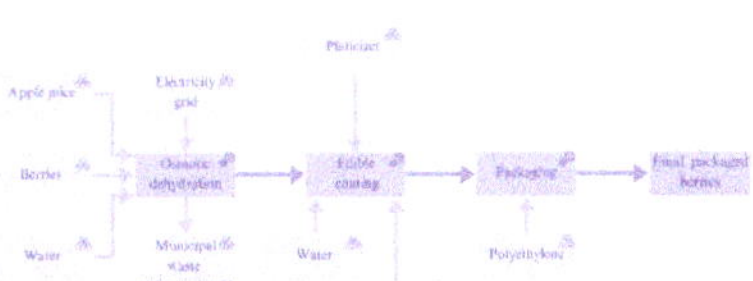

Figure 1. Flow chart of the innovative production of berries.

2.1.2. Process Analysis

Figure 2 provides a graphical representation of the production chain, illustrating the process from start to finish. It begins with the addition of freshly harvested berries into the osmotic solution, followed by their immersion in the edible coating solution, and concludes with the packaging of the berries as the final product.

Figure 2. Process of the innovative production of berries.

Osmotic Dehydration

Osmotic dehydration is a natural and mild method of removing water from food using osmotic solutions. Based on the principle of osmosis, the process exploits the movement of water from a region of low solute concentration to a region of high solute concentration through a semipermeable membrane [18]. In the osmotic dehydration of berries, apple juice at 42°Brix has been selected as the optimum solution, after an optimization conducted in the laboratory regarding the dehydration kinetics, the quality characteristics of the final product, and the shelf life [25]. The duration of dehydration varies between the different types, with blueberries and raspberries at 360 min and strawberries at 200 min to achieve the optimal result. The process is carried out at 40 °C, maintaining the nutritional value and organoleptic characteristics of the berries and making it suitable for industrial application. This method helps preserve the natural texture, aroma, and flavor of the berries, as it does not require high temperatures or mechanical stresses that could damage the cell walls [26]. At the same time, osmotic dehydration reduces the need for preservatives and extensive processing, making the berries healthier and more appealing to consumers seeking naturally processed products. Finally, this method offers economic and environmental benefits, as it reduces energy and resource consumption, contributing to a sustainable and efficient production of products.

Edible Coating

Edible coating is an innovative technology applied in food processing, aimed at improving their quality and shelf life. This process involves coating the berries with an edible film made from natural materials such as polysaccharides, proteins, and lipids [27]. Edible coatings are designed to create a protective barrier that reduces moisture loss and prevents oxygen ingress while simultaneously protecting the berries from microbial contamination [28]. The edible coating was selected based on the evaluation of various coatings conducted by Mari et al. (2024) [29]. In their study, the coating derived from the protein of *Chlorella vulgaris* was identified as the most effective for enhancing the shelf life of berries. The system studied includes the process of receiving, processing, and packaging berries in an innovative production line, as shown in Figure 1. Specifically, compared to conventional berries that are received and directly packaged, processing methods of osmotic dehydration and edible coating have been added to the production phase.

Packaging

The berries were packaged using High-Density Polyethylene (HDPE) plastic, selected for its durability and protective qualities. The packaging was designed to minimize physical damage during handling, transportation, and storage. HDPE serves as an effective barrier against moisture, oxygen, and contaminants, helping to maintain the freshness, appearance, and quality of the berries over time. The material's structural integrity ensures that the berries remain intact, reducing the risk of bruising or other damage [11]. The packaging was sealed to prevent moisture loss and exposure to air, thereby slowing down the deterioration process. HDPE's transparent nature allows for easy visibility of the product, aiding in consumer decision-making while also providing a stable surface for labeling and branding. Furthermore, the choice of HDPE was based on its cost-effectiveness and widely recognized performance in extending the shelf life for perishable items [28–31].This packaging method is commonly used for fresh produce due to its effectiveness in protecting the product and facilitating handling and transport. However, the environmental impact of HDPE packaging was also considered, with a focus on exploring more sustainable alternatives for future applications.

2.1.3. Functional Unit

The functional unit for the berry systems is defined as 1 kg (1 kg), while an analysis was also conducted using 1 euro (EUR) of revenue as a functional unit.

2.1.4. Assumptions and Limitations

The data used for berry production in both cases is derived from experimental studies and supplemented with literature reviews to ensure accuracy and represent the current industry conditions. This study primarily aims to assess the environmental footprint of the proposed methods and evaluate their feasibility for managing solid and liquid waste. Additionally, energy consumption is carefully considered, particularly in the osmotic dehydration process, where the highest energy losses occur.

2.1.5. Data Requirements

For the collection of data and the establishment of the inventory, values were obtained from experiments conducted by our research team between 2021 and 2024. These experimental data were combined with relevant literature data, and all figures were appropriately adjusted and verified through direct communication within the team.

2.2. *Life Cycle Inventory*

The Life Cycle Inventory (LCI) links processes with quantitative data based on the selected functional unit (1 kg of final packaged berries). Tables 1–3 present the input and output data for each process involved in berry processing, as depicted in Figure 2, for each berry. Literature and experimental data were used as a reference for data collection and inventory establishment, with appropriate adjustments made based on the specific context. These numbers were verified through careful review and consultation with relevant sources.

Table 1. Life Cycle Inventory (LCI) for innovative blueberries, expressed on a unit basis (1 kg of final packaged berries) of the final blueberry product.

Process	Flow	Quantity
Osmotic dehydration	[In] Blueberries (kg)	0.91
	[In] Apple juice (kg)	1.20
	[In] Water (kg)	1.02
	[In] Electricity (MJ)	0.25
	[Out] Blueberries (kg)	0.90
	[Out] Wastewater (kg)	1.72

Table 1. *Cont.*

Process	Flow	Quantity
Edible coating	[In] Berries (kg)	0.90
	[In] Protein (kg)	0.01
	[In] Tween 20 (kg)	3.58×10^{-4}
	[In] Glycerol (kg)	2.69×10^{-3}
	[In] Water (kg)	0.08
	[Out] Blueberries (kg)	0.99
Packaging	[In] Blueberries (kg)	0.99
	[In] HDPE (kg)	0.01
	[Out] Blueberries (kg)	1.00

Table 2. Life Cycle Inventory (LCI) for innovative raspberries, expressed on a unit basis (1 kg of final packaged berries) of the final raspberry product.

Process	Flow	Quantity
Osmotic dehydration	[In] Raspberries (kg)	1.17
	[In] Apple juice (kg)	1.52
	[In] Water (kg)	1.30
	[In] Electricity (MJ)	0.20
	[Out] Raspberries (kg)	0.90
	[Out] Wastewater (kg)	2.60
Edible coating	[In] Berries (kg)	0.90
	[In] Protein (kg)	0.01
	[In] Tween 20 (kg)	3.58×10^{-4}
	[In] Glycerol (kg)	2.69×10^{-3}
	[In] Water (kg)	0.08
	[Out] Raspberries (kg)	0.99
Packaging	[In] Raspberries (kg)	0.99
	[In] HDPE (kg)	0.01
	[Out] Raspberries (kg)	1.00

Table 3. Life Cycle Inventory (LCI) for innovative strawberries, expressed on a unit basis (1 kg of final packaged berries) of the final strawberry product.

Process	Flow	Quantity
Osmotic dehydration	[In] Strawberries (kg)	1.03
	[In] Apple juice (kg)	0.75
	[In] Water (kg)	1.15
	[In] Electricity (MJ)	0.25
	[Out] Strawberries (kg)	0.90
	[Out] Wastewater (kg)	3.94
Edible coating	[In] Berries (kg)	0.90
	[In] Protein (kg)	0.01
	[In] Tween 20 (kg)	3.58×10^{-4}
	[In] Glycerol (kg)	2.69×10^{-3}
	[In] Water (kg)	0.08
	[Out] Strawberries (kg)	0.99
Packaging	[In] Strawberries (kg)	0.99
	[In] HDPE (kg)	0.01
	[Out] Strawberries (kg)	1.00

3. Results and Discussion

3.1. Life Cycle Impact Assessment

The results of the environmental impact assessment throughout the life cycle of the production of innovative berries (blueberry, raspberry, and strawberry) with extended shelf life, per 1 kg of final packaged berries, are presented in Tables 4–6.

Table 4. Life cycle impact assessment results for the innovative blueberries (per 1 kg of final packaged berries) for the selected conventional and innovative midpoint impact categories.

Midpoint Impact Categories	Units	Osmotic Dehydration	Edible Coating	Packaging	Total
Climate change, default, excl biogenic carbon	kg CO_2 eq.	6.35E−01	1.27E−02	2.26E−02	6.70E−01
Climate change, incl biogenic carbon	kg CO_2 eq.	4.02E−01	4.21E−03	2.27E−02	4.29E−01
Fine Particulate Matter Formation	kg $PM_{2.5}$ eq.	6.42E−04	1.01E−05	7.88E−06	6.60E−04
Fossil depletion	kg oil eq.	1.20E−01	5.01E−03	2.28E−02	1.48E−01
Freshwater Consumption	m^3	6.92E−02	1.13E−04	9.86E−05	6.94E−02
Freshwater ecotoxicity	kg 1.4 DB eq.	1.03E−03	6.94E−06	8.24E−06	1.05E−03
Freshwater Eutrophication	kg P eq.	1.09E−05	1.08E−06	3.46E−08	1.20E−05
Human toxicity, cancer	kg 1.4-DB eq.	2.80E−04	3.79E−06	1.35E−05	2.97E−04
Human toxicity, non-cancer	kg 1.4-DB eq.	2.66E−02	1.27E−02	2.83E−03	4.21E−02
Ionizing Radiation	Bq C-60 eq. to air	1.29E−03	6.14E−04	7.77E−05	1.98E−03
Land use	Annual crop eq.·y	3.56E−01	1.31E−02	2.94E−04	3.69E−01
Marine ecotoxicity	kg 1.4-DB eq.	8.64E−04	1.11E−05	2.44E−05	9.00E−04
Marine Eutrophication	kg N eq.	4.29E−04	8.28E−06	2.86E−07	4.38E−04
Metal depletion	kg Cu eq.	5.35E−04	3.17E−04	1.63E−05	8.68E−04
Photochemical Ozone Formation, Ecosystems	kg NO_x eq.	2.99E−01	2.82E−02	1.62E−02	3.43E−01
Photochemical Ozone Formation, Human Health	kg NO_x eq.	1.87E−01	1.75E−02	1.00E−02	2.15E−01
Stratospheric Ozone Depletion	kg CFC-11 eq.	1.67E−06	2.18E−08	4.88E−09	1.70E−06
Terrestrial Acidification	kg SO_2 eq.	1.88E−03	4.35E−05	2.37E−05	1.95E−03
Terrestrial ecotoxicity	kg 1.4-DB eq.	7.14E−02	5.25E−03	3.12E−03	7.98E−02

Table 5. Life cycle impact assessment results for the innovative raspberries (per 1 kg of final packaged berries) for the selected conventional and innovative midpoint impact categories.

Midpoint Impact Categories	Units	Osmotic Dehydration	Edible Coating	Packaging	Total
Climate change, default, excl biogenic carbon	kg CO_2 eq.	1.36E+00	6.36E−02	2.26E−02	1.45E+00
Climate change, incl biogenic carbon	kg CO_2 eq.	8.35E−01	2.14E−02	2.26E−02	8.79E−01
Fine Particulate Matter Formation	kg $PM_{2.5}$ eq.	1.39E−03	5.05E−05	7.86E−06	1.45E−03
Fossil depletion	kg oil eq.	2.56E−01	2.51E−02	2.27E−02	3.04E−01
Freshwater Consumption	m^3	1.55E−01	5.67E−04	9.83E−05	1.56E−01
Freshwater ecotoxicity	kg 1.4 DB eq.	2.27E−03	3.45E−05	8.22E−06	2.31E−03
Freshwater Eutrophication	kg P eq.	2.24E−05	5.39E−06	3.45E−08	2.78E−05
Human toxicity, cancer	kg 1.4-DB eq.	5.53E−04	1.90E−05	1.35E−05	5.86E−04
Human toxicity, non-cancer	kg 1.4-DB eq.	5.67E−02	6.27E−02	2.82E−03	1.22E−01
Ionizing Radiation	Bq C-60 eq. to air	2.86E−03	3.07E−03	7.74E−05	6.01E−03
Land use	Annual crop eq.·y	7.92E−01	6.52E−02	2.93E−04	8.57E−01
Marine ecotoxicity	kg 1.4-DB eq.	1.89E−03	5.53E−05	2.44E−05	1.97E−03
Marine Eutrophication	kg N eq.	9.53E−04	4.12E−05	2.85E−07	9.94E−04
Metal depletion	kg Cu eq.	1.12E−03	1.58E−03	1.63E−05	2.72E−03
Photochemical Ozone Formation, Ecosystems	kg NO_x eq.	5.78E−01	1.41E−01	1.61E−02	7.35E−01
Photochemical Ozone Formation, Human Health	kg NO_x eq.	3.61E−01	8.77E−02	1.00E−02	4.59E−01
Stratospheric Ozone Depletion	kg CFC-11 eq.	3.71E−06	1.08E−07	4.86E−09	3.82E−06
Terrestrial Acidification	kg SO_2 eq.	4.09E−03	2.16E−04	2.36E−05	4.33E−03
Terrestrial ecotoxicity	kg 1.4-DB eq.	1.29E−01	2.61E−02	3.11E−03	1.58E−01

Table 6. Life cycle impact assessment results for the innovative strawberries (per 1 kg of final packaged berries) for the selected conventional and innovative midpoint impact categories.

Midpoint Impact Categories	Units	Osmotic Dehydration	Edible Coating	Packaging	Total
Climate change, default, excl biogenic carbon	kg CO_2 eq.	6.65E−01	2.55E−02	2.27E−02	7.13E−01
Climate change, incl biogenic carbon	kg CO_2 eq.	4.15E−01	8.51E−03	2.27E−02	4.46E−01
Fine Particulate Matter Formation	kg $PM_{2.5}$ eq.	6.75E−04	2.02E−05	7.90E−06	7.03E−04
Fossil depletion	kg oil eq.	1.23E−01	1.01E−02	2.29E−02	1.56E−01
Freshwater Consumption	m3	7.41E−02	2.27E−04	9.88E−05	7.44E−02
Freshwater ecotoxicity	kg 1.4 DB eq.	1.15E−03	1.39E−05	8.26E−06	1.17E−03
Freshwater Eutrophication	kg P eq.	1.51E−05	2.16E−06	3.47E−08	1.73E−05
Human toxicity, cancer	kg 1.4-DB eq.	4.03E−04	7.60E−06	1.36E−05	4.24E−04
Human toxicity, non-cancer	kg 1.4-DB eq.	2.81E−02	2.53E−02	2.84E−03	5.62E−02
Ionizing Radiation	Bq C-60 eq. to air	1.44E−03	1.23E−03	7.78E−05	2.75E−03
Land use	Annual crop eq.·y	3.92E−01	2.61E−02	2.94E−04	4.18E−01
Marine ecotoxicity	kg 1.4-DB eq.	9.60E−04	2.22E−05	2.45E−05	1.01E−03
Marine Eutrophication	kg N eq.	4.78E−04	1.65E−05	2.87E−07	4.95E−04
Metal depletion	kg Cu eq.	6.25E−04	6.33E−04	1.64E−05	1.27E−03
Photochemical Ozone Formation, Ecosystems	kg NO_x eq.	2.52E−01	5.65E−02	1.62E−02	3.25E−01
Photochemical Ozone Formation, Human Health	kg NO_x eq.	1.58E−01	3.51E−02	1.01E−02	2.03E−01
Stratospheric Ozone Depletion	kg CFC-11 eq.	1.83E−06	4.35E−08	4.89E−09	1.88E−06
Terrestrial Acidification	kg SO_2 eq.	1.98E−03	8.69E−05	2.37E−05	2.09E−03
Terrestrial ecotoxicity	kg 1.4-DB eq.	4.81E−02	1.05E−02	3.13E−03	6.17E−02

Subsequently, detailed diagrams are presented in Figures 3–5, illustrating the contribution of each individual process to the total footprint of the selected midpoint impact categories, providing a comprehensive analysis of the environmental burden.

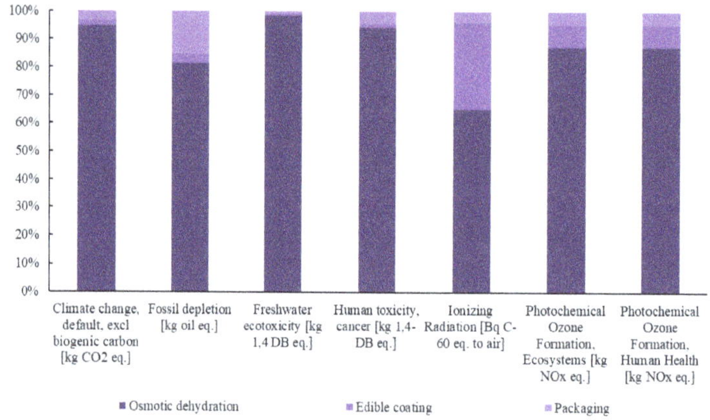

Figure 3. Contribution of each process to the final footprint of the selected midpoint impact categories for the production of innovative blueberries.

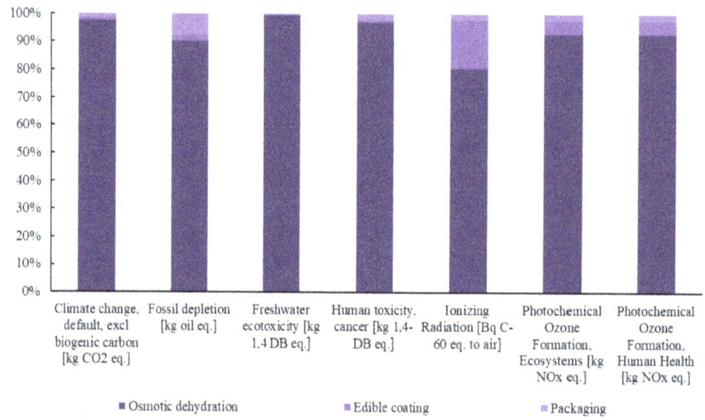

Figure 4. Contribution of each process to the final footprint of the selected midpoint impact categories for the production of innovative raspberries.

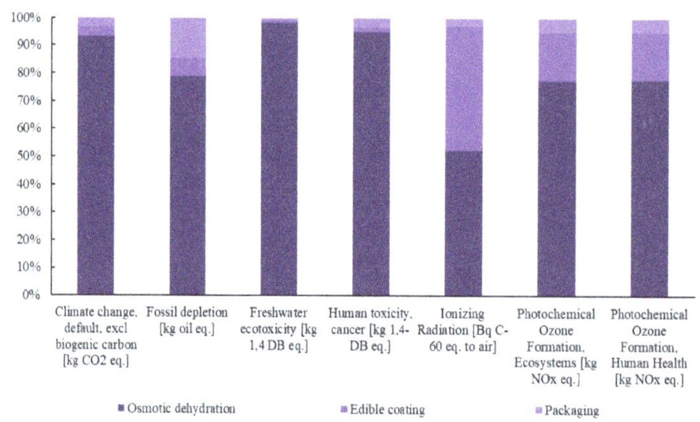

Figure 5. Contribution of each process to the final footprint of the selected midpoint impact categories for the production of innovative strawberries.

Figure 6 represents the total climate change, default, and biogenic carbon of the berry process. The impact assessment was conducted using two functional units: product mass (1 kg) and economic value (1 EUR). This dual approach aims to evaluate the environmental and economic trade-offs, ensuring a comprehensive and well-founded analysis. The study investigates the environmental and socio-economic implications in relation to profit margins to derive robust and data-driven conclusions. The pricing of berries was determined based on an analysis of the Greek market, with the following estimated values: blueberries at 24 EUR/kg, raspberries at 39 EUR/kg, and strawberries at 5 EUR/kg.

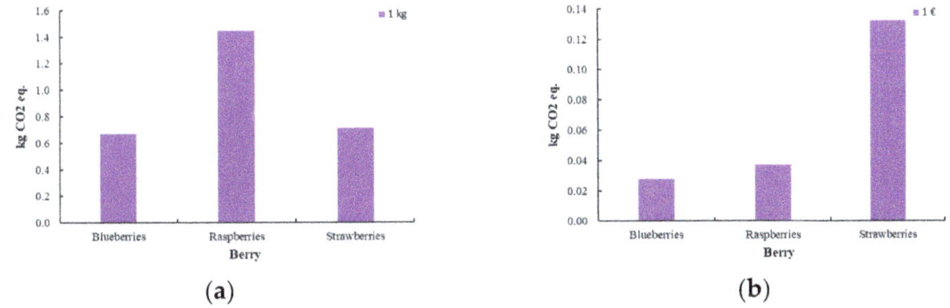

Figure 6. Climate change, default, and biogenic carbon [kg CO_2 eq.] for the production of innovative berries (blueberry, raspberry, strawberry) (**a**) per 1 kg of product and (**b**) per 1 EUR of product.

Figures 7–9 illustrate the endpoint impact categories, which describe the environmental consequences associated with the production of innovative berries. These impacts are classified into three main categories: effects on human health, ecosystem integrity, and resource availability. Human health impacts (DALY) quantify the potential risks arising from exposure to pollutants during berry processing. Ecosystem impacts (species.yr) assess potential biodiversity loss and ecosystem destabilization resulting from agricultural practices, while resource availability ($) reflects the depletion of natural resources and the corresponding economic costs associated with their use.

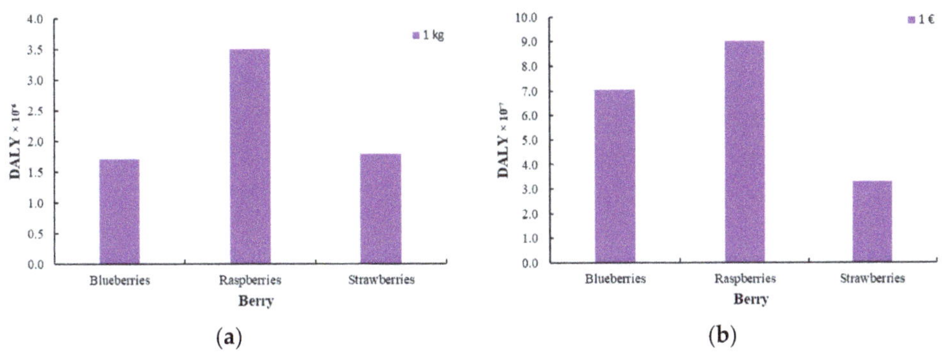

Figure 7. Damage to human health [DALY] for the production of innovative berries (blueberry, raspberry, and strawberry) (**a**) per 1 kg of product and (**b**) per 1 EUR of product.

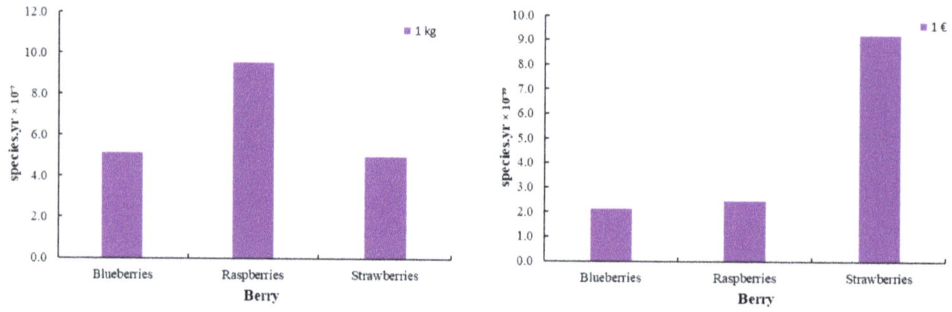

(a) (b)

Figure 8. Damage to ecosystems [species.yr] for the production of innovative berries (blueberry, raspberry, and strawberry) (**a**) per 1 kg of product and (**b**) per 1 EUR of product.

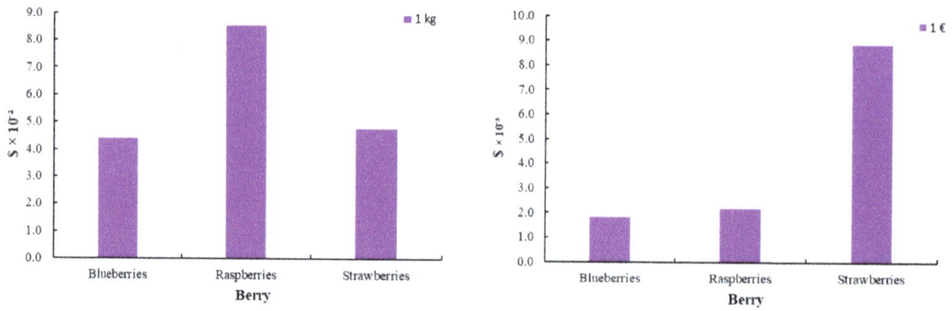

(a) (b)

Figure 9. Damage to resource availability [$] for the production of innovative berries (blueberry, raspberry, and strawberry) (**a**) per 1 kg of product and (**b**) per 1 EUR of product.

The impact assessment was carried out using two functional units to ensure a comprehensive evaluation: one based on product mass (1 kg) and the other on economic value (1 EUR). This dual approach allows for a more detailed analysis, considering both the physical quantity of the product and its financial significance in the market.

3.2. Interpretation

The Life Cycle Assessment (LCA) of the innovative berry production system, which integrates osmotic dehydration and edible coating techniques, offers critical insights into the environmental impact of these processing methods. This analysis highlights that osmotic dehydration is the primary driver of the increased environmental burden, contributing significantly to the environmental footprint across nearly all impact categories. Specifically, its contribution to climate change and land use is notably higher when compared to edible coating and packaging processes.

The use of osmotic dehydration significantly increases the environmental burden due to the use of apple juice as the osmotic agent, which has a substantial environmental footprint due to the cultivation and processing requirements of apples [32]. Furthermore, osmotic dehydration has been found to have detrimental environmental effects in other cases, particularly when alternative solvents such as glycerol are used. These solvents also contribute to heightened environmental impacts due to the energy required for their production, processing, and the waste generated during the dehydration process [33].

The environmental footprint of osmotic dehydration is not primarily attributed to electricity consumption. This indicates that, in terms of direct energy usage, the method is relatively energy-efficient compared to other dehydration techniques. However, the overall environmental impact stems mainly from the production and processing of osmotic agents rather than from the energy required to operate the dehydration process itself.

Despite these challenges, osmotic dehydration is widely used to extend the shelf life of fruit products, reducing spoilage and food waste [14,34,35]. A comprehensive assessment of its environmental benefits would require a full life cycle analysis, particularly evaluating its impact at the end-of-life stage. However, such studies have not yet been conducted due to a lack of sufficient data.

Opportunities exist for reducing the environmental footprint of osmotic dehydration. Recycling and reusing the osmotic solution through reverse osmosis technology is an important strategy for minimizing the environmental impact. Additionally, reducing the total volume of solvent required and optimizing the ratio of fruit to osmotic solution can further improve sustainability. Performing the process at temperatures close to ambient conditions can also yield significant energy savings, improving both energy efficiency and the overall sustainability of the method [36].

The edible coating process, on the other hand, does not contribute significantly to the environmental footprint, primarily because of the simplicity and sustainability of its production process. The raw materials used are biodegradable and plant-based, and the process requires minimal energy, particularly when performed at low temperatures. This helps reduce greenhouse gas emissions. Furthermore, the waste generated by edible coatings is minimal, as the residues are biodegradable and can be safely integrated into the environment. In addition to its low environmental impact, edible coating effectively extends the shelf life of food products by reducing moisture loss and slowing down microbial growth. Its environmental impact is almost insignificant in most categories, making it a highly sustainable food preservation technology and an environmentally friendly alternative to other food preservation methods [37,38].

Packaging, while still having a relatively low environmental impact, contributes more significantly to issues such as fossil fuel depletion and human toxicity, primarily due to the production and disposal of packaging materials [39]. In this process, the packaging remains the same as that used for fresh products, offering no additional environmental burden. However, sustainable and innovative packaging solutions, such as biodegradable or recyclable materials, could further enhance sustainability and should be explored in future research [40].

Despite the increased environmental burden, the innovative production method offers several advantages, most notably in extending the shelf life and improving the nutritional value of berries. These benefits, however, come with a higher environmental footprint compared to fresh berries, primarily due to the additional processing steps required. The overall environmental burden, particularly in terms of human health, resource availability, and ecosystem impacts, is higher for the processed berries, with osmotic dehydration and edible coating being the main contributors. Among the berries, raspberries exhibit the highest environmental footprint across all categories, which is attributed to the increased solvent use during osmotic dehydration, as optimized in the process. In contrast, the environmental impact of edible coating is consistent across all berry types, with no significant variations.

When the environmental footprint is normalized based on the selling price (EUR), significant differences in the impact of the preservation methods are observed. The environmental footprint of raspberries decreases and becomes more comparable to that of blueberries, while the footprint for strawberries increases substantially. This shift can be attributed to the higher commercial value of blueberries and raspberries, which justifies their relatively higher environmental impact despite the lower footprint observed in the analysis. The increased cost of strawberries, in comparison, amplifies their environmental footprint when normalized to price, reflecting the complexities of balancing economic value with environmental sustainability. On the other hand, the lower price of strawberries is insufficient to offset the environmental burden of the processing method, rendering it less efficient from both an environmental and economic perspective. However, if the extended shelf life of the product is taken into account, the overall environmental impact could be reconsidered, as reduced food waste may contribute to sustainability. Nevertheless, given that strawberry prices are not consistently low, the economic advantage remains limited, and the positive impact of the process is not significant.

While the environmental footprint of the innovative processing method is higher, the extended shelf life of the berries provides several economic and logistical benefits. The improved shelf life reduces losses during transportation and storage, enhancing supply chain efficiency. Furthermore, the ability to sell the berries at a higher price due to their extended shelf life increases profit margins and allows for distribution to high-demand markets and expanded geographical areas, thus improving commercial prospects.

Despite the increased environmental impact, the economic benefits derived from enhanced product preservation and reduced waste may mitigate some of the environmental costs. To further minimize the ecological footprint, it is essential to promote more environmentally friendly production practices while maintaining the commercial advantages offered by the extended shelf life of innovative berries.

4. Conclusions

In conclusion, the Life Cycle Assessment (LCA) of the innovative berry production system reveals that, while osmotic dehydration significantly contributes to the environmental footprint, the overall environmental impact can be mitigated through various strategies. The primary environmental burden stems from the use of apple juice as the osmotic agent, with its cultivation and processing contributing to high environmental costs. However, opportunities for reducing these impacts exist, such as optimizing the use of solvents, implementing reverse osmosis for solution recycling, and conducting the process at lower temperatures to save energy. On the other hand, edible coating and packaging demonstrate relatively low environmental footprints, with edible coating in particular proving to be a highly sustainable preservation method due to its low energy demand and minimal waste generation. To reduce the overall environmental burden, it is essential to refine the osmotic dehydration process and further explore sustainable packaging solutions.

Despite the higher environmental footprint compared to fresh berries, the innovative processing methods offer substantial economic and logistical benefits. The extended shelf life achieved through osmotic dehydration and edible coating reduces losses during transportation and storage, enhancing the supply chain efficiency and allowing berries to be sold at higher prices. This not only increases profit margins but also enables distribution to more markets, expanding commercial opportunities. Therefore, while the environmental challenges associated with these processing techniques should not be overlooked, the economic advantages, along with strategies to minimize ecological impacts, offer a pathway toward more sustainable and profitable berry production. Emphasizing environmentally friendly production practices and optimizing the entire production process will be key to ensuring the sustainability of this innovative berry preservation approach.

Author Contributions: Conceptualization, A.M., T.K., C.B. and M.K.; methodology, A.M., T.K. and C.B.; software, A.M., T.K. and C.B.; validation, A.M., T.K. and C.B.; formal analysis, A.M., T.K. and C.B.; investigation, A.M., T.K. and C.B.; resources, A.M.; data curation, A.M.; writing—original draft preparation, A.M.; writing—review and editing, A.M., T.K. and C.B.; visualization, M.K.; supervision, C.B. and M.K.; project administration, M.K.; funding acquisition, M.K. All authors have read and agreed to the published version of the manuscript.

Funding: This project has received funding from the European Union's Horizon 2020 research and innovation programme under the Marie Skłodowska-Curie grant agreement No 101007783.

Institutional Review Board Statement: Not applicable.

Informed Consent Statement: Not applicable.

Data Availability Statement: The original contributions presented in this study are included in the article. Further inquiries can be directed to the corresponding author.

Conflicts of Interest: The authors declare no conflicts of interest.

Abbreviations

LCA—Life Cycle Assessment, ISO—International Organization for Standardization, EU-28—European Union (28 member states), HDPE—High-Density Polyethylene, MJ—Megajoule, CO_2 eq.—Carbon Dioxide Equivalent, PM2.5 eq.—Particulate Matter (diameter $\leq$ 2.5 μm) Equivalent, DB eq.—1,4-Dichlorobenzene Equivalent (used in toxicity assessments), P eq.—Phosphorus Equivalent, N eq.—Nitrogen Equivalent, Cu eq.—Copper Equivalent, NOx eq.—Nitrogen Oxides Equivalent, CFC-11 eq.—Trichlorofluoromethane Equivalent (Ozone Depleting Substance), SO_2 eq.—Sulfur Dioxide Equivalent, Bq C-60 eq. to air—Becquerel of Cobalt-60 Equivalent to Air (Ionizing Radiation Indicator), DALY—Disability-Adjusted Life Years (used for Human Health Impact), GaBi—Software used for Life Cycle Analysis (by Sphera Solutions), ReCiPe—A life cycle impact assessment method, FAO—Food and Agriculture Organization, CE—Circular Economy

References

1. Kussmann, M.; Abe Cunha, D.H.; Berciano, S. Bioactive Compounds for Human and Planetary Health. *Front. Nutr.* **2023**, *10*, 1193848. [CrossRef]
2. Beattie, J.; Crozier, A.; Duthie, G. Potential Health Benefits of Berries. *Curr. Nutr. Food Sci.* **2005**, *1*, 71–86. [CrossRef]
3. Devirgiliis, C.; Guberti, E.; Mistura, L.; Raffo, A. Effect of Fruit and Vegetable Consumption on Human Health: An Update of the Literature. *Foods* **2024**, *13*, 3149. [CrossRef] [PubMed]
4. Slavin, J.L.; Lloyd, B. Health Benefits of Fruits and Vegetables. *Adv. Nutr.* **2012**, *3*, 506–516. [CrossRef]
5. Shah, H.M.S.; Singh, Z.; Kaur, J.; Hasan, M.U.; Woodward, A.; Afrifa-Yamoah, E. Trends in Maintaining Postharvest Freshness and Quality of *Rubus* Berries. *Compr. Rev. Food Sci. Food Saf.* **2023**, *22*, 4600–4643. [CrossRef]
6. Martini, D.; Marino, M.; Angelino, D.; Del Bo', C.; Del Rio, D.; Riso, P.; Porrini, M. Role of Berries in Vascular Function: A Systematic Review of Human Intervention Studies. *Nutr. Rev.* **2020**, *78*, 189–206. [CrossRef]
7. Dhall, R.K. Advances in Edible Coatings for Fresh Fruits and Vegetables: A Review. *Crit. Rev. Food Sci. Nutr.* **2013**, *53*, 435–450. [CrossRef]
8. Nowacka, M.; Dadan, M.; Tylewicz, U. Current Applications of Ultrasound in Fruit and Vegetables Osmotic Dehydration Processes. *Appl. Sci.* **2021**, *11*, 1269. [CrossRef]
9. Menéndez-Cañamares, S.; Blázquez, A.; Albertos, I.; Poveda, J.; Díez-Méndez, A. Probiotic Bacillus Subtilis SB8 and Edible Coatings for Sustainable Fungal Disease Management in Strawberry. *Biol. Control* **2024**, *196*, 105572. [CrossRef]
10. Ma, M.; Sun, Q.J.; Li, M.; Zhu, K.X. Deterioration Mechanisms of High-Moisture Wheat-Based Food—A Review from Physico-chemical, Structural, and Molecular Perspectives. *Food Chem.* **2020**, *318*, 126495. [CrossRef]
11. Karanth, S.; Feng, S.; Patra, D.; Pradhan, A.K. Linking Microbial Contamination to Food Spoilage and Food Waste: The Role of Smart Packaging, Spoilage Risk Assessments, and Date Labeling. *Front. Microbiol.* **2023**, *14*, 1198124. [CrossRef]
12. Dey, T.; Bhattacharjee, T.; Nag, P.; Ritika; Ghati, A.; Kuila, A. Valorization of Agro-Waste into Value Added Products for Sustainable Development. *Bioresour. Technol. Rep.* **2021**, *16*, 100834. [CrossRef]

13. Seberini, A. Economic, Social and Environmental World Impacts of Food Waste on Society and Zero Waste as a Global Approach to Their Elimination. *SHS Web Conf.* **2020**, *74*, 03010. [CrossRef]
14. Gribova, N.A.; Perov, V.I.; Eliseeva, L.G.; Berketova, L.V.; Nikolayeva, M.A.; Soltaeva, N.L. Innovative Technology of Processing Berries by Osmotic Dehydration. In *IOP Conference Series: Earth and Environmental Science, Proceedings of the International Conference on World Technological Trends in Agribusiness, Omsk City, Russia, 4–5 July 2020*; IOP Publishing Ltd.: Bristol, UK, 2021; Volume 624.
15. Pateiro, M.; Vargas-Ramella, M.; Franco, D.; Gomes da Cruz, A.; Zengin, G.; Kumar, M.; Dhama, K.; Lorenzo, J.M. The Role of Emerging Technologies in the Dehydration of Berries: Quality, Bioactive Compounds, and Shelf Life. *Food Chem. X* **2022**, *16*, 100465. [CrossRef]
16. Márquez-Villacorta, L.; Pretell-Vásquez, C.; Hayayumi-Valdivia, M. Optimization of Edible Coating with Essential Oils in Blueberries. *Ciência Agrotecnologia* **2022**, *46*, 198. [CrossRef]
17. César de Albuquerque Sousa, T.; de Lima Costa, I.H.; Gandra, E.A.; Meinhart, A.D. Use of Edible Coatings as a New Sustainable Alternative to Extend the Shelf Life of Strawberries (*Fragaria ananassa*): A Review. *J. Stored Prod. Res.* **2024**, *108*, 102375. [CrossRef]
18. Mari, A.; Parisouli, D.N.; Krokida, M. Exploring Osmotic Dehydration for Food Preservation: Methods, Modelling, and Modern Applications. *Foods* **2024**, *13*, 2783. [CrossRef]
19. Pié-Amill, A.; Colás-Medà, P.; Viñas, I.; Falcó, I.; Alegre, I. Efficacy of an Edible Coating with Carvacrol and Citral in Frozen Strawberries and Blueberries to Control Foodborne Pathogens. *Foods* **2024**, *13*, 3167. [CrossRef]
20. Peña, C.; Civit, B.; Gallego-Schmid, A.; Druckman, A.; Pires, A.C.-; Weidema, B.; Mieras, E.; Wang, F.; Fava, J.; Canals, L.M.i; et al. Using Life Cycle Assessment to Achieve a Circular Economy. *Int. J. Life Cycle Assess.* **2021**, *26*, 215–220. [CrossRef]
21. Cacace, F.; Bottani, E.; Rizzi, A.; Vignali, G. Evaluation of the Economic and Environmental Sustainability of High Pressure Processing of Foods. *Innov. Food Sci. Emerg. Technol.* **2020**, *60*, 102281. [CrossRef]
22. Smol, M.; Mejia, A.; Gastaldi, M.; D'Adamo, I. Environmental Indicators for Assessment of Circular Economy (CE) Implementation in the Water and Wastewater Sector. *Bus. Strategy Environ.* **2025**, *34*, 1993–2011. [CrossRef]
23. Stramarkou, M.; Boukouvalas, C.; Koskinakis, S.E.; Serifi, O.; Bekiris, V.; Tsamis, C.; Krokida, M. Life Cycle Assessment and Preliminary Cost Evaluation of a Smart Packaging System. *Sustainability* **2022**, *14*, 7080. [CrossRef]
24. Boukouvalas, C.; Kekes, T.; Oikonomopoulou, V.; Krokida, M. Life Cycle Assessment of Energy Production from Solid Waste Valorization and Wastewater Purification: A Case Study of Meat Processing Industry. *Energies* **2024**, *17*, 487. [CrossRef]
25. Akharume, F.U.; Singh, K.; Sivanandan, L. Characteristics of Apple Juice and Sugar Infused Fresh and Frozen Blueberries. *LWT* **2016**, *73*, 448–457. [CrossRef]
26. Tsakiri-Mantzorou, Z.; Drosou, C.; Mari, A.; Stramarkou, M.; Laina, K.T.; Krokida, M. Edible Coating with Encapsulated Antimicrobial and Antibrowning Agents via the Emerging Electrospinning Process and the Conventional Spray Drying: Effect on Quality and Shelf Life of Fresh-Cut Potatoes. *Potato Res.* **2024**. [CrossRef]
27. Mari, A.; Fafalis, C.; Krokida, M. Extension of Blueberry Shelf-Life with Edible Coatings from Chlorella Vulgaris. *Chem. Eng. Trans.* **2024**, *110*, 79.
28. Verghese, K.; Lewis, H.; Lockrey, S.; Williams, H. Packaging's Role in Minimizing Food Loss and Waste Across the Supply Chain. *Packag. Technol. Sci.* **2015**, *28*, 603–620. [CrossRef]
29. Boz, Z.; Korhonen, V.; Koelsch Sand, C. Consumer Considerations for the Implementation of Sustainable Packaging: A Review. *Sustainability* **2020**, *12*, 2192. [CrossRef]
30. Hussain, S.; Akhter, R.; Maktedar, S.S. Advancements in Sustainable Food Packaging: From Eco-Friendly Materials to Innovative Technologies. *Sustain. Food Technol.* **2024**, *2*, 1297–1364. [CrossRef]
31. Bou-Mitri, C.; Abdessater, M.; Zgheib, H.; Akiki, Z. Food Packaging Design and Consumer Perception of the Product Quality, Safety, Healthiness and Preference. *Nutr. Food Sci.* **2021**, *51*, 71–86. [CrossRef]
32. Khanali, M.; Kokei, D.; Aghbashlo, M.; Nasab, F.K.; Hosseinzadeh-Bandbafha, H.; Tabatabaei, M. Energy Flow Modeling and Life Cycle Assessment of Apple Juice Production: Recommendations for Renewable Energies Implementation and Climate Change Mitigation. *J. Clean. Prod.* **2020**, *246*, 118997. [CrossRef]
33. Mari, A.; Kekes, T.; Boukouvalas, C.; Drosou, C.; Krokida, M.; Tsartsaris, C. Evaluating the Environmental and Economic Benefits of New Technologies in Low-Salt Olive Fermentation. *Agriculture* **2024**, *14*, 2077. [CrossRef]
34. Nsonzi, F.; Ramaswamy, H.S. Osmotic Dehydration Kinetics of Blueberries. *Dry. Technol.* **1998**, *16*, 725–741. [CrossRef]
35. Panagiotou, N.M.; Karathanos, V.T.; Maroulis, Z.B. Effect of osmotic agent on osmotic dehydration of fruits. *Dry. Technol.* **1999**, *17*, 175–189. [CrossRef]
36. Ciurzyńska, A.; Kowalska, H.; Czajkowska, K.; Lenart, A. Osmotic Dehydration in Production of Sustainable and Healthy Food. *Trends Food Sci. Technol.* **2016**, *50*, 186–192. [CrossRef]
37. Rajendran, N.; Runge, T.; Bergman, R.D.; Nepal, P.; Pottackal, N.T.; Rahman, M.M. Economic and Environmental Analysis of Producing Soy Protein-Cellulose-Based Bionanocomposite Fruit Coating. *Ind. Crops Prod.* **2024**, *211*, 118213. [CrossRef]

38. Lecart, B.; Baumsteiger, C.; Monie, F.; Di Maria, A.; Detrembleur, C.; Richel, A.; Vanderschuren, H. Towards Green Chemicals and Edible Coatings from Barks and Peels with near Critical Extraction of Suberin. *Green Chem.* **2023**, *25*, 9282–9291. [CrossRef]
39. Zhao, X.; Wang, Y.; Chen, X.; Yu, X.; Li, W.; Zhang, S.; Meng, X.; Zhao, Z.M.; Dong, T.; Anderson, A.; et al. Sustainable Bioplastics Derived from Renewable Natural Resources for Food Packaging. *Matter* **2023**, *6*, 97–127. [CrossRef]
40. Arfelli, F.; Roguszewska, M.; Torta, G.; Iurlo, M.; Cespi, D.; Ciacci, L.; Passarini, F. Environmental Impacts of Food Packaging: Is It All a Matter of Raw Materials? *Sustain. Prod. Consum.* **2024**, *49*, 318–328. [CrossRef]

foods

MDPI

Article

Effect of Pulsed Electric Fields and Osmotic Dehydration on the Quality of Modified-Atmosphere-Packaged Fresh-Cut and Fried Potatoes

Efimia Dermesonlouoglou *, George Seretis, Maria Katsouli, Alexandros Katsimichas, Petros Taoukis and Maria Giannakourou *

Laboratory of Food Chemistry and Technology, School of Chemical Engineering, National Technical University of Athens (NTUA), 9, Iroon Polytechniou Str, 15772 Zografou, Greece; gseretis123@gmail.com (G.S.); mkatsouli@chemeng.ntua.gr (M.K.); alkats@chemeng.ntua.gr (A.K.); taoukis@chemeng.ntua.gr (P.T.)

* Correspondence: efider@chemeng.ntua.gr (E.D.); mgian@chemeng.ntua.gr (M.G.)

Abstract: The aim of this research was to study the effect of osmotic dehydration (OD) and/or pulsed electric field (PEF) on the quality of MAP-packed potatoes, both as raw materials and after deep frying. Fresh-cut potato strips (from Naxos island) were osmotically dehydrated using a solution of 20% glycerol, 5% sodium chloride, and 1% ascorbic acid (wt) at a 5:1 liquid-to-food ratio at 35 °C for 120 min. OD-treated and untreated samples were packaged at MAP (0.2% O_2 + 12% CO_2) and stored at 4, 8, and 12 °C. Color (Browning Index, *BI*), texture (hardness, F_{max}), sensory characteristics (including total sensory quality), and microbial stability (total aerobic and anaerobic counts, *Pseudomonas*, Entrobacteriaceae, and yeasts/molds) were monitored during storage. After package opening, samples were deep-fried at 180 °C for up to 8 min, and the oil content of fried samples was quantified. Sensory evaluation of raw and fried samples was conducted. Untreated fresh-cut potatoes were characterized by detrimental color degradation starting from the third day of storage at 4 °C and presented microbial growth (total viable counts: 6 log (CFU)/g) on the sixth day, whereas pre-treated potato samples retained their color and microbiological stability after 15 and 18 days of cold storage, respectively. OD pre-treatment reduced the oil uptake during frying (up to 30%).

Keywords: *Solanum tuberosum*; osmotic dehydration; pulsed electric fields; browning; texture; quality; deep drying; storage

1. Introduction

Potatoes are highly popular due to their rich content of nutrients, minerals, and dietary fibers and their versatility in various food applications, such as potato fries [1–3]. However, during the fresh-cutting process, the mechanical damage to the tissue makes them susceptible to enzymatic browning and reduces their shelf life [4–6]. The browning of fresh-cut fruits and vegetables occurs when phenolic compounds undergo enzymatic oxidation, catalyzed by polyphenol oxidase, leading to the formation of quinones, which then condense into colored pigments [4,5]. Blanching (at 55–75 °C for a minimum of 10 min) is commonly used in the production of par-fried potatoes to deactivate polyphenol oxidase. However, blanching requires significant amounts of water and is not energy efficient [7]. Additionally, during blanching, a colorless complex may form between potato iron and chlorogenic acid. Upon oxidation, this complex can create ferri-chlorogenic acid, resulting in an undesirable dark color. To mitigate this, anti-browning agents, such as sulfites, are often added to the blanching solution [8]. Previous studies have also highlighted the positive effect of osmotic dehydration (as a single treatment or in combination with other technologies) in preserving the color (browning) of fresh-cut fruits and vegetables [9–16].

Osmotic dehydration (OD) is a process where food is immersed in a hypertonic solution—containing carbohydrates, salts, and other specific ingredients such as antioxidants and antimicrobials—at mild temperatures. The key parameters of this process include temperature, time, food-to-osmotic solution ratio, and the osmotic solution formulation [9,10]. For each food product, the optimal processing conditions must be chosen to ensure efficient dehydration within a reasonable timeframe while preserving food quality. The influence of different parameters, particularly temperature, time duration, and the composition of the osmotic solution, on mass transfer during OD has been widely studied [10,12]. Osmotic dehydration has been extensively explored as a pre-treatment method, particularly before drying or freezing foods with a high water content. Studies have shown that OD pre-treatment can increase the quality and stability of fresh-cut fruits and vegetables [9,10,13–17]. Additionally, the impact of OD on frying kinetics and the quality of potato fries (French type) has been explored by Krokida et al. [18], Gupta et al. [19], and Kwaw et al. [20]. The studies found that pre-frying dehydration reduced oil absorption in potato fries while significantly enhancing their color and structural properties. Osmotic dehydration also helped prevent discoloration that occurred during subsequent frying.

Recently, there has been increasing interest in innovative technologies that offer significant potential for reducing energy consumption and shortening processing times [16,21,22]. Among these, pulsed electric field (PEF) has emerged as a promising process for various food applications owing to its low power requirements [16,21,22]. PEF is recognized as an effective pre-treatment for enhancing mass transfer processes, such as dehydration (air drying, freeze drying, and osmotic dehydration) and extraction [23–27]. The process involves the use of high-voltage pulses on semi-liquid or liquid foods placed between two electrodes. This electric field induces electroporation, which creates pores in the cell membranes and walls. As a result, cell permeability increases, allowing intracellular water to move into the extracellular matrix [23,28]. This movement of water accelerates dehydration but also leads to a loss of turgor pressure, causing tissue softening [23,24,29,30]. PEF technology is now used in the potato industry to reduce cutting forces [30,31]. Research has also explored the potential of PEF pre-treatment on potato tubers for French fries production [30–37]. PEF treatment has been shown to enhance the permeability of potato cell membranes, reduce oil and acrylamide content in fried potatoes, and improve their texture [30,36,37]. However, studies on the combined use of PEF (as a pre-treatment for osmotic dehydration) and osmotic dehydration before frying are still relatively limited.

The objective of this study is twofold: (1) to assess and model the impact of pulsed electric field (PEF) pre-treatment conditions on the osmotic dehydration (OD) parameters (including mass transfer) and product quality characteristics (water activity, texture, and color) of fresh-cut potatoes and (2) to evaluate the quality and shelf-life stability (with regard to microbial growth) of potatoes treated with OD and PEF + OD and packaged under modified atmosphere during cold storage. The quality of OD-treated and PEF + OD-treated potatoes during/after frying was also determined (oil uptake during frying, texture/color, and sensory properties of the fried product). For this study, potatoes with protected geographical indication certifications from Naxos Island (Greece) were used. Preserving the unique qualities of "Naxos" potatoes at their freshest while extending their shelf life is essential from a quality standpoint. This will ensure that they can be successfully distributed to both domestic and foreign markets.

2. Materials and Methods

2.1. Materials

Potatoes (of the Spunta variety) were purchased from a local potato producer (SKLIRAKIS, J.,-JACOB GIAKOUMIS O.E., Naxos, Greece) at the maturity stage (based on crop physiology and the presence of a thick, firm skin) and stored at room temperature in the dark for 7 days (maximum). The potatoes with similar color, size, and maturity and no obvious disease, insect, and mechanical damage were peeled (1–2 mm) and cut along the longitudinal direction (in the middle of the potato) to obtain potato strips of weight 5 g and length 5 cm. The water content of the raw material was 0.874 $\pm$ 0.024 g water/g sample weight.

2.2. Pulsed Electric Field (PEF) Treatment of Potatoes

Potato strips were treated in water (tap, with an electrical conductivity of 800 μS/cm) at an electric field strength of 0.5 kV/cm for 200 pulses, resulting in a specific energy input equal to 0.02 kJ/kg [38]. The treatment was performed in a stainless-steel parallel plate batch electrode chamber with an electrode spacing of 8 cm and a total volume of 300 mL. Bipolar pulses of near-rectangular shape were delivered to the treatment chamber using an Elcrack-5 kW unit (DIL, Quackenbrück, Germany). In brief, approximately 80 g of potato strips were transferred to the treatment chamber for each PEF treatment and covered with tap water. The pulse width was set to 15 μs, the amplitude of electric pulses was 4 kV, and the pulse frequency was maintained at 20 Hz. The initial temperature of potato strips was approximately 25 °C, and the post-treatment temperature remained practically unchanged. Sample temperature was measured by a digital thermometer (General Tools & Instruments, Secaucus, NJ, USA) before and after each PEF treatment. After each treatment, selected quality parameters (texture/color) of the potatoes were evaluated.

2.3. Osmotic Dehydration (OD) of Untreated and PEF-Pre-Treated Potatoes

Potato strips (untreated and PEF-pre-treated) were first weighed and placed in cylindrical glass containers, which were then filled with preheated osmotic medium of 20% glycerol, 5% sodium chloride, and 1% ascorbic acid at 35 °C to achieve a food-to-liquid sample ratio of 1:5. The containers were immersed in a thermostatic water bath with continuous agitation at 240 rpm (Grant GL5400 Linear Shaking Water Bath, Royston, UK). Samples were taken at 0, 30, 60, 120, and 180 min, rinsed with water to remove excess solution, and gently blotted with absorbent paper [32]. Water loss (*WL*), solid gain (*SG*), water activity (a_w), and the quality properties of texture (firmness) and color (browning) change were determined throughout the OD process.

Osmotic dehydration of potatoes for the shelf-life study was performed in pilot-scale equipment consisting of a cylindrical (Ø28 × 60 cm) stainless steel 100 L tank fitted with spigots, which allowed draining of the solution. Stirring was also mildly applied (set at a rotational speed of 120 rpm) through an electrical motor paddle-type stirrer, carefully submerged into the solution. A heating option was also provided, where required, with a complementary stainless steel helical tube, which could be heated via a continuous circulation of water from a 3 kW water bath. In our work, gentle heating was employed (temperature set at 35 °C, monitored with a thermocouple throughout the whole process) to ensure homogeneous dissolution of the osmotic medium constituents. The solution required for osmotic dehydration was prepared in situ within the stainless-steel tank by dosing the ingredients using a high-capacity electronic balance. Potatoes were manually loaded and unloaded in the tank (batch process). After OD treatment, the potatoes were rinsed with centrifugal household vegetable equipment and placed for 30 min on a perforated rack to drain off excess osmotic solution. The scaling-up of the OD process up to the 100 kg scale was deemed successful, and the measured parameters agreed with their counterparts obtained after experiments in the lab scale.

2.4. Quality Monitoring of Untreated and PEF-Pre-Treated OD Potatoes During Cold Storage

OD potato strips, untreated and PEF-treated, were packaged in PA (polyamide)/PE (polyethylene) bags (250 × 175 × 0.04 mm; ten potato strips from each repetition) in modified atmosphere (12.2% CO_2-0.188% O_2) (Boss NT42N, Bad Homburg, Germany) and stored in controlled temperature cabinets at 4, 8, and 12 °C (±0.2 °C) (Sanyo MIR 153, Sanyo Electric, Osaka, Japan). The temperature in the cabinets was continuously monitored using temperature data loggers (COX TRACER, Belmont, NC, USA). At predetermined sampling storage times, three bags per processing with one bag per repetition were selected for the shelf-life study. Quality (texture/color) and microbial stability of samples were measured. Gas headspace of packages was analyzed (CheckMate 9900 O_2/CO_2, PBI Dansensor, Ringsted, Denmark).

OD potato strips, untreated and PEF-treated, were fried in sunflower oil at 180 °C (±1 °C) for 0, 2, 4, 6, 8, and 10 min (Tefal Visialis Deep Fryer, Groupe SEB, Paris, France). The fryer was filled with 2 L of oil in a potato strips-to-oil ratio of 1:50 *w*/*v*. Fried potato strips were drained for a few seconds on paper.

2.5. Analytical Protocols and Mathematical Modeling

2.5.1. Physicochemical Parameters

Water activity (a_w) was determined by a water activity meter (Aqua lab 4TEV, Decagon Devices, Pullman, WA, USA). The water (moisture) content was measured by drying at 105 °C for 18 h (WTB BINDER 7200, Type C53, Tuttlingen, Germany) [39]. pH was measured using a pH meter (338, Amel Instruments, Milano, Italy). Ten grams of homogenized potato sample was diluted in 90 mL of sterilized Ringer solution [40].

Mass Transfer During Osmotic Dehydration

Water loss (*WL*) and solid gain (*SG*) values were calculated using Equations (1) and (2) [41], respectively.

$$WL = \frac{m_0 - m_0 \cdot DW_{wb} - \left(m_{wet} - m_{dry}\right)}{m_0 \cdot DW_{wb}} \quad (1)$$

$$SG = \frac{m_{dry} - DW_{wb} \cdot m_0}{DW_{wb} \cdot m_0} \quad (2)$$

where m_0 represents the initial wet weight of the sample prior to immersion in the osmotic solution, DW_{wb} is the dry weight of the untreated sample (g of water/g of sample wet base), m_{wet} is the wet weight of the treated sample (g), and m_{dry} is the dry weight of the OD-pre-treated sample (g).

According to our kinetic approach, the penetration model was used to describe water loss and solid gain, where the ratio of WL/SG is proportional to the square root of the contact time, according to Equation (3) [41].

$$WL \text{ or } SG = k_X \cdot \sqrt{t_{OD}} \quad (3)$$

where *WL* is in $\frac{\text{g water}}{\text{g DW}}$ or *SG* is in $\frac{\text{g solid}}{\text{g DW}}$ at time t_{OD}; k_X is the water loss constant (k_{WL}) in $\frac{\text{g water}}{\text{g DW}\cdot\text{s}^{\frac{1}{2}}}$ or the solid gain constant (k_{SG}) in $\frac{\text{g solids}}{\text{g DW}\cdot\text{s}^{\frac{1}{2}}}$; and *t* is the osmotic dehydration time in min.

2.5.2. Potato Product Composition

The glycerol content of the potato strip product was measured according to the assay proposed by Kuhn et al. [42] with minor modifications. The final measurement was executed in a microplate reader at 410 nm over a period of 25 min. The final glycerol content was calculated based on a glycerol standard curve and expressed as mg glycerol/g potato. The salt content of the potato strip product was measured titrimetrically by the Mohr method [39]. Vitamin C (mg L (ascorbic acid)/100 g potato) of the potato strip product was determined by a high-performance liquid chromatography method [43]. For all measurements, two replicates were used.

2.5.3. Potato Product Firmness

Potato strips were compressed to a deformation of 20% by a knife attached to a texture analyzer (TA-XT2i, Stable Micro Systems, Godalming, UK) (test speed of 1.00 mm/s; distance of 4.00 mm). The knife was used to cut the potatoes along the longitudinal direction in the middle of each potato sample. The firmness was recorded as the maximum compression force (F_{max}, N) [11]. A minimum of five replicates were performed for each measurement.

2.5.4. Potato Product Color

An Xrite-i1 portable digital colorimeter (Gretag-Macbeth, Grand Rapids, MI, USA) was used to assess the objective color on the potato surface, expressed in the CIE-Lab color scale. At least five replicates were taken by measuring the color at three different evenly distributed locations on the surface of each potato. The degree of browning was quantified using the Browning Index (*BI*) calculated using Equation (4) [44]. In the present study, *BI* is used (as the most representative parameter) to indicate color changes in the processed potato strips during OD processing as well as during cold storage [1–3,45].

$$BI = \frac{100\left(\frac{(a + 1.75 \times L)}{(5.645 \times L + a - 0.3012 \times b)} - 0.31\right)}{0.17} \tag{4}$$

where L, a, and b represent the measured CIE-Lab color parameters.

2.5.5. Oil Content

Five grams of smashed potato strips were weighed and placed in a 250 mL beaker together with hexane at a 1:10 sample–solvent ratio. The mixture was allowed to stir with a magnetic stirrer for about 40 min, and the liquid solvent–oil phase was then placed in pre-weighed 100 mL round-bottom flasks by filtration. The vacuum evaporator (Heidolph G1, Germany) was then used to evaporate the hexane at a temperature of 40 °C, with gentle rotation of the flask. Finally, the spherical bottles were re-weighed, and the gross bottle-oil weight was measured, with the fat content calculated through the weight and the initial amount.

For the prediction of the oil uptake during the frying process, a first-order kinetic model proposed in the literature was used [32] (Equation (5)).

$$Y = Y_e\left[1 - \exp\left(-k_{OIL} \times t_{fr}\right)\right] \tag{5}$$

where Y is the oil content (kg/kg db), Y_e is the equilibrium value of Y (kg/kg db), k_{OIL} is the rate constant (min^{-1}), and t_{fr} is the frying time.

2.5.6. Sensory Analysis

The sensory evaluation was performed by eight to ten trained adult assessors [46]. Participation in the assessments was voluntary, and an informed written consent was obtained by all participants prior to the sensory evaluation study, which could be withdrawn at any point of the evaluation. The organizer of the sensory test provided a detailed description of the test to all participants to inform them about the food samples that would be assessed. A questionnaire was also distributed where they were obliged to report any indispositions or allergies, and if such was the case, the particular volunteer was excluded from the tests.

Sensory attributes of fried potato strips (including the appearance and odor of the raw samples) were assessed, with scores recorded on appropriate forms to reflect the organoleptic changes indicating quality deterioration (provided as a Supplementary Materials). The fried potato strips were described in terms of appearance (oil detection and color browning), odor, texture (crispiness, firmness, and gumminess), taste (overall, sweet taste, salty taste, and other), flavor, aftertaste, and total sensory quality. Each sensory parameter, as well as the total sensory quality, was scored separately on a 1–9 scale (where 9 represented high quality and 1 represented poor quality). A sensory score of 5 was considered the threshold for minimum acceptability.

2.5.7. Microbial Growth

The growth of key microorganisms associated with the spoilage of fresh-cut potatoes was monitored throughout storage. For microbiological analysis, a 10 g sample was transferred to a sterile stomacher bag containing 90 mL of sterilized Ringer solution (Merck, Darmstadt, Germany) and homogenized for 60 s using a Stomacher (Bag Mixer, Interscience, Saint Nom la Bretêche, France). Mold and yeast counts were performed using Rose Bengal Chloramphenicol (RBC, Merck, Darmstadt, Germany) after incubation at 25 °C for 120 h. Total aerobic mesophiles were counted using the standard plate count method with plate count agar (PCA, Merck, Darmstadt, Germany) incubated at 25 °C for 72 h. Total anaerobic bacteria were enumerated on sulfite-polymyxinsulfadiazin-A (SPS) agar, with plates incubated at 35 ± 0.5 °C for 24–36 h in anaerobic jars with an atmosphere generation system (Oxoid, Basingstoke, Hampshire, UK). *Pseudomonas* spp. was measured on cetrimide agar (CFC, Merck, Darmstadt, Germany) after incubation at 25 °C for 48 h. The pour plate method was used for Enterobacteriaceae enumeration using violet red bile glucose agar (VRBG, Merck, Darmstadt, Germany) after incubation at 37 °C for 24–48 h. Two replicates from at least three appropriate dilutions were enumerated. Microbial counts were expressed as log CFU per gram of tissue. Microbial growth was modeled using the Baranyi growth model. The DMFit program (available at www.combase.cc accessed on 24 July 2024) was used for curve fitting and calculation of the kinetic parameters, lag phase (λ in d), and microbial growth rate (k_{micr} in d^{-1}).

2.6. Statistical Analysis

Analyses were conducted on at least three duplicate samples, and treatments were conducted at least twice. The mean value ± SD is used to report the results. The significance level for analyses of variance (ANOVA) was set at $p < 0.05$ using Statista 7.0 (StatSoft, Inc., Tulsa, OK, USA). As a post hoc analysis, Duncan's multiple range test was used to separate means with significant differences.

3. Results and Discussion

3.1. PEF and OD

3.1.1. Effect of PEF and OD on Mass Transfer and Water Activity of Potatoes

For both untreated and PEF-pre-treated potato strips, water loss increased and water activity rapidly decreased at the start of the OD process, reaching stable values after

120 min, as shown in Figure 1a,c. PEF pre-treatment significantly affected water loss up to 180 min of OD ($p < 0.05$). The increase in water loss after PEF processing was also noted by the increase in the water loss constant (k_{WL}) (Equation (3); Table 1). The effect of PEF on solid gain values (under the applied processing conditions) was not significant ($p > 0.05$)(k_{SG}) (Figure 1b; Table 1). Dimopoulos et al. (2024) reported that PEF pre-treatment of osmotically dehydrated spinach (glycerol concentration in the osmotic solution: 60%) led to a significant *WL* as well as *SG* increase [41]. Dermesonlouoglou et al. (2019) reported that the increase in glycerol concentration (from 40 to 60%) in the osmotic solution significantly affected both the water and the solids diffusion of OD peach fruit [47]. In this study, the glycerol concentration was kept low (at 20%), aiming at a minimum alteration of potato sensory attributes.

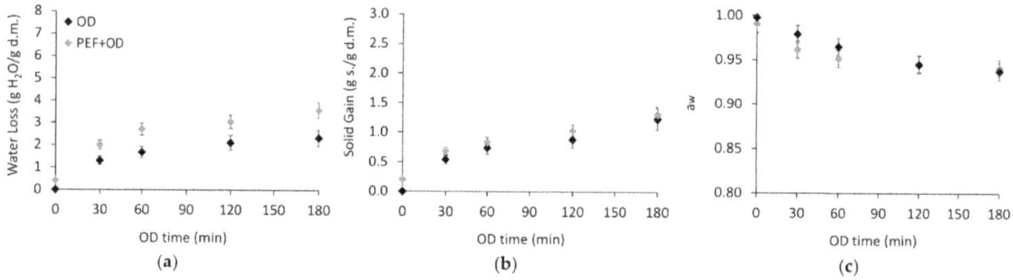

Figure 1. Mass transfer: (**a**) water loss (*WL*, g water/g d.m.), (**b**) solid gain (*SG*, g s./g d.m.), and (**c**) water activity (a_w) during osmotic dehydration (OD) of OD-treated and PEF + OD-treated potato strips. (values: mean ± standard error).

Table 1. Water loss (*WL*) and solid gain (*SG*) constants for osmotic dehydration and/or pulsed electric fields (OD and PEF + OD) (Equation (3)).

	Water Loss Constant $k_{WL}\left(\frac{\text{g water}}{\text{g DW}\cdot\text{s}^{\frac{1}{2}}}\right)$	R^2	**Solid Gain Constant** $k_{SG}\left(\frac{\text{g solids}}{\text{g DW}\cdot\text{s}^{\frac{1}{2}}}\right)$	R^2
OD-treated	0.1913 ± 0.0095^{a}	0.9885	0.0893 ± 0.0088^{a}	0.9948
PEF + OD-treated	0.2320 ± 0.0151^{b}	0.9708	0.0798 ± 0.0068^{a}	0.9927

Different superscript letters indicate significant differences between means k_{WL} and k_{SG} ± standard deviation as calculated by Duncan's multiple range test for a significance level of $p = 0.05$.

Pre-treatment of OD potatoes with PEF induced slight changes in the product water activity ($a_{w,PEF/OD\ sample}$: 0.9375; $a_{w,OD\ sample}$: 0.9406) ($p > 0.05$). Significantly different a_w values for OD-treated and PEF + OD-treated potato strips were observed for osmotic dehydration times up to 60 min ($p < 0.05$). According to Katsouli et al. (2024) [38], although the mass transfer and water activity values are significantly lower at more intense process conditions (PEF application as a pre-osmotic dehydration step), the temporal evolution (rate of a_w decrease) is not affected by PEF treatments. This observation suggests that electroporation leads to an immediate release of free water right after the treatment. A rapid immersion in the osmotic solution is adequate to eliminate this released free water and reduce water activity.

3.1.2. PEF and OD Effect on Color and Texture of Potato Strips

PEF pre-treatment influenced the quality of the product, specifically in terms of color degradation and texture firmness during osmotic dehydration (OD). The degradation of color in treated potatoes as a function of the duration of osmotic dehydration was quantified by the Browning Index (*BI*) (Figure 2a). The *BI* values for samples subjected to PEF pre-treatment were found to be higher compared to those of the OD-treated samples. Furthermore, the application of OD to PEF-treated samples resulted in a decrease in *BI* values. Throughout the OD process, *BI* values for both standard OD-treated and PEF + OD-treated samples exhibited a decline. It was observed that after 120 min of OD, the *BI* values approached 14.0. For OD times > 120 min and *BI* values < 14, the potato strips were characterized as «white-colorless». Katsouli et al. (2024) [38] similarly noted that the *BI* values of potato strips subjected to PEF treatment prior to osmotic dehydration (OD) were lower than those of untreated samples. The firmness of potatoes post-PEF treatment is a significant aspect, as it affects the sensory attributes of the final product. Figure 2b illustrates the variation in firmness of the samples in relation to the duration of osmotic dehydration. A softening effect was observed in the potatoes following both PEF and OD treatments, with firmness decreasing from approximately 8.15 N in the fresh-cut samples to 6.43 N and 5.74 N after PEF and PEF + OD treatments, respectively. This finding is consistent with the observations made by Katsouli et al. (2024) [38] regarding potato strips treated with PEF and OD. Conversely, during the osmotic dehydration process, an increase in firmness values was recorded. Specifically, a 120 min duration of OD resulted in firmness values that approached those of fresh-cut potatoes, i.e., approximately 9.00 N. The OD-treated and PEF + OD-treated potato product characteristics for OD time of 120 min (selected as the appropriate OD treatment time) are shown in Table 2. The PEF + OD-treated potato strips were characterized by significantly lower water activity and pH values compared to the values of the untreated potato strips. PEF + OD-treated potato strips presented different color and texture parameters. The OD-treated potato strips were positively evaluated during the sensory evaluation testing. Regarding the composition of the OD-modified and PEF + OD-modified potato tissue, glycerol, salt, and ascorbic acid (vitamin C) content were calculated by analytical methods.

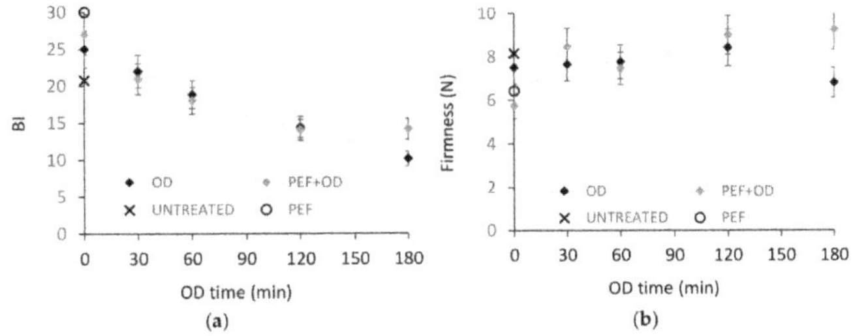

Figure 2. (**a**) Browning index (*BI*) and (**b**) firmness (F_{max}, N) values of raw (untreated), OD-treated, PEF-treated, and PEF + OD-treated potato strips during osmotic dehydration (0, 30, 60, 120, and 180 min) (values: mean ± standard error).

Table 2. Potato product characteristics (untreated, OD-treated, and PEF + OD-treated).

Sample	Water Content	a_w	pH	Vitamin C (g asc.acid/g)	Glycerol (g /g)	Salt (g NaCl/g)	Color (*L*, *a*, *b*, *BI*)	Texture (Firmness, F_{max}, N)	Sensory Score (1–9)
Untreated	0.8572	0.9927	6.036	0.0013	_	0.004	53.77; −0.57; 20.48	5.76	7.9
OD-treated	0.7014	0.9462	4.798	0.011	0.372	0.054	54.22; −0.98; 8.10	8.41	7.1
PEF + OD-treated	0.6821	0.9457	4.468	0.010	0.356	0.049	52.99; −1.17; 7.88	9.55	5.7

3.2. *Quality of OD-Treated and PEF + OD-Treated Potatoes After Frying*

3.2.1. Oil Uptake During Frying of OD-Treated and PEF + OD-Treated Potato Strips

The oil content of potatoes during frying up to 10 min is presented in Figure 3. OD, as well as the combined PEF and OD application, significantly decreased the oil uptake during frying ($p < 0.05$). For all sample categories, the data for oil uptake showed an exponential increase in oil uptake with the increase in frying time. In Table 3, the oil content constants for different pre-treatments (OD, PEF, PEF + OD) are reported as calculated by Equation (5) after appropriate data fitting.

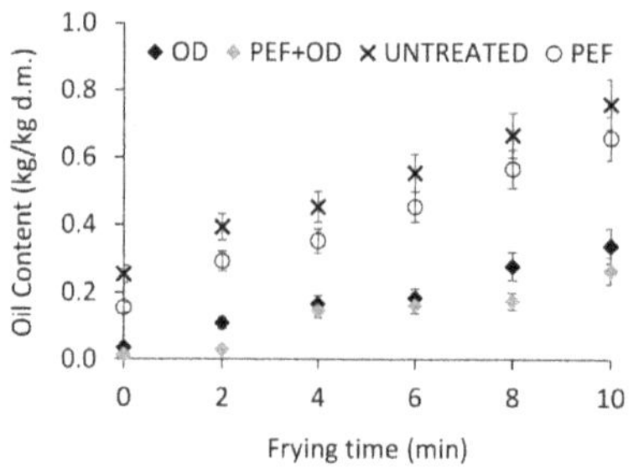

Figure 3. Oil content (kg/kg d.m.) during frying (values: mean ± standard error).

Table 3. Oil uptake constant (k_{OIL}) values for untreated, OD-treated, PEF-treated, and PEF + OD-treated potato strips as calculated by Equation (5) after fitting to the experimental data.

	Oil Uptake Constant k_{OIL} (min^{-1})	R^2
Untreated	0.199 [a]	0.9483
PEF-treated	0.198 [a]	0.9875
OD-treated	0.162 [b]	0.9302
PEF + OD-treated	0.142 [b]	0.8961

Different superscript letters indicate significant differences between means k_{OIL} ± standard deviation as calculated by Duncan's multiple range test for a significance level of $p = 0.05$.

Pre-drying procedures lower the crust's permeability, moisture content, and oil absorption of fried products. When potatoes are dried, a skin layer forms on their surface, which lowers porosity and lessens the amount of vapor that passes through the surface. There is less vacant space for oil to remain after frying when porosity decreases. Because the dry matter level of potatoes contributes to the undesirable mealy texture of fries, manufacturers try to increase the dry matter of potatoes by pre-drying them before frying. Chips made from tubers with a high dry matter content absorb less oil when fried. Oladejo et al. (2017) investigated the impact of osmotic dehydration on the quality attributes of French fries [48]. They found that samples that were pre-treated in osmotic solutions had much less oil and moisture compared to the untreated samples [48]. They reported that the increased moisture movement inside the food product's structure during frying, which results in high vapor pressure within the food structure and reduces oil absorption, could be the cause of the reduced oil content of the osmotically pre-treated samples. Furthermore, a crust formed on the surface of the fried samples treated with OD, which may have stopped the oil from being absorbed during the cooling process. This outcome matches the findings of Krokida et al. (2001) [32], Karizaki et al. (2013) [33], Dehghannya et al. (2015) [49], and Barani et al. (2020) [50]. Krokida et al. (2001) reported that the oil content reduced as drying time increased [32].

The positive effects of electroporation were also highlighted by Liu et al. (2017, 2018) [51,52], who observed a reduction in oil absorption during the frying of potato discs. Their research found that PEF treatment was more effective than a blanching-induced gel layer in reducing oil uptake in French fries. Specifically, PEF-treated chips showed a 38.7% reduction in oil uptake, while blanching only resulted in a 3.8% decrease [51]. PEF treatment reduces oil absorption by promoting water diffusion from the potato core to the surface, which forms a thicker water vapor layer that minimizes both dehydration and oil absorption during frying. Additionally, the smoother surface of PEF-treated potatoes enhances oil drainage after frying. Regardless of the processing conditions, PEF treatments increase the moisture content and softens the potato cubes without altering their fresh appearance. More recently, it has been shown that combining PEF treatment with convective air drying significantly shortens frying times and reduces oil absorption in fried potatoes [52]. Furthermore, the efficiency of potato dehydration was notably improved when PEF treatment was combined with vacuum drying at sub-atmospheric pressure [53].

3.2.2. Quality of OD-Treated and PEF + OD-Treated Fried Potato Strips

The organoleptic quality of fried potato strips (frying time 10 min) can be described in terms of appearance (oil detection and color browning), odor, texture (crispiness, firmness, and gumminess), taste (overall, sweet taste, salty taste, and other), flavor, aftertaste, and total sensory quality. In Figure 4, average scores (scale 1–9) for the characteristics (axis 1–13) of untreated, OD-treated, and PEF + OD-treated potato strips are demonstrated.

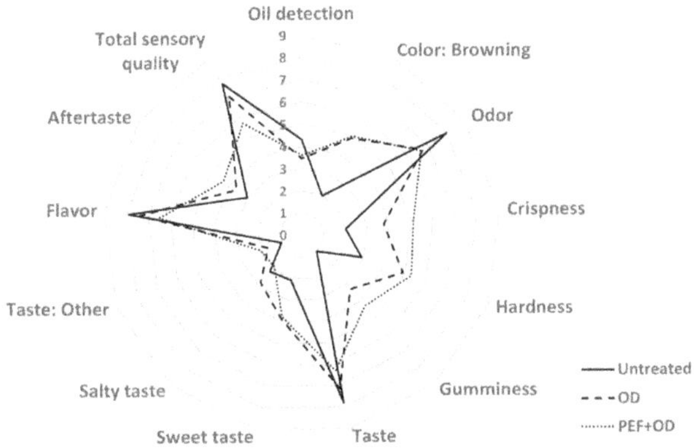

Figure 4. Average scores (scale 1–9) for sensory characteristics of untreated (——), OD-treated (- - -) and PEF + OD-treated (·········) fried potato strips (180 °C, 10 min): oil detection, color browning, odor, crispiness, firmness, gumminess, taste overall, sweet taste, salty taste, other taste, flavor, aftertaste, and total sensory quality.

Fresh-cut potatoes and/or minimally processed potatoes are convenient but highly perishable products. Unlike most fresh-cut vegetables, which are ready to eat, fresh-cut potatoes must be cooked before consumption. Therefore, in addition to the safety (chemical and microbiological), quality, and sensory characteristics of raw fresh-cut potatoes, the same requirements should be applied for cooked potatoes [54]. In the present study, the sensory evaluation was conducted on raw potatoes as well as fried potatoes. Sensory scores for the individual sensory/quality parameters as well as the overall sensory quality of these samples were presented and mathematically modeled. Score for overall sensory quality encompasses the panelists' general opinion about the examined samples and is a good indicator of the evolution of a product's quality.

From the sensory scores, the OD-treated fried potato strips received a high score for the total sensory quality (7.1/9.0 compared to 7.7/9 for the untreated fried potato sample). PEF pre-treatment negatively affected the sensory quality of the OD-treated fried samples. PEF + OD-treated fried potato strips received the lowest score for total sensory quality (6.0/9.0), mainly due to their increased firmness. PEF-pre-treated potato strips presented the value of 10 N for firmness compared to 8 N of the OD-treated sample and 5.5 N of the untreated sample. The water content of untreated, OD-treated, and PEF + OD-treated samples was calculated as 0.413, 0.249, and 0.298 g/g, respectively. All treated samples had a sticky texture due to glycerol content compared to the untreated sample.

French fries have two distinct textures: a "crispy" crust, similar in physical characteristics to potato chips (or crisps) [55], and a "firm-mealy" core, which shares some of the textural properties of boiled potatoes. The key structural parameters that influence the crispness of French fries include moisture content, oil absorption, and the starch content and distribution within the potato tissues. Other important parameters affecting crispness are related to the manufacturing process, such as pre-drying and par-frying conditions [56,57].

PEF-treated samples received lower scores for aroma, taste, and oil content. Both OD-treated and PEF + OD-treated samples were accepted in terms of appearance and presented improved sensory characteristics. Previous studies of Abedpour and Dehghannya (2016) [58] and da Costa Ribeiro et al. (2016) [59] on osmotic dehydration pre-treatment of potato preceding frying revealed that OD pre-treatment improved sensory characteristics of the fried potato.

The OD-treated and PEF + OD-treated samples exhibited a dark yellow color, in contrast to the light yellow color of the untreated sample (Scheme 1). Among these, the OD-treated samples were considered the most visually appealing. The golden yellow hue of the fried crust is an important factor in consumer preference, often influencing their choice even before tasting. This color results from the Maillard reaction, which leads to the caramelization of sugars when fried at high temperatures [37,59]. The intensity of the color is influenced by the presence of reducing sugars, amino acids, and proteins on the surface, as well as the frying temperature, edible coating, and pre-treatment method [19,28,60]. Zhang et al. (2018) reported that combining PEF treatment with blanching significantly enhanced the brightness of French fries, giving them a golden yellow appearance [61]. The discoloration observed during frying was primarily caused by the Maillard reaction, with the depth of color depending on the levels of reducing sugars, amino acids, or proteins on the surface of the potato strips, as well as frying temperature and duration [60]. Blanching likely resulted in the loss of reducing sugars and amino acids due to electrolyte leakage from the potato strips. As a result, the PEF–blanching combined pre-treatment slowed the Maillard reaction and significantly reduced browning during frying [61,62].

Scheme 1. Photos of (**a**) untreated, (**b**) OD-treated, and (**c**) PEF + OD-treated potato strips after deep frying at 180 °C for 10 min (day 0 of the storage).

3.3. *Shelf-Life Study of OD and PEF + OD Raw Potatoes*

3.3.1. Texture and Color Change During Storage

The texture (firmness) and color (browning) of OD raw potato strips were retained during 18 days of storage. In Figure 5, the firmness change of OD-treated, PEF + OD-treated, and untreated potato strips during storage are presented. PEF-treated samples presented the highest firmness values compared to the untreated and OD-treated samples. During the 18 days of storage, the firmness of OD-treated and PEF + OD-treated samples increased, whereas the firmness of untreated samples remained stable during their shorter 5 days of storage. The firmness of OD-treated samples presented a 10%, 22%, and 31% increase at the end of their storage at 4, 8, and 12 °C, respectively. The firmness values of PEF + OD-treated samples significantly decreased at a storage temperature of 12 °C (12%). Firmness in potato tubers is influenced by three primary changes that occur due to chemical, physical, and structural transformations during the manufacturing process [63]. First, starch undergoes gelatinization; second, cell walls weaken, resulting in increased permeability; and third, the adhesion between adjacent cells diminishes [64]. The extent of these changes is determined by thermal conditions, including processing temperature and treatment duration.

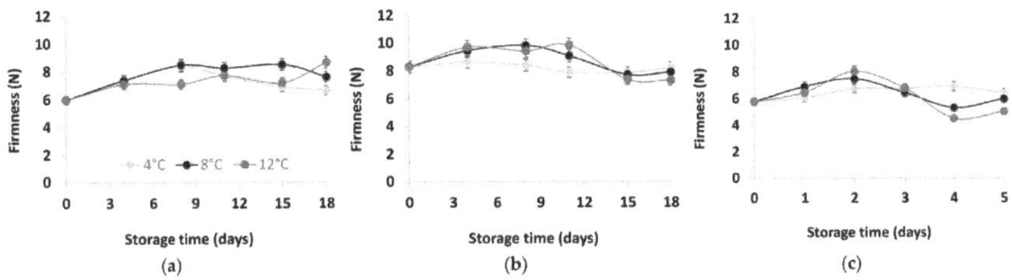

Figure 5. Evolution of texture (firmness, N) of (**a**) OD-treated, (**b**) PEF + OD-treated, and (**c**) untreated potato strips during their storage at 4, 8, and 12 °C.

Browning index values were significantly lower for OD-treated (66%) and PEF + OD-treated (71%) potato strips (at day 1 and throughout storage) compared to untreated ones, showing that OD pre-treatment protected potato color, which is a major problem of fresh-cut potatoes ($p < 0.05$). In Figure 6, the *BI* changes of OD-treated, PEF + OD-treated, and untreated potato strips during storage are presented. The *BI* values of untreated samples were 20.48 (day 1) and 23.89 (day 5 at 12 °C) compared to *BI* values of OD-treated and PEF + OD-treated samples, which were 6.81 and 5.76 (day 1) and 10.66 and 8.45 (day 5 at 12 °C), respectively. *BI* values of all samples slightly increased during storage. The same trend was observed by Katsouli et al. [38] for PEF + OD-treated and high-pressure (HP) + OD-treated potato strips, which presented lower *BI* values (and lighter color) compared to untreated potato strips. More specifically, HP and PEF reduced the *BI* by 12.5% and 27.9%, respectively, compared to the samples that had only been osmo-dehydrated.

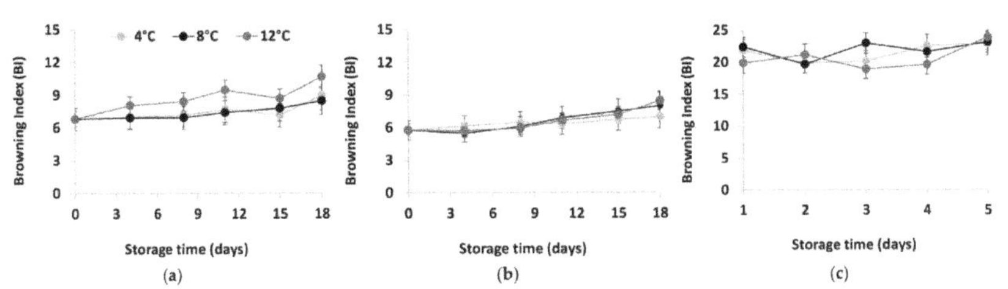

Figure 6. Evolution of color (Browning Index, *BI*) of (**a**) OD-treated, (**b**) PEF + OD-treated, and (**c**) untreated potato strips during their storage at 4, 8, and 12 °C.

The potato variety Spunta, commonly cultivated for the fresh market in southern Europe, the Middle East, and North Africa, is among the most susceptible to browning [65]. Therefore, for such delicate tissue, the milder color (PEF and OD) could be considered commercially significant. However, despite the stability of the *BI* value of the untreated sample during the 5 days of storage, untreated samples were rejected at day 3 by the sensory panelists. This could be due to the discolorations on the surface of the potato strips. Although color measurements were taken at three representative points of the potato tissue, the *BI* failed to accurately reflect the color differences. Therefore, visual inspection proves to be a more reliable method for monitoring processing. In a study on enzymatic browning in potatoes, Cantos et al. (2002) assessed color development using both the *BI* and organoleptic evaluation given the unique characteristics of the tissue and observed discrepancies between the two methods [65].

3.3.2. Evolution of Microbial Load During Storage

In the present study, the primary focus was on the quality changes incurred by spoilage, microbial growth, and/or enzymatic activity. The evolution of the total aerobic bacteria, Enterobacteriaceae, and anaerobic microorganisms (no significant growth of other microorganisms was observed) of untreated, OD-treated, and PEF + OD-treated potato strips is shown in Figure 7. OD pre-treatment significantly increased the microbial stability of potato strips. The microbial counts of OD-treated and PEF + OD-treated potatoes presented an extended lag period before reaching about 10^4–10^5 CFU/g at 18 days of storage, as also reported by Dimopoulos et al. (2024) [41]. The microbial counts of untreated potatoes reached 10^5 (4 °C)–10^8 (12 °C) CFU/g (total anaerobic) after 5 days of storage. The initial total viable load (TVC) of untreated potato strips was 4 log(CFU)/g. The total viable counts for pre-treated with OD and PEF + OD were <6 log(CFU)/g for 18 days of chill storage. Enterobacteriaceae were at levels below the limit of detection (<10 CFU/g). In the case of the untreated sample, Enterobacteriaceae growth was observed after 5 days of storage. The kinetic parameters of the microbial growth model, microbial growth rate (*k*), and lag phase (λ) are presented in Table 4. The effect of OD treatment regarding microbial growth inhibition can be attributed to the difficulty of microorganism penetration into cellular spaces, which are filled with the concentrated osmotic solution [66]. Glycerol, chosen as the primary agent for reducing water activity in the osmotic medium, also demonstrates benefits as a microbial protectant. The combination of water activity levels, pH values, and packaging conditions in the developed osmo-dehydrated potatoes ensures their microbial stability.

Table 4. Baranyi model parameters for microbial growth (total viable counts (TVC), Enterobacteriaceae, and total anaerobic count) for untreated potato strips. The dependence of the rate constant μ_m on storage temperature was modeled with the Arrhenius equation.

Storage Temperature (°C)	$\mu_m(\mathrm{d}^{-1})$	$logN_o\left(\frac{logCFU}{g}\right)$	$logN_o\left(\frac{logCFU}{g}\right)$	R^2	$E_{a,\mu m}\left(\frac{\mathrm{kJ}}{\mathrm{mol}}\right)$	$\mu_m, ref(\mathrm{d}^{-1})$	R^2
			Total Viable Count (TVC)				
4	0.483 ± 0.14	3.977 ± 0.318	0.801 ± 1.159	0.839			
8	0.718 ± 0.053	3.998 ± 0.083	0.293 ± 0.219	0.995	118.96	0.839	0.9317
12	2.062 ± 0.326	4.085 ± 0.142	1.267 ± 0.201	0.990			
			Enterobacteriaceae				
4	0.734 ± 0.206	3.508 ± 0.157	2.611 ± 0.559	0.876			
8	0.834 ± 0.105	3.670 ± 0.156	0.668 ± 0.362	0.984	33.61	0.335	0.9513
12	1.106 ± 0.058	3.643 ± 0.111	–	0.994			
			Total Anaerobic Count				
4	0.523 ± 0.153	3.758 ± 0.309	–	0.858			
8	0.975 ± 0.243	3.972 ± 0.319	0.703 ± 0.640	0.944	104.88	0.518	0.9995
12	1.876 ± 0.145	4.084 ± 0.122	0.982 ± 0.147	0.996			

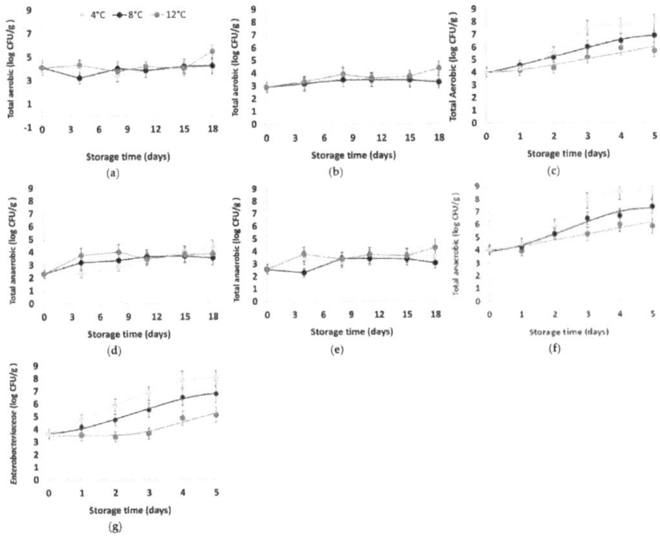

Figure 7. Microbial growth: total viable counts (TVC) of (**a**) OD-treated, (**b**) PEF + OD-treated, and (**c**) untreated samples; total anaerobic count of (**d**) OD-treated, (**e**) PEF + OD-treated, and (**f**) untreated samples; and (**g**) Enterobacteriaceae of untreated potato strips. (**a**,**d**,**g**) Experimental values and (**a**–**c**) experimental values and data fit to the Baranyi model. Error bars represent the standard deviation of three replicates.

3.3.3. Evolution of Sensory Quality Loss During Storage

The quality of fresh-cut potatoes (slices, strips, or cubes) is defined by their bright appearance, firm texture, slightly moist surface, and flesh color characteristic of the variety, free from signs of darkening or dehydration. To preserve these qualities, a combination of anti-browning agents, such as sodium bisulfite or organic acids, is commonly used alongside modified atmosphere packaging or vacuum packaging to maintain very low oxygen levels. Key parameters influencing quality include the potato cultivar, harvesting and handling practices, tuber maturity, and the peeling method employed. In the present study, the modified atmosphere of 12.2% CO_2–87.6% N_2 (0.188% O_2) used for the package kept the microbial counts at low levels for 5 and 18 days of storage at 4 °C (untreated, OD-treated, and PEF + OD-treated samples, respectively). During storage, the O_2 and CO_2 concentration levels decreased for all temperatures studied. For example, the O_2 and CO_2 concentrations of untreated potatoes stored at 4 °C reduced from 0.188 and 12.2 to 0.139 and 9.0, respectively. The same was observed for OD-treated and PEF + OD-treated samples, to a lesser extent. The respective values for OD-treated and PEF + OD-treated samples were 0.122 and 10.9 (OD) and 0.121 and 11.4 (PEF + OD). Ma et al. (2010) demonstrated that potato slices could be preserved for 8 days at 5 °C under a controlled atmosphere of 12% CO_2 and 3% O_2, combined with a dip in 0.025% sodium bisulfite [67]. Similarly, a shelf life exceeding 7 days at 5 °C was achieved when bisulfite was replaced with citric and ascorbic acids and packed under a modified atmosphere of 20% CO_2 and 80% N_2 [68]. Angos et al. (2008) reported that a super-atmospheric oxygen composition of 80% O_2 and 10–20% CO_2 effectively controlled browning and suppressed the respiration rate compared to a controlled atmosphere with 2.5% O2 and 10% CO2 over 14 days of storage at 4 °C [69]. Additionally, an active modified atmosphere with initial O_2 levels of 0.5–1.0% using a 35 µm thick polypropylene (PP) film preserved the appearance of fresh-cut potatoes for 10 days at 4 °C, though slight off-odors developed [70]. Modified atmosphere packaging (MAP) with low oxygen levels (1–3%) at 0–5 °C can reduce cut surface browning but is insufficient to completely prevent it [71]. A shelf life of up to 3 weeks at 2 °C can be achieved by applying an active MAP with an initial nitrogen flush, resulting in a gas composition of 2–5% O_2 and 3–5% CO_2 during storage [72]. These gas levels do not impact microbial populations compared to storage in air [73]. Beltran et al. (2005) concluded that vacuum packaging was the most effective method for preserving the sensory quality of fresh-cut potatoes for up to 14 days at 4 °C. Under MAP conditions, browning was inhibited only with sodium sulfite, but this treatment led to the development of off-odors after 14 days at 4 °C [74].

In the present study, the OD-treated potato samples were rejected on day 15 due to off-odor development. The untreated fried potato samples were rejected on day 3 due to discoloration (based on sensory deterioration as well as microbial growth). As reported, (sole) MAP packaging could not maintain the color of fresh-cut potatoes [74]. The OD-treated and PEF + OD-treated fried potato strips were sensorially evaluated up to day 15 for the three temperatures studied. The average sensory scores for OD-treated potato fries were 6.5, 6.3, and 6.0 for the storage temperatures of 4, 8, and 12 °C, respectively, showing that the effect of temperature was not statistically significant. The same was observed for PEF + OD-treated samples. PEF + OD-treated samples received 6.5 (4 °C), 6.4 (8 °C), and 6.3 (12 °C) on storage day 15. The shelf life of OD-treated as well as PEF + OD-treated fried potato strips was limited to 15 days at 4–12 °C by the sensory rejection of (raw, after package opening) potato strips. Both OD-treated and PEF + OD-treated potato samples were microbially stable, as shown in Section 3.3.2.

4. Conclusions

The scope of this study was to explore the application of PEF and OD for effective treatment of potatoes with the aim of retaining or improving the quality and shelf life of modified-atmosphere-packaged raw potato strips as well as reducing oil uptake during subsequent frying. OD (35 °C, 120 min) with/without PEF (0.5 kV/cm, 200 pulses) pre-treatment was found to promote the retention of the overall quality (texture and color) and increase the shelf-life stability of the fresh-cut potato strips. Additionally, the OD-treated and PEF + OD-treated samples resulted in high-quality potato strips with reduced browning during chilled storage. Raw OD-pre-treated potatoes were microbiologically stable for 18 days at 4 °C; however, their shelf life was estimated at 15 days based on sensory acceptance. The untreated potato strips were rejected at 3 days (based on sensory deterioration as well as microbial growth). OD pre-treatment reduced the oil uptake during frying (up to 30%) and resulted in potato fries judged positively by sensory panelists. OD in combination with MAP can extend the shelf life and improve the commercial value of fresh-cut potatoes (especially concerning the "Naxos" variety, which has a special added value), while OD pre-treatment can be utilized as an alternative method of producing low-fat potato fries.

A point to consider regarding PEF and OD feasibility in the food industry is related to existing technical and economic challenges and scalability but also to the need for optimization of these emerging technologies. Taking also into account each technology's limitations, their successful combined use may assist in delivering safer and environmentally friendly fresh-cut potato products with prolonged shelf life. More specifically, OD, while retaining sensory qualities, cannot provide the required microbial stability and consumes large volumes of osmotic solution, increasing operational costs and rendering scalability difficult. On the other hand, PEF, although energy efficient, is not yet fully explored regarding the optimization of processing parameters, and validation of results on a larger scale remains a point of investigation. Nonetheless, PEF has been extensively explored and recently implemented in various fresh-cut product treatments, which makes it a promising technology for broader commercial application in the fresh-cut potato industry. A strong competitive advantage of PEF is that it has already been used in the potato industry. In this context, future research should focus on addressing energy efficiency issues for smaller-scale operations and designing an appropriate optimization study to ensure consistency and scalability, especially in terms of texture and sensory quality. Future studies should also investigate the effect of these technologies on different potato cultivars and processing conditions to retain the initial superior quality of raw material and maximize the shelf life of the end product [16].

Supplementary Materials: The following supporting information can be downloaded at https://www.mdpi.com/article/10.3390/foods14030420/s1, Table S1: Tested sensory characteristics for raw and fried potatoes.

Author Contributions: Conceptualization, P.T. and M.G.; methodology, E.D. and M.K.; validation, G.S., M.K. and A.K.; investigation, E.D. and M.K.; data curation, G.S. and A.K.; writing—original draft preparation, E.D. and G.S.; writing—review and editing, E.D., M.G. and P.T.; visualization, E.D.; supervision, M.K. and E.D.; project administration, M.G. and P.T. All authors have read and agreed to the published version of the manuscript.

Funding: This study was funded by the European Union-Next Generation EU in the framework of the National Recovery and Resilience Plan (Greece 2.0) under call ID 16971 "RESEARCH-CREATE—INNOVATE" [project code T2EDK-03121—project acronym: Fresh4ever].

Institutional Review Board Statement: Institutional Review Board approval for human subject study (IRB) is not applicable in this case. Our Institution, National Technical University of Athens (NTUA) is a Technical University in Athens, Greece. The Laboratory of Food Chemistry and Technology of the School of Chemical Engineering of NTUA consists of a chemical laboratory and a sensory laboratory. The chemical laboratory (well equipped with basic and specialized instruments for food analyses) facilities have a total area of 250 m^2 and the sensory laboratory a total area of 50 m^2. The sensory laboratory was accredited in 2010 according to the ISO 17025 standard. The preparation of food samples was conducted in the preparation area under hygienic conditions according to the principles of Good Laboratory Practice (GLP) (suitably separated from the testing room, equipped for preparing and serving food samples), and the sensory testing was conducted in sensory booths by trained assessors. All methods were performed in accordance with the relevant guidelines and regulations included in the International Organization for Standardization (ISO) concerning sensory analysis.

Informed Consent Statement: Regarding the ethics statement during sensory testing, the following procedure was considered. Only adults participated in the recruitment to the sensory team. Participation in the tests and assessments was voluntary. Informed written consent was obtained from the participants in the sensory evaluation study. Each of them could withdraw their consent without providing any justification. Each participant also consented to the processing of their personal data in accordance with Article 6 of Regulation (EU) 2016/679 of the European Parliament and of the Council of 27 April 2016 on the protection of natural persons regarding the processing of personal data and on the free movement of such data, and repealing Directive 95/46/EC (General Data Protection Regulation). All participants obtained a detailed description of the test and were informed about the food samples that would be assessed. Each of the assessors was obliged to report any indispositions and allergies and if such was the case, the subject did not participate in the tests.

Data Availability Statement: The data presented in this study are available on request from the corresponding author. The data are not publicly available due to project grant confidentiality restrictions.

Conflicts of Interest: The authors declare no conflicts of interest.

References

1. Dourado, C.; Pinto, C.; Barba, F.J.; Lorenzo, J.M.; Delgadillo, I.; Saraiva, J.A. Review Innovative non-thermal technologies affecting potato tuber and fried potato quality. *Trends Food Sci. Technol.* **2019**, *88*, 274–289. [CrossRef]

2. Qiao, L.; Han, X.; Wang, H.; Gao, M.; Tian, J.; Lu, L.; Liu, X. Novel alternative for controlling enzymatic browning: Catalase and its application in fresh-cut potatoes. *J. Food Sci.* **2021**, *86*, 3529–3539. [CrossRef] [PubMed]
3. Tian, J.; Chen, J.; Ye, X.; Chen, X. Health benefits of the potato affected by domestic cooking: A review. *Food Chem.* **2016**, *202*, 165–175. [CrossRef] [PubMed]
4. Li, L.; Wu, M.; Zhao, M.; Guo, M.; Liu, H. Enzymatic properties on browning of fresh-cut potato. *Mater. Sci. Eng.* **2018**, *397*, 012116. [CrossRef]
5. Hou, Z.; Feng, Y.; Wei, S.; Wang, Q. Effects of curing treatment on the browning of fresh-cut potatoes. *Am. J. Potato Res.* **2014**, *91*, 655–662. [CrossRef]
6. Liu, P.; Xu, N.; Liu, R.; Liu, J.; Peng, Y.; Wang, Q. Exogenous proline treatment inhibiting enzymatic browning of fresh-cut potatoes during cold storage. *Postharvest Biol. Technol.* **2022**, *184*, 111754. [CrossRef]
7. Bingol, G.; Wang, B.; Zhang, A.; Pan, Z.; McHugh, T.H. Comparison of water and infrared blanching methods for processing performance and final product quality of French fries. *J. Food Eng.* **2014**, *121*, 135–142. [CrossRef]
8. Wang-Pruski, G.; Nowak, J. Potato after-cooking darkening. *Am. J. Potato Res.* **2004**, *81*, 7–16. [CrossRef]
9. Ahmed, I.; Qazi, I.M.; Jamal, S. Developments in osmotic dehydration technique for the preservation of fruits and vegetables. *Innov. Food Sci. Emerg. Technol.* **2016**, *34*, 29–43. [CrossRef]
10. Yadav, A.K.; Singh, S.V. Osmotic dehydration of fruits and vegetables: A review. *J. Food Sci. Technol.* **2014**, *51*, 1654–1673. [CrossRef]
11. Krokida, M.K.; Kiranoudis, C.T.; Maroulis, Z.B.; Marinos-Kouris, D. Effect of pretreatment on color of dehydrated products. *Dry. Technol.* **2000**, *18*, 1239–1250. [CrossRef]
12. Osae, R.; Adjonu, R.; Owusu-Ansah, P.; Apaliya, M.T.; Fauzia, A.S.; Engmann, F.N.; Otoo, G.S.; Alolga, R.N. Freeze-thawing and osmotic dehydration pretreatments on physicochemical properties and quality of orange-fleshed sweet potato slice during hot air drying. *Food Chem. Adv.* **2024**, *5*, 100843. [CrossRef]
13. Torreggiani, D.; Bertolo, G. Present and Future in Process Control and Optimization of Osmotic Dehydration: From Unit Operation to Innovative Combined Process: An Overview. *Adv. Food Nutr. Res.* **2004**, *48*, 173–238. [CrossRef] [PubMed]
14. Cheng, X.; Wang, S.; Iqbal, M.S.; Pan, L.; Hong, L. Effect of ultrasound-assisted osmotic dehydration on the drying kinetics, water state, and physicochemical properties of microwave vacuum-dried potato slices. *Ultrason. Sonochem.* **2023**, *99*, 106557. [CrossRef] [PubMed]
15. Mari, A.; Andriotis, P.; Drosou, C.; Krokida, M. Enhancing Shelf-life Stability of Refrigerated Potatoes through Osmotic Dehydration and Ohmic Heating Optimization: A Strategy to Mitigate Enzymatic Browning. *Eur. Potato J.* **2024**. [CrossRef]
16. Drosou, C.; Sklirakis, I.; Polyzou, E.; Yakoumis, I.; Boukouvalas, C.J.; Krokida, M. Processing Fresh-Cut Potatoes Using Non-Thermal Technologies and Edible Coatings. *Appl. Sci.* **2024**, *14*, 11039. [CrossRef]
17. Akbarian, M.; Ghasemkhani, N.; Moayedi, F. Osmotic dehydration of fruits in food industrial: A review. *Int. J. Biosci.* **2013**, *3*, 1–16.
18. Krokida, M.K.; Oreopoulou, V.; Maroulis, Z.B.; Marinos-Kouris, D. Effect of osmotic dedydration pretreatment on quality of french fries. *J. Food Eng.* **2001**, *49*, 339–345. [CrossRef]
19. Gupta, P.; Shivhare, U.S.; Bawa, A.S. Studies on frying kinetics and quality of French fries. *Dry. Technol.* **2000**, *8*, 311–321. [CrossRef]
20. Kwaw, E.; Osae, R.; Apaliya, M.T.; Sackey, A.S.; Alolga, R.N.; Kaburi, S.A.; Hinson, M.; Bediako, G.; Botwe, A.K.; Pitcher, V.M. Effect of different pre-treatments on the physical properties, frying kinetics and organoleptic physiognomies of fried sweet potato (*Ipomoea batatas*) chips. *Food Human.* **2024**, *3*, 100351. [CrossRef]
21. Nowosad, K.; Sujka, M.; Pankiewicz, U.; Kowalski, R. The application of PEF technology in food processing and human nutrition. *J. Food Sci. Technol.* **2020**, *58*, 397–411. [CrossRef] [PubMed]
22. Tiwari, B.; O'Donell, C.; Cullen, P. New challenges in food science and technology: An industrial perspective. *Trends Food Sci. Technol.* **2009**, *20*, 180–181. [CrossRef]
23. Donsì, F.; Ferrari, G.; Pataro, G. Applications of Pulsed Electric Field Treatments for the Enhancement of Mass Transfer from Vegetable Tissue. *Food Eng. Rev.* **2010**, *2*, 109–130. [CrossRef]
24. González Pérez, J.E.; Ramírez Corona, N.; López Malo, A. Mass Transfer During Osmotic Dehydration of Fruits and Vegetables: Process Factors and Non-Thermal Methods. *Food Eng. Rev.* **2021**, *13*, 344–374. [CrossRef]
25. Soliva-Fortuny, R.; Balasa, A.; Knorr, D.; Martín-Belloso, O. Effects of pulsed electric fields on bioactive compounds in foods: A review. *Trends Food Sci Technol.* **2009**, *20*, 544–556. [CrossRef]
26. Wiktor, A.; Lammerskitten, A.; Barba, G.; Michalski, M.; Toepfl, S.; Parniakov, O. Drying Processes Assisted by PEF for Plant-Based Materials. In *Innovative Food Processing Technologies*; Elsevier: Amsterdam, The Netherlands, 2021; pp. 271–280. [CrossRef]
27. Wiktor, A.; Singh, A.P.; Parniakov, O.; Mykhailyk, V.; Mandal, R.; Witrowa-Rajchert, D. PEF as an alternative tool to prevent thermolabile compound degradation during dehydration processes. In *Pulsed Electric Fields to Obtain Healthier and Sustainable Food for Tomorrow*; Barba, F.J., Parniakov, O., Wiktor, A., Eds.; Academic Press: Cambridge, MA, USA, 2020; pp. 155–202. [CrossRef]

28. Barba, F.J.; Parniakov, O.; Pereira, S.A.; Wiktor, A.; Grimi, N.; Boussetta, N.; Saraiva, J.; Raso, J.; Martin-Belloso, O.; Witrowa-Rajchert, D.; et al. Current applications and new opportunities for the use of pulsed electric fields in food science and industry. *Food Res. Int.* **2015**, *77*, 773–798. [CrossRef]
29. Boussetta, N.; Grimi, N.; Lebovka, N.I.; Vorobiev, E. "Cold" electroporation in potato tissue induced by pulsed electric field. *J. Food Eng.* **2013**, *115*, 232–236. [CrossRef]
30. Janositz, A.; Noack, A.; Knorr, D. Pulsed electric fields and their impact on the diffusion characteristics of potato slices. *Food Sci. Technol.* **2011**, *44*, 1939–1945. [CrossRef]
31. Faustera, T.; Schlossnikl, D.; Ostermeier, R.R.; Teufel, F.; Toepfl, S.; Jaeger, H. Impact of pulsed electric field (PEF) pretreatment on process performance of industrial French fries production. *J. Food Eng.* **2018**, *235*, 16–22. [CrossRef]
32. Krokida, M.K.; Oreopoulou, V.; Maroulis, Z.B.; Marinos-Kouris, D. Effect of pre-treatment on viscoelastic behaviour of potato strips. *J. Food Eng.* **2001**, *50*, 11–17. [CrossRef]
33. Karizaki, V.M.; Sahin, S.; Sumnu, G.; Mosavian, M.T.H.; Luca, A. Effect of ultrasound-assisted osmotic dehydration as a pretreatment on deep fat frying of potatoes. *Food Bioprocess Technol.* **2013**, *6*, 3554–3563. [CrossRef]
34. Liu, C.; Deng, H.; Lv, M.; Du, H.; Li, B.; Grimi, N.; Liu, Y.; Chen, W. Impact of pulsed electric fields and ultrasound on the frying characteristics of sweet potato chips. *Food Bioprod. Process.* **2025**, *149*, 49–57. [CrossRef]
35. Su, Y.; Zhang, M.; Chitrakar, B.; Zhang, W. Reduction of oil uptake with osmotic dehydration and coating pre-treatment in microwave-assisted vacuum fried potato chips. *Food Biosci.* **2021**, *39*, 100825. [CrossRef]
36. Ostermeier, R.; Hill, K.; Dingis, A.; Töpfl, S.; Jäger, H. Influence of pulsed electric field (PEF) and ultrasound treatment on the frying behavior and quality of potato chips. *Innov. Food Sci. Emerg. Technol.* **2021**, *67*, 102553. [CrossRef]
37. Schouten, M.A.; Genovese, J.; Tappi, S.; Di Francesco, A.; Baraldi, E.; Cortese, M.; Capriolid, G.; Angelonid, S.; Vittori, S.; Rocculi, P.; et al. Effect of innovative pre-treatments on the mitigation of acrylamide formation in potato chips. *Innov. Food Sci. Emerg. Technol.* **2020**, *64*, 102397. [CrossRef]
38. Katsouli, M.; Dermesonlouoglou, E.; Dimopoulos, G.; Karafantalou, E.; Giannakourou, M.; Taoukis, P. Shelf-Life Enhancement Applying Pulsed Electric Field and High-Pressure Treatments Prior to Osmotic Dehydration of Fresh-Cut Potatoes. *Foods* **2024**, *13*, 171. [CrossRef]
39. AOAC. *Official Methods of Analysis*, 15th ed.; Association of Official Analytical Chemists: Rockville, MD, USA, 1990.
40. Garcia, E.; Barrett, D.M. Preservative Treatments for Fresh-cut Fruits and Vegetables. In *Fresh-Cut Fruits and Vegetables*, 1st ed.; Lamikanra, O., Ed.; CRC Press: Boca Raton, FL, USA, 2002. [CrossRef]
41. Dimopoulos, G.; Katsimichas, A.; Balachtsis, K.; Dermesonlouoglou, E.; Taoukis, P. Effect of pulsed electric fields on the shelf stability and sensory acceptability of osmotically dehydrated spinach: A mathematical modeling approach. *Foods* **2024**, *13*, 1410. [CrossRef]
42. Kuhn, J.; Müller, H.; Salzig, D.; Czermak, P. A rapid method for an offline glycerol determination during microbial fermentation. *Electron. J. Biotechnol.* **2015**, *18*, 252–255. [CrossRef]
43. Giannakourou, M.C.; Taoukis, P.S. Kinetic modelling of vitamin c loss in frozen green vegetables under variable storage conditions. *Food Chem.* **2003**, *83*, 33–41. [CrossRef]
44. Aguirre-Joya, J.A.; Ventura-Sobrevilla, J.; Martínez-Vazquez, G.; Ruelas-Chacón, X.; Rojas, R.; Rodríguez-Herrera, R.; Aguilar, C.N. Effects of a natural bioactive coating on the quality and shelf life prolongation at different storage conditions of avocado (*Persea americana* Mill.) cv. Hass. *Food Pack. Shelf Life* **2017**, *14*, 102–107. [CrossRef]
45. Wong, J.X.; Ramli, S.; Desa, S.; Chen, S.N. Use of *Centella Asiatica* Extract in Reducing Microbial Contamination and Browning. Effect in Fresh Cut Fruits and Vegetables during Storage: A Potential Alternative of Synthetic Preservatives. *LWT* **2021**, *151*, 112229. [CrossRef]
46. *ISO 8586-1*; Sensory Analysis—General Guidance for the Selection, Training and Monitoring of Assessors, Part 1: Selected As-sessors. International Organization for Standardization: Geneva, Switzerland, 1993.
47. Dermesonlouoglou, E.K.; Giannakourou, M.C. Evaluation and modelling of osmotic pre-treatment of peach using alternative agents in a multiple-component solution. *J. Sci. Food Agric.* **2019**, *99*, 1240–1249. [CrossRef] [PubMed]
48. Oladejo, A.O.; Ma, H.; Qu, W.; Zhou, C.; Wu, B. Effects of ultrasound on mass transfer kinetics, structure, carotenoid and vitamin c content of osmodehydrated sweet potato (*Ipomea batatas*). *Food Bioproc. Tech.* **2017**, *10*, 1162–1172. [CrossRef]
49. Dehghannya, J.; Gorbani, R.; Ghanbarzadeh, B. Effect of ultrasound-assisted osmotic dehydration pretreatment on drying kinetics and effective moisture diffusivity of mirabelle plum. *J Food Proc. Preserv.* **2015**, *39*, 2710–2717. [CrossRef]
50. Barani, Y.H.; Zhang, M.; Wang, B.; Devahastin, S. Influences of four pretreatments on anthocyanins content, color and flavor characteristics of hot-air dried rose flower. *Dry. Technol.* **2020**, *38*, 1988–1995. [CrossRef]
51. Liu, T.; Dodds, E.; Leong, S.Y.; Eyres, G.T.; Burritt, D.J.; Oey, I. Effect of pulsed electric fields on the structure and frying quality of "kumara" sweet potato tubers. *Innov. Food Sci. Emerg. Technol.* **2017**, *39*, 197–208. [CrossRef]

52. Liu, C.; Grimi, N.; Lebovka, N.; Vorobiev, E. Effects of pulsed electric fields treatment on vacuum drying of potato tissue. *LWT* **2018**, *95*, 289–294. [CrossRef]
53. Liu, C.; Grimi, N.; Lebovka, N.; Vorobiev, E. Effects of preliminary treatment by pulsed electric fields and convective air-drying on characteristics of fried potato. *Innov. Food Sci. Emerg. Technol.* **2018**, *47*, 454–460. [CrossRef]
54. Levaj, B.; Pelaic, Z.; Galic, K.; Hunjek, D.D.; Pedisic, S.; Kurek, M.; Šcetar, M.; Poljak, M.; Balbino, S.; Cošic, Z.; et al. Maintaining the Quality and Safety of Fresh-Cut Potatoes (*Solanum tuberosum*): Overview of Recent Findings and Approaches. *Agronomy* **2023**, *13*, 2002. [CrossRef]
55. Pedreschi, F.; Aguilera, J.; Pyle, L. Textural characterization and kinetics of potato strips during frying. *J. Food Sci.* **2002**, *66*, 314–318. [CrossRef]
56. Pedreschi, F.; Moyano, P.; Santis, N.; Pedreschi, R. Physical properties of pre-treated potato chips. *J. Food Eng.* **2007**, *79*, 1474–1482. [CrossRef]
57. Kita, A. The influence of potato chemical composition on crisp texture. *Food Chem.* **2002**, *76*, 173e179. [CrossRef]
58. Abedpour, L.; Dehghannya, J. Investigation of oil uptake during potato strips deep-fat frying pretreated with ultrasound and osmotic dehydration. *J. Food Sci. Technol.* **2016**, *12*, 79–94.
59. Da Costa Ribeiro, A.S.; Aguiar-Oliveira, E.; Maldonado, R.R. Optimization of osmotic dehydration of pear followed by conventional drying and their sensory quality. *LWT-Food Sci. Technol.* **2016**, *72*, 407–415. [CrossRef]
60. Graham-Acquaah, S.; Ayernor, G.S.; Bediako-Amoa, B.; Saalia, F.S.; Afoakwa, E.O.; Abbey, L. Effect of blanching and frying on textural profile and appearance of yam (*Dioscorea rotundata*) french fries. *J. Food Process. Preserv.* **2015**, *39*, 19–29. [CrossRef]
61. Zhang, Y.; Kahl, D.H.W.; Bizimungu, B.; Lu, Z.X. Effects of blanching treatments on acrylamide, asparagine, reducing sugars and colour in potato chips. *J. Food Sci. Technol.* **2018**, *55*, 4028–4041. [CrossRef]
62. Zhang, G.; Rei, K.; Lyu, X.; Yang, R. Effects of combined pulsed electric field and blanching pretreatment on the physiochemical properties of French fries. *Innov. Food Sci. Emerg. Technol.* **2021**, *67*, 102561. [CrossRef]
63. Thybo, A.K.; Christiansen, J.; Kaack, K.; Petersen, M.A. Effect of cultivars, wound healing and storage on sensory quality and chemical components in pre-peeled potatoes. *LWT Food Sci. Technol.* **2006**, *39*, 166–176. [CrossRef]
64. Van Marle, J.T.; Stolle-Smits, T.; Donkers, J.; van Dijk, C.; Voragen, A.G.J.; Recourt, K. Chemical and microscopic characterization of potato (solanum tuberosum l.) cell walls during cooking. *J. Agric. Food Chem.* **1997**, *45*, 50–58. [CrossRef]
65. Cantos, E.; Tudela, J.A.; Gil, M.I.; Espín, J.C. Phenolic compounds and related enzymes are not rate-limiting in browning development of fresh-cut potatoes. *J. Agric. Food Chem.* **2002**, *50*, 3015–3023. [CrossRef]
66. Castelló, M.L.; Fito, P.J.; Chiralt, A. Changes in respiration rate and physical properties of strawberries due to osmotic dehydration and storage. *J. Food Eng.* **2010**, *97*, 64–71. [CrossRef]
67. Ma, Y.; Wang, Q.; Hong, G.; Cantwell, M. Reassessment of treatments to retard browning of fresh-cut Russet potato with emphasis on controlled atmospheres and low concentrations of bisulphite. *Int. J. Food Sci. Technol.* **2010**, *45*, 1486–1494. [CrossRef]
68. Laurila, E.K.; Hurme, E.U.; Ahvenainen, R.T. Shelf life of sliced raw potatoes of various cultivar varieties—Substitution of bisulfites. *J. Food Prot.* **1998**, *61*, 1363–1371. [CrossRef] [PubMed]
69. Angos, I.; Virseda, P.; Fernandez, T. Control of respiration and color modification on minimally processed potatoes by means of low and high O_2/CO_2 atmospheres. *Postharvest Biol. Technol.* **2008**, *48*, 422–430. [CrossRef]
70. Tudela, J.A.; Gil, M.I. Tubers: Fresh-cut potatoes Group on Quality and Safety. In *Controlled and Modified Atmospheres for Fresh and Fresh-Cut Produce*; Department of Food Science and Technology, CEBAS-CSIC: Murcia, Spain, 2020; Chapter 25.1; pp. 625–628. [CrossRef]
71. Gorny, J.R. A summary of CA and MA requirements and recommendations for fresh-cut (minimally processed) fruits and vegetables. *Acta Hortic.* **2003**, *600*, 609–614. [CrossRef]
72. Gunes, G.; Lee, C.Y. Color of minimally processed potatoes as affected by modified atmosphere packaging and ant browning agents. *J. Food Sci.* **1997**, *62*, 572–576. [CrossRef]
73. Gunes, G.; Splittstoesser, D.F.; Lee, C.Y. Microbial quality of fresh potatoes: Effect of minimal processing. *J. Food Prot.* **1997**, *60*, 863–866. [CrossRef]
74. Beltran, D.; Selma, M.V.; Tudela, J.A.; Gil, M. Effect of different sanitizers on microbial and sensory quality of fresh-cut potato strips stored under modified atmosphere or vacuum packaging. *Postharv. Biol. Technol.* **2005**, *37*, 37–46. [CrossRef]

foods

MDPI

Article

Edible Coating Combining Liquid Smoke from Oil Palm Empty Fruit Bunches and Turmeric Extract to Prolong the Shelf Life of Mackerel

Muhammad Faisal [1,2,3,4], Asri Gani [1,4], Murna Muzaifa [5], M. Bagas Heriansyah [5], Hera Desvita [6,*], Suraiya Kamaruzzaman [1,2], Ahmad Sauqi [1] and Daru Ardiansa [1]

1 Department of Chemical Engineering, Faculty of Engineering, Universitas Syiah Kuala, Banda Aceh 23111, Indonesia; mfaisal@usk.ac.id (M.F.); asri_gani@usk.ac.id (A.G.); suraiya.k@usk.ac.id (S.K.); asauqi154@gmail.com (A.S.); daruardiansa@gmail.com (D.A.)
2 Climate Change Research Center, Universitas Syiah Kuala, Banda Aceh 23111, Indonesia
3 Halal Research Center, Universitas Syiah Kuala, Banda Aceh 23111, Indonesia
4 Oil Palm and Coconut Research Center, Universitas Syiah Kuala, Banda Aceh 23111, Indonesia
5 Department of Agriculture Product Technology, Faculty of Agriculture, Universitas Syiah Kuala, Banda Aceh 23111, Indonesia; murnamuzaifa@usk.ac.id (M.M.); mbagasheriansyah@gmail.com (M.B.H.)
6 Research Center for Chemistry, National Research and Innovation Agency, B.J. Habibie Science and Techno Park, Serpong, South Tangerang 15314, Indonesia
* Correspondence: hera005@brin.go.id

Abstract: This research aimed to evaluate the use of edible coating from a combination of liquid smoke and turmeric extract as a preservative for mackerel at room temperature. Liquid smoke was obtained from the pyrolysis of oil palm empty fruit bunches (OPEFB) at a temperature of 380 °C and purified by distillation at 190 °C. Liquid smoke with a concentration of 3% was combined with turmeric extract at a ratio of 2, 4, 6, and 8 g/L (CLS 2:1, CLS 4:1, CLS 6:1 and CLS 8:1). TVB-N testing showed that the mixture of liquid smoke and turmeric at a ratio of CLS 6: 1 and CLS 8: 1 maintains the freshness of fish for 48 h. Meanwhile, organoleptic testing reports that the best mixture was CLS 8:1. The number of colonies in the CLS 2:1, CLS 4:1, CLS 6:1, and CLS 8:1 mixtures were 4.92, 4.92, 4.16, and 4×10^5 colonies/g after 44 h of soaking. The MPN test result at 48 h of soaking is 1.1×10^3 MPN/g. Generally, mackerel preserved with a mixture of turmeric extract and liquid smoke with a ratio of 8:1 can be consumed up to a shelf life of 48 h at room temperature storage.

Keywords: OPEFB; liquid smoke; turmeric; TVB-N; TPC; MPN; organoleptics

1. Introduction

Fish is reported to contain high protein, vitamins, omega-3 polyunsaturated fatty acids, docosahexaenoic acid, eicosapentaenoic acid, and high nutritional value beneficial for body health [1,2]. This food ingredient is easily damaged since the meat is an ideal substrate for the life and growth of spoilage microorganisms, especially bacteria [3,4]. The water content in fish is high, around 65–80% [5], allowing possible biochemical reactions by enzymes in the body of fresh fish. The presence of biochemical reactions and oxidation causes tissue to decompose and produces chemical changes such as the texture of the meat becoming soft and the appearance of basic compounds in the protein causing a rancid odor. The presence of these volatile chemical compounds from the decomposition of fish meat provides a strong indication of a decline in quality. Therefore, these compounds are often used as an index and fish can last about 8 h after being caught [6,7].

To overcome the problem of quality decline in fish, various preservation methods have been developed, including those with natural ingredients [8]. An effective preservation method to extend the shelf life of fish is needed. People often use hazardous additives as a fish preservative, and this can have a negative impact on consumer health. The use of formalin has become a controversial issue and is prohibited in many countries due to the potential health risks. Several countries have implemented a ban on the use of the product to preserve fish. This compound is classified as a carcinogen and causes toxic effects in humans, such as irritation of the digestive tract, nausea, vomiting, and other digestive disorders. Therefore, the availability of safe and inexpensive natural preservatives is needed.

Natural preservatives used include red algae [9], peppermint essential oil [10], rosemary essential oil [11], clove essential oil [12], cinnamon oil [13], 'Gabsi' pomegranate peel extracts [14], terminalia ferdinandiana [15], thyme oil [16], curcumin [17], and liquid smoke [18]. Liquid smoke has bioactive compounds such as phenol, carbonyl, and organic acids functioning as antibacterials to maintain the quality of fish [19–21]. Liquid smoke can be produced from various biomass wastes such as coconut shells [22], terminalia cattapa wood [23], rice husks [24], and oil palm empty fruit bunches (OPEFB) [25]. However, the use as a preservative is still less popular because liquid smoke produces a fairly pungent odor. To reduce the odor, the concentration needs to be reduced and additional compounds from other natural ingredients such as turmeric extract are needed hence the antimicrobial properties are maintained.

Turmeric rhizome (*Curcuma domestical val*) is a biomaterial containing ingredients that can function as antioxidants and has received attention in food packaging [26,27]. This biomaterial is also widely used in herbal medicine and functional food, acting as anti-carcinogenic activities, antiangiogenesis, and antidiabetic [28–31]. Different bioactives are due to the presence of curcumin compounds [32]. Turmeric rhizomes can maintain fish quality due to curcumin compounds and essential oils. According to Pasaraeng et al. [33], the higher the concentration of curcumin, the lower the Total Volatile Base Nitrogen (TVB-N) value of fish. This shows that the inhibitory power of curcumin on bacterial growth is better. The combination of liquid smoke with turmeric, which contains active compounds such as curcumin, is expected to provide a synergistic effect in inhibiting the growth of microorganisms and slowing down the oxidation process in fish meat. This method is effective in extending shelf life as a sustainable solution for the fish processing industry. Therefore, this research aimed to evaluate natural preservatives from a combination of liquid smoke from OPEFB and curcumin extract to preserve mackerel. The ability of the natural preservatives was analyzed through Total Volatile Base (TVB) Testing, total plate count (TPC), most probable number (MPN), and organoleptic tests.

2. Materials and Methods

2.1. Materials

The materials used include OPEFB, turmeric extract, mackerel, distilled water (H_2O), Sodium Hydroxide (NaOH; Merck, Darmstadt, Germany), Potassium Carbonate (K_2CO_3, Merck, Darmstadt, Germany), Trichloroacetate (TCA; Merck, Darmstadt, Germany), Ethanol (Merck, Darmstadt, Germany), Hydrochloric Acid (HCl; Merck, Darmstadt, Germany), Phenolptelain Indicator (PP; Merck, Darmstadt, Germany), Peptonwasser Buffering (BPW; Merck, Darmstadt, Germany), and Sodium Chloride (NaCl; Merck, Darmstadt, Germany).

2.2. Sample Preparation

OPEFB was obtained from a palm oil mill in Cot Girek village, North Aceh district, Aceh, Indonesia, and cut into 5–8 cm pieces and dried by being exposed to the sun for about three days until they were completely dry. Additionally, the 5% water content in the sample was tested. The turmeric was cleaned and dried to reduce the water content before grounding into powder.

2.3. Making Preservatives from a Combination of Liquid Smoke with Turmeric

A total of 3 kg of OPEFB was put into the reactor and pyrolyzed at a temperature of 380 °C. The resulting steam flowed to the condenser through a pipe connected to the reactor lid and produced condensate in the form of grade 3 liquid smoke and tar. The complete process of making liquid smoke was consistent with previous research [24], as reported in Figure 1.

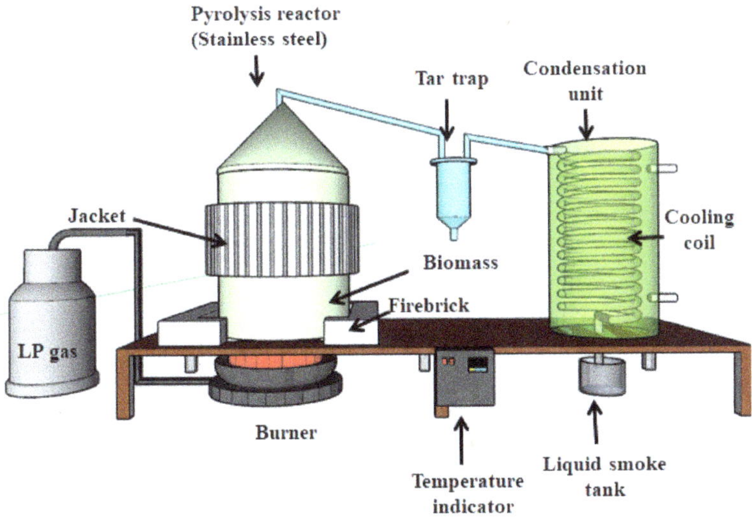

Figure 1. Schematic of the pyrolysis process.

The liquid smoke is purified using distillation at a temperature of 190 °C and diluted to 3% with distilled water. Furthermore, 3% (v/v) liquid smoke is added with turmeric powder at 2, 4, 6, and 8 g/L to obtain 4 combinations of CLS 2:1, CLS 4:1, CLS 6:1, and CLS 8:1, respectively.

2.4. Testing of a Mixture of Liquid Smoke with Curcumin on Mackerel Fish (Scomberomorus Commerson)

The test was conducted by dipping the mackerel into a mixture of liquid smoke with curcumin and analyzed every 4 h for 48 h. The mackerel (*Scomberomorus commerson*) samples were transferred to a sealed container and stored at room temperature during the preservation process. Mackerel that had not received any treatment were used as the control sample. The tests conducted were TVB-N followed the procedures outlined in SNI 2354.8:2009 [34]. The TVB-N value is calculated using the following formula:

$$TVBN = \frac{Vs - Vb \times NHCl \times 14.007 \times 100}{Sample\ weight}$$

where TVB-N: Total Volatile Base (mgN/100 g), Vs: Sample volume (mL), Vb: Volume of solution without sample (mL).

The TPC test was conducted in accordance with SNI 02-2725-1992 [35]. Nutrient Agar (NA) media in Petri dishes was used for the total plate count (TPC) test. The diluted mackerel sample was carefully transferred onto the surface of the NA media in the Petri dishes under aseptic conditions. After that, the plates were covered with plastic wrap and incubated at 37 °C for 24 to 48 h. A colony counter was used to count the bacterial colonies following incubation in order to calculate the total plate count.

The MPN test was carried out in following SNI 2897:2008 guidance [36]. The test consists of two phases: presumptive testing and confirmation testing, which confirms the results from the presumptive phase. A positive result is indicated by the production of gas or bubbles in the Durham tube.

The organoleptic tests such as Kamaruzzaman et al. [37]. Previous research calculated the average organoleptic value using the following equation [7].

$$AverageValue : x = \frac{\sum xi}{n} \tag{1}$$

Description: x = average score, xi = organoleptic value of panelist i, n = number of panelists. Organoleptic testing involves using 37 panelist senses to test food, such as the hands to feel texture, the eyes to see color, and the nose to detect aroma. Seven of them were standard panelists, which are people who have knowledge and expertise in determining the grade of fish that is still fit for consumption, as well as good aptitude and sensitivity to product quality. The remaining thirty panelists were non-standard panelists, meaning they lacked the necessary training to perform organoleptic evaluations. The panelists were trained, educated, and given information on how to do the organoleptic evaluations prior to the test. To ascertain the degree of preference for mackerel fish preserved with different liquid smoke combinations and the duration of preservation, an organoleptic test was performed. Different ratios of liquid smoke mixture (CL 2:1, CL 4:1, CL 6:1, and 8:1) were used to soak the mackerel.

2.5. Statistical Analysis

All the experiments were performed with three replications. The data of the analyses were pooled, averaged, and standard deviation were calculated using MS-Excel software 2010.

3. Results

3.1. Total Volatile Base Nitrogen (TVB-N) Test

TVB-N value is an important parameter used to determine the quality of food ingredients. In this context, a food ingredient will be considered unfit to eat with a TVB-N value that exceeds the acceptance limit. Pearson [38] stated that the acceptance limit for TVB-N value was 20–30 mg N/100 g, as reported in Figure 2.

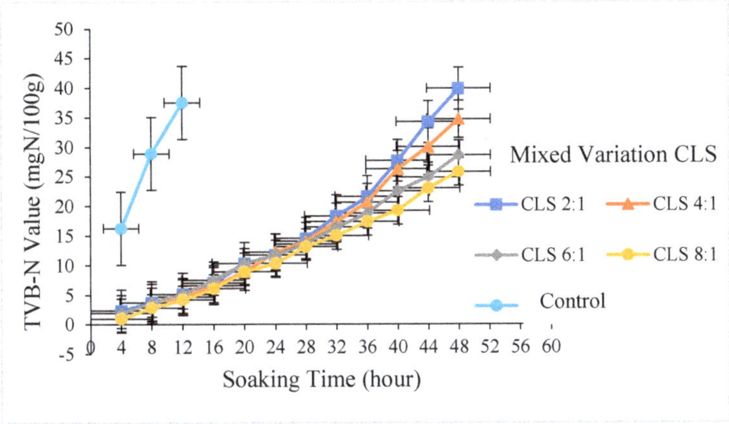

Figure 2. TVB-N value in mackerel fish samples coated with different concentrations of liquid smoke and turmeric.

The TVB-N values of mackerel fish shown in Figure 2 on the first day in each mixture were 11.725, 12.194, 11.725, and 10.318 mgN/100 g. These values show that at the beginning of storage, the fish was categorized as fresh with a TVB-N value range of 10–20 mgN/100 g [39]. The addition of liquid smoke from OPEFB with a combination of turmeric can reduce TVB-N value in mackerel fish compared to control samples. The TVB-N value after soaking for 8 h was 28.901 mgN/100 g without preservative treatment from a mixture of liquid smoke with turmeric. This value is close to the maximum TVB-N value which is not good for consumption. Meanwhile, the TVB-N value obtained was 37.501 mgN/100 g at 12 h. Fish soaked using liquid smoke and CLS in 2:1 and 8:1 mixtures of turmeric and liquid smoke lasted for 40 and 48 h, respectively.

The TVB-N value continues to increase during storage time, indicating a decline in the quality of mackerel. The soft texture and high protein content of fish causes protein degradation, peptide compounds, and amino acid content to produce volatile base compounds [40]. Bekhit et al. [41] stated that the degradation of enzymes produced volatile base compounds. The TVB-N value is influenced by the amount of non-protein nitrogen in the fish [42]. The highest value of mackerel in the CLS 2:1 sample was obtained with a TVB-N value of 39.86 mgN/100 g at a storage time of 48 h, and the value exceeded the limit for consumption. Fadhli et al. [39] stated that fish were included in the fresh category with a maximum TVB-N value of 30 mgN/100 g. According to Izza et al. [43], the TVB-N value of tofu preserved using liquid smoke derived from teak and pine wood indicates a shelf life of 4 days, maintaining 24.51 mgN/100 g. However, the TVB- value exceeds the safe consumption limit (>35 mgN/100 g) on day 5. The presence of curcumin compounds in turmeric can inhibit the growth of bacteria damaging fish meat [44]. Therefore, the preservative from CLS preserves mackerel for 48 h of storage.

3.2. Organoleptic Test

3.2.1. Taste

Sensory evaluation of preserved mackerel, treated with a combination of liquid smoke and curcumin, was conducted through taste testing by a panel of assessors. Before testing, the mackerel was steamed for 10 min, as reported in Figure 3.

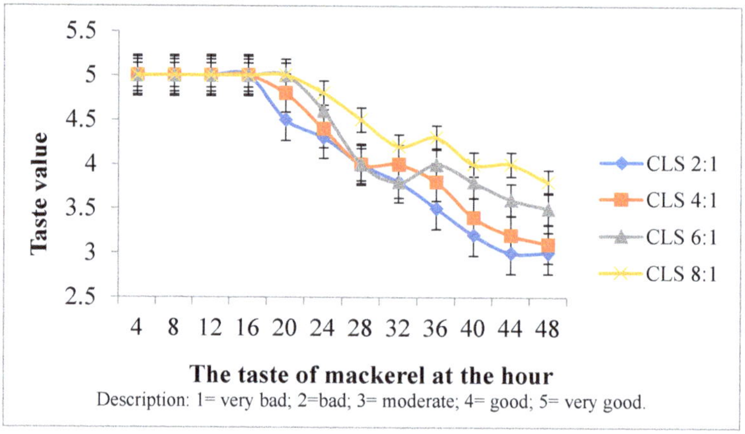

Figure 3. Taste testing of preserved mackerel in various CLS.

Figure 3 shows that the addition of liquid smoke with a high concentration in preservation can slow down the aroma and taste of mackerel and last up to 30 h. In terms of taste, fish with CLS variations of 20 h of immersion are still delicious. However, the best is CLS 8:1 mixture as reported by the different variations. This mixture has a delicious taste compared to others at a soaking time of 48 h because the concentration of curcumin inhibits bacterial activity. Therefore, the taste is delicious and the addition of liquid smoke to the soaking water causes the fish to be dominated by a slightly smoky taste. The taste changes faster in less than 12 h when compared to mackerel. For CLS 8:1 at 20 h, the percentage of panelist assessment was 100% by giving a score of 5. At 48 h, the panelists gave more scores of 4, resulting in a percentage of 68%, while others reported 3. The results are appropriate to Elshehawy and Farag [45], where smoked chicken with 1% and 2% received higher acceptance and value for taste, texture, color, and aroma. Faisal et al. [7] showed that 2–3% liquid smoke had a taste acceptable to the panelists for 48 h of preservation time.

3.2.2. Aroma

Aroma testing was conducted using the nose of each panelist. Meanwhile, the fish was steamed for 10 min before testing. In terms of aroma in mackerel (Figure 4) with CLS 2:1 and CLS 4:1 mixtures, the fish still had a good aroma at 20 h of soaking time. In contrast to the CLS 8:1 and CLS 6:1 mixtures of 32 h, the aroma of the fish still tasted good. After 10 h without treatment with the CLS preservative mixture, the aroma of the mackerel developed signs of spoilage. The smoky aroma produced was absorbed into the fish layer. The aroma became pungent due to the gradual reduction in acetic acid content. An unpleasant aroma could also be used as an indication of product damage caused by an oxidation reaction. Additionally, the occurrence of fat oxidation leads to an undesirable odor in fish [46]. In comparison to other variations, CLS 8:1 mixture showed an improved profile due to the ability to increase the amount of turmeric and increase the odor associated with liquid smoke. This is caused by the substances contained in turmeric, namely curcumin content which gives a distinctive aroma to the preservative mixture [47]. The preference value for preserved mackerel in the CLS 8:1 variation with a soaking time of 32 h was 65% at a score of 5, while others were 3 and 4. Similar results with the use of 3% liquid smoke from durian skin can still be accepted at 48 h [7].

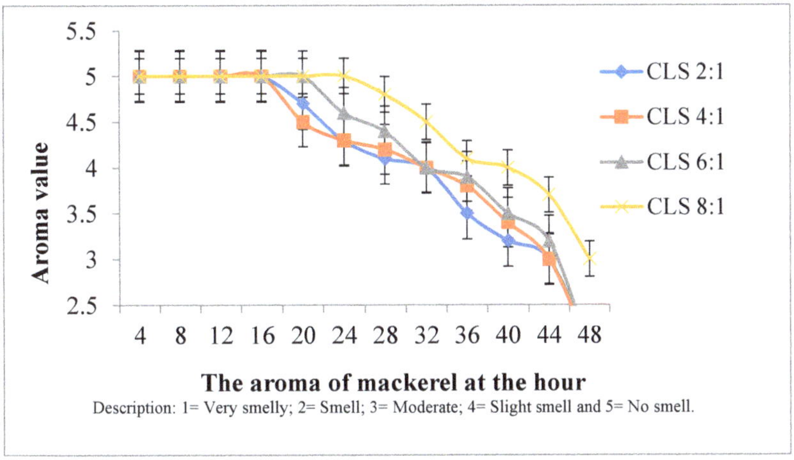

Figure 4. Aroma testing on preserved mackerel in various CLS.

3.2.3. Texture

Based on Figure 5, the best texture of mackerel is in fish with the addition of liquid smoke in CLS 8:1 and CLS 6:1 mixture variations. In the CLS 2:1 and CLS 4:1 variations, the texture of the fish begins to change to soft at 32 h. Generally, the texture of the fish soaked in water without preservatives begins to change at 10 h because of increased bacterial growth. Fish experience a decline in quality when the texture of the meat becomes soft due to the enzymatic process in muscle tissue, such as cathepsin and collagenase. The cathepsin enzyme causes the texture to become soft due to protein degradation, while the collagenase breaks down polypeptide bonds [48]. For the presentation of preference at CLS 8:1 with a time of 32 h, 73% of panelists gave a score of 4 while others scored 3 and 5. At 48 h, the score given by all panelists was 3. According to Syarif et al. [49], pyrolysis of ironwood at a temperature of 400 °C produced liquid smoke used to preserve mackerel. This was achieved by using 5% liquid smoke to maintain a shelf life of 3 days with a texture value lower than fresh fish.

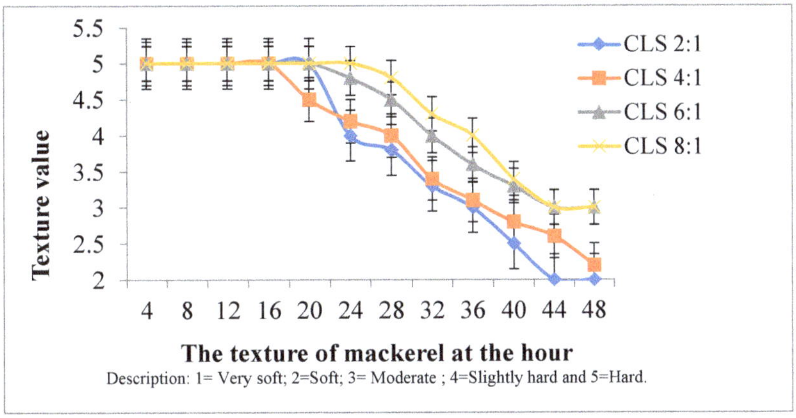

Figure 5. Texture testing on preserved mackerel in various CLS.

3.2.4. Color

The color evaluation of the mackerel in the organoleptic test was based on the color of its flesh. Figure 6 describes the color score values utilized in the organoleptic test.

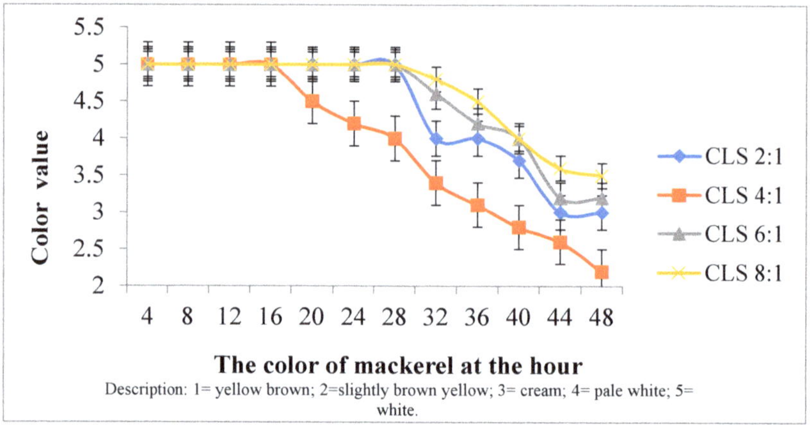

Figure 6. Color testing on preserved mackerel in various CLS.

The mackerel soaked with various mixtures of liquid smoke and turmeric gave different color change effects to the soaking time (Figure 6). According to Joesidawati [50], carbonyl compounds (aldehydes and ketones) have a major influence because the product changes due to the interaction between carbonyl and amino groups, and the color of the fish remains white. The rate at which the turmeric is absorbed into the fish tissue or the way the turmeric extract reacts with the other chemicals in the liquid smoke mixture. Turmeric's structure is changed throughout the extraction and purification processes, making it incapable of being absorbed [51]. In the variation in CLS mixture 8:1, the color is yellowish due to the influence of high turmeric. Meanwhile, the color of the fish soaked without CLS mixture began to change to brownish at 12 h of soaking. A type of food product with high nutritional value, taste, and texture but without good color can reduce the demand of consumers. The preference value for mackerel at CLS 8:1 was 65%, giving a score of 3. The use of liquid smoke from durian skin with a concentration of 0.5–3% maintains the value during storage and the color begins to change after 42 h due to the high content of phenol and acetic acid [7]. The intensity of the change in the sample is due to the Maillard reaction occurring between the carbonyl and amino acid group or protein [52].

3.3. Total Plate Count (TPC) Test

The bacterial content in a product is among the microbiological parameters in determining the suitability of a product for consumption [53]. In this context, the microbial contamination of fishery products occurs during handling, distribution, as well as storage, and processing. Analysis of the number of bacteria determines the growth rate during storage and the results of the TPC analysis of mackerel are presented in Table 1.

Table 1. Analysis data of total plate count test of mackerel with various CLS.

Storage Time (Hours)	Number of Colonies in CLS Variation ($\times 10^5$ Colonies/g)			
	CLS 2:1	**CLS 4:1**	**CLS 6:1**	**CLS 8:1**
4	1.2	0.68	0.48	0.16
8	1.8	1.44	1.2	0.48
12	2.52	2.28	1.72	1.24
16	2.88	2.8	2.68	2.56
20	2.08	1.88	1.76	1.56
24	1.88	1.36	1.24	1.08
28	1.6	1.2	1.08	0.92
32	2.28	1.84	1.48	1.4
36	3.36	3.4	3.12	3.08
40	4.08	3.8	3.56	3.44
44	4.92	4.92	4.16	4
48	5.64	5.44	5.08	4.96

Note: CLS = Turmeric: liquid smoke combination (g/L).

The number of microbes increased from the initial state in the CLS 2:1 mixture, which was 1.2×10^5 colonies/g to 1.88×10^5 colonies/g. In this context, the microbes at 24 h immersion are in the safe consumption zone of 1×10^5 colonies/g based on SNI 02-2725-1992 [35]. At 20 h to 28 h immersion time, the number of microbial colonies decreased to 2.08×10^5 colonies/g, 1.88×10^5 colonies/g and 1.6×10^5 colonies/g. This is due to high bacteriostatic properties, affecting the reproducibility of bacteria [54]. According to Sasongko et al. [55] the average bacterial colony/g of smoked rabbit meat using immersion with 0% contained 27.4×10^5 colonies/g of bacteria. Meanwhile, the treatment of rabbit meat soaked using 1% liquid smoke contained 18.4×10^5 colonies/g of bacteria. This shows the activity of coconut shells in inhibiting and killing bacteria in smoked rabbit meat. Based on other research, the TPC value of fish soaked with liquid smoke with a concentration of 2.5% experienced the greatest microbial growth on the 4th day of storage of 3.1×10^5 colonies/g. During storage until the 20th day, the TPC value of fish boiled using 2.5% liquid smoke experienced a decrease of 3.1×10^5 colonies/g [56]. According to Xin et al. [57], soaking green mussels can reduce the TPC value from 3.65 to 3.03 log CFU/g. At the time of soaking 36 h, the number of microbes continues to increase with the combination of mixtures. The media supporting adaptation and environmental conditions suitable for microbial growth are important factors influencing the increase in microbes. However, the use of liquid smoke with a combination of turmeric on mackerel can inhibit bacterial growth due to the content of phenol, carbonyl, and acid as well as the derivatives [58].

3.4. Most Probable Number of *Escherichia coli*

MPN test on mackerel preservation against *Escherichia coli* bacteria can be seen in Table 2. Bacterial growth increased after 12 h of preservation, where the value was more than 0.3 MPN/g. The number of *E. coli* in mackerel with the treatment ranged from 3.0 MPN/g to 3.6 MPN/g at 4 h of observation and decreased to 0.3 MPN/g at 8 h. The presence of acid, phenol, and phenolic compounds in liquid smoke interferes with the growth of *E. coli*. Liquid smoke is a strong bactericide that can stop the growth of *E. coli* and other pathogens [59]. Previous research showed that smoked fish products given liquid smoke from sawdust could inhibit the growth of coliform or *E. coli* [60]. In addition, turmeric has antioxidant properties, which maintain fat stability and extend the shelf life of products [61]. The use of the extract can give a natural yellow color to mackerel which adds to the visual appeal. In the CLS2:1 treatment at a storage time of 32 h, the MPN value exceeded the maximum limit of microbial contamination at 1×10^5 MPN/g [36]. Based on the observations, mackerel given liquid smoke with a higher concentration of turmeric showed better resistance to microbial contamination compared to those coated with a lower concentration of the extract. Each edible coating CLS 4:1, CLS6:1, and CLS8:1 can extend shelf life by 36, 40, and 48 h, respectively.

Table 2. Most probable number of *E. coli* data of mackerel in various CLS.

Storage Time (Hours)	MPN/g			
	CLS 2:1	**CLS 4:1**	**CLS 6:1**	**CLS 8:1**
4	3.6×10^1	3×10^1	3×10^1	3×10^1
8	$<0.3 \times 10^1$	$<0.3 \times 10^1$	$<0.3 \times 10^1$	$<0.3 \times 10^1$
12	0.3×10^1	0.3×10^1	$<0.3 \times 10^1$	$<0.3 \times 10^1$
16	2.9×10^1	1.5×10^1	1.5×10^1	0.7×10^1
20	2.1×10^2	2.7×10^1	2.3×10^1	1.4×10^1
24	2.9×10^2	2.8×10^1	2.3×10^1	1.2×10^2
28	4.6×10^2	1.6×10^2	1.2×10^2	1.5×10^2
32	$>1.1 \times 10^3$	2.9×10^2	1.5×10^2	2.4×10^2
36	$>1.1 \times 10^3$	1.1×10^3	2.4×10^2	2.9×10^2
40	$>1.1 \times 10^3$	$>1.1 \times 10^3$	2.9×10^2	4.6×10^2
44	$>1.1 \times 10^3$	$>1.1 \times 10^3$	$>1.1 \times 10^3$	1.1×10^3
48	$>1.1 \times 10^3$	$>1.1 \times 10^3$	$>1.1 \times 10^3$	1.1×10^3

Note: CLS = Turmeric: liquid smoke combination (g/L).

4. Conclusions

In conclusion, the ratio of liquid smoke and turmeric concentrations affected the preservation ability of edible coating. Meanwhile, TVB, TPC, and MPN values decreased with the increasing ratio. At 48 h of storage time and CLS 8:1, the TVB, TPC, and MPN values were 25.795 mgN/100 g, 4.96×10^5 colonies/g, and 1.1×10^3 MPN/g, respectively. This showed that the condition of the fish was still suitable for consumption within the permitted limit. For CLS 2:1, the TVB value increased to 39.865 mgN/100 g since the condition of the fish was no longer suitable for consumption. The results of the organoleptic test for taste on the CLS 8:1 edible coating showed that the fish lasted up to 48 h with a value of 3.8. For texture and aroma, the mixture of CLS 8:1 was quite good since the mackerel was in good condition. Therefore, a mixture of edible coating liquid smoke from OPEFB and turmeric could be an effective natural preservative to extend the shelf life of mackerel with suitable packing techniques like vacuum sealing or modified atmosphere packaging. Future research could be explored further in future studies.

Author Contributions: Conceptualization, M.F., A.G. and H.D.; methodology, M.B.H., A.S. and D.A.; software, M.M. and S.K.; validation, M.F., A.G. and H.D.; formal analysis, H.D. and M.M.; investigation, S.K., A.S. and D.A.; resources, M.F.; data curation, M.F., H.D. and M.M.; writing—original draft preparation, M.F. and H.D.; writing—review and editing, M.F. and H.D.; visualization, A.G., S.K. and M.M.; supervision, M.F.; project administration, M.B.H.; funding acquisition, M.F. All authors have read and agreed to the published version of the manuscript.

Funding: The authors are grateful to Syiah Kuala University for the research grant provided (Grant No. 101/UN11.2.1/PG.01.03/SPK/PTNBH/2024).

Institutional Review Board Statement: This study followed the sensory evaluation methodologies provided in the Standard Nasional Indonesia (SNI) for sensory analysis. According to the established rules, the sensory evaluation process does not require formal ethical approval for this type of study. The study was conducted in accordance with the ethical principles and standards that were applicable at the time of its execution. Throughout the study, there was no involvement of human body or animal testing that would necessitate ethical clearance beyond the scope of sensory evaluation prescribed by national regulations.

Informed Consent Statement: Informed consent was obtained from all subjects involved in the study.

Data Availability Statement: The original contributions presented in the study are included in the article, further inquiries can be directed to the corresponding author.

Acknowledgments: The authors express gratitude to the Research Centre for Chemistry and the National Research and Innovation Agency for their invaluable support with facilities and analysis.

Conflicts of Interest: The authors declare no conflicts of interest.

References

1. Mei, J.; Ma, X.; Xie, J. Review on natural preservatives for extending fish shelf life. *Foods* **2019**, *8*, 490. [CrossRef] [PubMed]
2. Lilly, T.T.; Immaculate, J.K.; Jamila, P. Macro and micronutrients of selected marine fishes in Tuticorin, South East coast of India. *Int. Food Res. J.* **2017**, *24*, 191–201.
3. Zhou, P.; Chu, Y.; Lv, Y.; Xie, J. Quality of frozen mackerel during storage as processed by different freezing methods. *Int. J. Food Prop.* **2022**, *25*, 593–607. [CrossRef]
4. Jiang, D.; Liu, Y.; Jiang, H.; Rao, S.; Fang, W.; Wu, M.; Fang, W. A novel screen-printed mast cell-based electrochemical sensor for detecting spoilage bacterial quorum signaling molecules (N-acyl-homoserine-lactones) in freshwater fish. *Biosens. Bioelectron.* **2018**, *102*, 396–402. [CrossRef]
5. Balami, S.; Sharma, A.; Karn, R. Significance of nutritional value of fish for human health. *Malays. J. Halal Res.* **2019**, *2*, 32–34. [CrossRef]
6. Riyantono, R.; Abida, I.W.; Farid, A. Tingkat ketahanan kesegaran ikan mas (*Cyprinus carpio*) menggunakan asap cair. *J. Kelaut. Indones. J. Mar. Sci. Technol.* **2009**, *2*, 66–72.
7. Faisal, M.; Gani, A.; Mulana, F. Preliminary assessment of the utilization of durian peel liquid smoke as a natural preservative for mackerel. *F1000Research* **2019**, *8*, 240. [CrossRef]
8. Ariestya, D.I.; Swastawati, F.; Susanto, E. Antimicrobial activity of microencapsulation liquid smoke on tilapia [Oreochromis niloticus (*Linnaeus*, 1758)] meat for preservatives in cold storage (±5 C). *Aquat. Procedia* **2016**, *7*, 19–27. [CrossRef]
9. Sari, A.P.; Nurdin, G.M.; Manguntungi, B.; Mustopa, A.Z. Potential of Red, Brown, and Green Macroalgae from Dato Beach, Majene, Indonesia as Natural Food Preservative. *Philipp J. Sci.* **2023**, *152*, 1483–1493. [CrossRef]
10. Kang, J.; Jin, W.; Wang, J.; Sun, Y.; Wu, X.; Liu, L. Antibacterial and anti-biofilm activities of peppermint essential oil against *Staphylococcus aureus*. *LWT* **2019**, *101*, 639–645. [CrossRef]
11. Yang, J.; Goksen, G.; Zhang, W. Rosemary essential oil: Chemical and biological properties, with emphasis on its delivery systems for food preservation. *Food Control* **2023**, *154*, 110003. [CrossRef]
12. Sirena, J.T.; Dal Magro, J.; Junges, A.; Steffens, C.; Cansian, R.L.; Paroul, N. Characterization of free and encapsulated cinnamon and clove essential oils for enhancing fresh sausage quality: A natural substitute for synthetic preservatives. *Food Biosci.* **2024**, *61*, 104649. [CrossRef]
13. Yitbarek, R.M.; Admassu, H.; Idris, F.M.; Fentie, E.G. Optimizing the extraction of essential oil from cinnamon leaf (*Cinnamomum verum*) for use as a potential preservative for minced beef. *Appl. Biol. Chem.* **2023**, *66*, 47. [CrossRef]
14. Kharchoufi, S.; Licciardello, F.; Siracusa, L.; Muratore, G.; Hamdi, M.; Restuccia, C. Antimicrobial and antioxidant features of 'Gabsi' pomegranate peel extracts. *Ind. Crop. Prod.* **2018**, *111*, 345–352. [CrossRef]
15. Akter, S.; Netzel, M.E.; Tinggi, U.; Osborne, S.A.; Fletcher, M.T.; Sultanbawa, Y. Antioxidant rich extracts of *Terminalia ferdinandiana* inhibit the growth of foodborne bacteria. *Foods* **2019**, *8*, 281. [CrossRef] [PubMed]
16. Wan, J.; Zhong, S.; Schwarz, P.; Chen, B.; Rao, J. Enhancement of antifungal and mycotoxin inhibitory activities of food-grade thyme oil nanoemulsions with natural emulsifiers. *Food Control* **2019**, *106*, 106709. [CrossRef]
17. Morsy, M.K.; Al-Dalain, S.Y.; Haddad, M.A.; Diab, M.; Abd-Elaaty, E.M.; Abdeen, A.; Elsabagh, R. Curcumin nanoparticles as a natural antioxidant and antimicrobial preservative against foodborne pathogens in processed chicken fingers. *Front. Sustain. Food Syst.* **2023**, *7*, 1267075. [CrossRef]
18. Faisal, M.; Djuned, F.M.; Abubakar, Y.; Desvita, H. Chikuwa preservation by edible coating from a combination of young coconut shell liquid smoke and chitosan. *S. Afr. J. Chem. Eng.* **2024**, *50*, 135–142. [CrossRef]
19. Racioppo, A.; Speranza, B.; Pilone, V.; Stasi, A.; Mocerino, E.; Scognamiglio, G.; Corbo, M.R. Optimizing liquid smoke conditions for the production and preservation of innovative fish products. *Food Biosci.* **2023**, *53*, 102712. [CrossRef]

20. Trigo-Gutierrez, J.K.; Vega-Chacón, Y.; Soares, A.B.; Mima, E.G.D.O. Antimicrobial activity of curcumin in nanoformulations: A comprehensive review. *Int. J. Mol. Sci.* **2021**, *22*, 7130. [CrossRef]
21. Hussain, Y.; Alam, W.; Ullah, H.; Dacrema, M.; Daglia, M.; Khan, H.; Arciola, C.R. Antimicrobial potential of curcumin: Therapeutic potential and challenges to clinical applications. *Antibiotics* **2022**, *11*, 322. [CrossRef] [PubMed]
22. Silaban, R.; Simanjuntak, J.P.; Tambunan, B.H.; Putra, A.N. Production and Characterization of Liquid Smoke from Coconut Shell Waste as an Effort to Reduce the Impact on Environmental Pollution. *Eng. Environ. Tech.* **2024**, *25*, 162–170. [CrossRef] [PubMed]
23. Oramahi, H.A.; Maurisa, T.; Darwati, H.; Rifanjani, S. Optimization and Characterization of Liquid Smoke Produced by Terminalia catappa Wood Pyrolysis and its In Vitro Antifungal Activity. *Sci. Technol. Indones.* **2024**, *9*, 207–214. [CrossRef]
24. Faisal, M.; Desvita, H.; Abubakar, Y.; Azwar. A Preliminary Study on the Use of Rice Husk-Based Smoke Powder for Meatball Preservatives. *J. Food Qual.* **2022**, *2022*, 7915258. [CrossRef]
25. Silaban, R.; Lubis, I.; Siregar, R.E.; Agus, P. Production of liquid smoke from the combination of coconut shell and empty fruit bunch through pyrolysis process. In Proceedings of the 4th International Conference on Innovation in Education, Science and Culture, ICIESC 2022, Medan, Indonesia, 11 October 2022. [CrossRef]
26. Roy, S.; Priyadarshi, R.; Ezati, P.; Rhim, J.W. Curcumin and its uses in active and smart food packaging applications-a comprehensive review. *Food Chem.* **2022**, *375*, 131885. [CrossRef]
27. Aliabbasi, N.; Fathi, M.; Emam-Djomeh, Z. Curcumin: A promising bioactive agent for application in food packaging systems. *J. Environ. Chem. Eng.* **2021**, *9*, 105520. [CrossRef]
28. Xu, P.; Wang, T.; He, J.; Xiong, W.; Ren, J.; Feng, W.; Wang, R. Antibacterial rice protein nanoparticles with a high curcumin loading for fruit preservation. *Food Biosci.* **2024**, *61*, 104935. [CrossRef]
29. Lin, J.T.; Chiang, Y.C.; Li, P.H.; Chiang, P.Y. Structural and Release Properties of Combined Curcumin Controlled-Release Tablets Formulated with Chitosan/Sodium Alginate/HPMC. *Foods* **2024**, *13*, 2022. [CrossRef]
30. Wu, H.; Liu, Z.; Zhang, Y.; Gao, B.; Li, Y.; He, X.; Yu, L. Chemical Composition of Turmeric (*Curcuma longa* L.) Ethanol Extract and Its Antimicrobial Activities and Free Radical Scavenging Capacities. *Foods* **2024**, *13*, 1550. [CrossRef]
31. Kalaycıoğlu, Z.; Torlak, E.; Akın-Evingür, G.; Özen, İ.; Erim, F.B. Antimicrobial and physical properties of chitosan films incorporated with turmeric extract. *Int. J. Biol. Macromol.* **2017**, *101*, 882–888. [CrossRef]
32. Tao, R.; Zhang, F.; Tang, Q.J.; Xu, C.S.; Ni, Z.J.; Meng, X.H. Effects of curcumin-based photodynamic treatment on the storage quality of fresh-cut apples. *Food Chem.* **2019**, *274*, 415–421. [CrossRef] [PubMed]
33. Pasaraeng, E.; Abidjulu, J.; Runtuwene, M.R. Pemanfaatan rimpang kunyit (*Curcuma domestica* val.) dalam upaya mempertahankan mutu ikan layang (*Decapterus* sp.). *J. MIPA* **2013**, *2*, 84–87. [CrossRef]
34. *Standar Nasional Indonesia 2354.8*; Penentuan Kadar Total Volatil Base Nitrogen (TVB-N) Dan Trimetil Amin Nitrogen (TMA-N) Pada Produk Perikanan. Dewan Standarisasi Nasional: Jakarta, Indonesia, 2009.
35. *Standar Nasional Indonesia 02-2725*; Badan Minimum Cemaran Mikroba Pada Daging. Dewan Standarisasi Nasional: Jakarta, Indonesia, 1992.
36. *Standar Nasional Indonesia 2897:2008*; Metode Pengujian Cemaran Mikroba Dalam Daging, Telur dan Susu, Serta Hasil Olahannya. Dewan Standarisasi Nasional: Jakarta, Indonesia, 2008.
37. Kamaruzzaman, S.; Faisal, M.; Mukhlishien, M.; Hidayat, T.; Illahi, M.D.A.; Desvita, H. The Organoleptic Evaluation of Chicken Meatball Preservation by Liquid Smoke Powder from Durian Rinds (*Durio ziberthinus* murr.). In *E3S Web of Conferences EDP Sciences*; EDP Sciences: Les Ulis, France, 2024; Volume 503, p. 50001. [CrossRef]
38. Pearson, D. *The Chemical Analysis of Foods*, 7th ed.; Churchill Livingstone: Edinburgh, London, UK, 1976.
39. Fadhli, I.; Dewi, E.N.; Fahmi, A.S. Aplikasi methyl red sebagai label indikator kesegaran ikan bandeng (*Chanos chanos*) pada suhu penyimpanan dingin yang berbeda. *Jurnal Ilmu dan Teknologi Perikanan* **2022**, *4*, 15–23. [CrossRef]
40. Milijašević, J.B.; Milijašević, M.; Đinović-Stojanović, J.; Vranić, D. Effect of modified atmosphere and vacuum packaging on TVB-N production of rainbow trout (*Oncorhynchus mykiss*) and carp (*Cyprinus carpio*) cuts. In *IOP Conference Series: Earth and Environmental Science*; IOP Publishing: Bristol, UK, 2017; Volume 85, p. 12036. [CrossRef]
41. Bekhit, A.E.D.A.; Giteru, S.G.; Holman, B.W.; Hopkins, D.L. Total volatile basic nitrogen and trimethylamine in muscle foods: Potential formation pathways and effects on human health. *Compr. Rev. Food Sci. Food Saf.* **2021**, *20*, 3620–3666. [CrossRef] [PubMed]
42. Nguyen, H.T.; Hilmarsdóttir, G.S.; Tómasson, T.; Arason, S.; Gudjónsdóttir, M. Changes in protein and non-protein nitrogen compounds during fishmeal processing—Identification of unoptimized processing steps. *Processes* **2022**, *10*, 621. [CrossRef]
43. Izza, N.; Rihayat, T.; Astuti, R.D.D.; Aida, A.; Izzati, I.A.; Aidy, N.; Safitri, A. Comparison of raw materials for making liquid smoke with pyrolysis method as an alternative to formalin and borax in food. In *6th FIRST 2022 International Conference (FIRST-ESCSI-22)*; Atlantis Press: Paris, France, 2023; pp. 113–127. [CrossRef]

44. Chen, X.; Lan, W.; Xie, J. Natural phenolic compounds: Antimicrobial properties, antimicrobial mechanisms, and potential utilization in the preservation of aquatic products. *Food Chem.* **2023**, *440*, 138198. [CrossRef]
45. El Shehawy, S.M.; Farag, Z.S. Biochemical Characteristics of Refrigerated Smoked Chicken Luncheon as Affected by Liquid Smoke. *Egypt. J. Food Sci.* **2024**, *52*, 17–29. [CrossRef]
46. Liu, L.; Zhao, Y.; Zeng, M.; Xu, X. Research progress of fishy odor in aquatic products: From substance identification, formation mechanism, to elimination pathway. *Food Res. Int.* **2024**, *178*, 113914. [CrossRef]
47. Maring, M.; Nandi, S. Aromatic Plants as Potential Resources to Combat Osteoarthritis. *Comb. Chem. High Throughput Screen.* **2024**, *27*, 1434–1465. [CrossRef]
48. Naiu, A.S. Perkembangan terkini perubahan selama penurunan mutu ikan basah. *J. Saintek* **2011**, *6*, 1–12.
49. Syarif, T.; Aladin, A.; Modding, B.; Wiyani, L.; Dewi, F.C. Application of liquid smoke from pyrolysis byproducts of ulin wood sawdust (*Eusideroxylon zwageri*) as a preservative of mackerel (Rastrelliger). In *AIP Conference Proceedings*; AIP Publishing: Melville, NY, USA, 2023; Volume 2596. [CrossRef]
50. Joesidawati, I.M. Mutu Ikan Cucut (*Centrophorus squamosus*) Asap Dengan Metode Pengasapan Dan Lama Penyimpanan Yang Berbeda. *Jurnal Ilmiah Unirow Tuban* **2012**, *2*, 118–122.
51. Nasef, N.A.; Loveday, S.M.; Golding, M.; Martins, R.N.; Shah, T.M.; Clarke, M.; Singh, H. Food matrix and co-presence of turmeric compounds influence bioavailability of curcumin in healthy humans. *Food Funct.* **2019**, *10*, 4584–4592. [CrossRef] [PubMed]
52. Salsabila, N.; Rosyidi, D.; Susilo, A. Physico-chemical and sensory quality of Pekin duck jerky sonicated with coconut shell liquid smoke and stored for different period. *Online J. Anim. Feed. Res.* **2023**, *13*, 30–33. [CrossRef]
53. Arifan, F.; Winarni, S.; Wahyuningsih, W.; Pudjihastuti, I.; Broto, R.W. Total plate count (TPC) analysis of processed ginger on Tlogowungu Village. In *International Conference on Maritime and Archipelago (ICoMA 2018)*; Atlantis Press: Paris, France, 2019; pp. 377–379. [CrossRef]
54. Dien, H.A.; Montolalu, R.I.; Berhimpon, S. Liquid smoke inhibits growth of pathogenic and histamine forming bacteria on skipjack fillets. In *IOP Conference Series: Earth and Environmental Science*; IOP Publishing: Bristol, UK, 2019; Volume 278. [CrossRef]
55. Sasongko, P.; Mushollaeni, W.; Herman, H. Antibacterial activity of liquid smoke from coconut shell waste on smoked rabbit meat. *Buana Sains* **2014**, *14*, 193–197.
56. Zuraida, I.; Hasbullah, R.; Budijanto, S.; Prabawati, S. Aktivitas antibakteri asap cair dan daya awetnya terhadap bakso ikan. *J. Ilmu Pertan. Indones.* **2009**, *14*, 41–49.
57. Xin, X.; Bissett, A.; Wang, J.; Gan, A.; Dell, K.; Baroutian, S. Production of liquid smoke using fluidised-bed fast pyrolysis and its application to green lipped mussel meat. *Food Control* **2021**, *124*, 107874. [CrossRef]
58. Brustolin, A.P.; Soares, J.M.; Muraro, K.; Schwert, R.; Steffens, C.; Cansian, R.L.; Valduga, E. Investigating antimicrobial and antioxidant activity of liquid smoke and physical-chemical stability of bacon subjected to liquid smoke and conventional smoking. *J. Food Sci.* **2024**, *89*, 7217–7227. [CrossRef]
59. Dien, H.A.; Montolalu, R.I.; Mentang, F.; Berhimpon, S.; Nurkolis, F. Inhibition of microencapsulated liquid smoke on the food-borne pathogens and histamine-forming bacterias' growth in tuna loin sashimi: Inhibition of liquid smoke microencapsulation. *Open Access Maced. J. Med. Sci.* **2022**, *10*, 1200–1206. [CrossRef]
60. Amin, H.; Abouzied, A.S. Production and characterization of new hot and cold smoked mussel (*Brachidontes pharaonis*) meat products using sawdust and liquid smoke. *Aquat. Sci. Fish Resour. (ASFR)* **2024**, *5*, 1–11. [CrossRef]
61. Augustyńska-Prejsnar, A.; Topczewska, J.; Ormian, M.; Saletnik, A.; Sokołowicz, Z.; Lechowska, J. The effect of the addition turmeric on selected quality characteristics of duck burgers stored under refrigeration. *Appl. Sci.* **2022**, *12*, 805. [CrossRef]

foods

MDPI

Article

Microbial and Sensory Quality Changes in Broiler Chicken Breast Meat During Refrigerated Storage

Anna Augustyńska-Prejsnar [1], Miroslava Kačániová [2,3], Paweł Hanus [4,*], Zofia Sokołowicz [1] and Mirosław Słowiński [5]

1. Department of Animal Production and Poultry Products Evaluation, Institute of Food and Nutrition Technology, University of Rzeszow, 35-959 Rzeszow, Poland; aaugustynska@ur.edu.pl (A.A.-P.); zsokolowicz@ur.edu.pl (Z.S.)
2. Institute of Horticulture, Faculty of Horticulture and Landscape Engineering, Slovak University of Agri-Culture, Trieda Andreja Hlinku 2, 949 76 Nitra, Slovakia; miroslava.kacaniova@gmail.com or m.kacaniova@vizja.pl
3. School of Medical & Health Sciences, University of Economics and Human Sciences in Warsaw, Okopowa 59, 01-043 Warszawa, Poland
4. Department of Food Technology and Human Nutrition, Institute of Food and Nutrition Technology, University of Rzeszow, 35-959 Rzeszow, Poland
5. Department of Food Technology and Assessment, Institute of Food Sciences, Warsaw University of Life Sciences, Nowoursynowska 159, 02-787 Warsaw, Poland; miroslaw_slowinski@sggw.edu.pl

* Correspondence: phanus@ur.edu.pl

Abstract: The aim of the study was to assess the bacterial flora of broiler chicken breast meat using the MALDI method, as well as its sensory evaluation while stored refrigerated at a stable temperature (0.5 °C+/−0.5 °C). Bacterial identification based on peptidic spectra obtained by matrix-assisted laser desorption ionisation time-of-flight (MALDI-TOF MS) mass spectrometry is a rapid, inexpensive, and accurate method for identifying isolates that belong to certain bacterial phyla. The microbiological and sensory quality was assessed on the 1st and 3rd, 5th, 7th, 8th, 9th, 10th, 11th, and 12th day of refrigerated storage. The study identified psychrophilic bacteria to be the dominant microflora during the entire period of refrigerated storage. The species profile of the bacteria, however, varied in the subsequent days of storage. From the 8th day of storage, the profile was dominated by bacteria of the genus *Pseudomonas*. The proportionate content of *Pseudomonas* bacteria ranged from 89% on day 8 to 95% on day 11th of storage. The majority of the unfavourable microflora (*Aeromonas* species, *Alcaligenes* spp., *Klebsiella* spp., and *Yersinia* spp.) were observed on the 11th day of storage, which indicates that meat spoilage processes had commenced. The quality of breast meat from broiler chickens stored at 0.5 °C+/−0.5 °C was sensorially acceptable up to the 8th day of storage.

Keywords: poultry meat; refrigerated storage; microbiological quality; sensory quality

1. Introduction

Broiler chicken breast meat is characterised by a high protein content (20–24%) and a significant water content (70–76%), thus making it a perishable product that quickly loses freshness during storage [1,2]. Refrigeration is a common method of preserving poultry meat [3]. The basic condition for obtaining satisfactory effects of cold storage on broiler chicken meat is to maintain a high hygienic standard during the entire pre- and post-slaughter process, high microbiological quality of carcasses, and compliance with the principle of the so-called continuity of the cold chain at the time of slaughter [4,5]. The most frequently detected microorganisms in poultry carcasses belong to the genera *Acinetobacter*, *Pseudomonas*, and the Enterobactericeae family, while *Aeromonas* spp., Micrococcaceae and Lactobacillaceae are less common [4,6,7].

The loss of freshness of broiler chicken meat stored in refrigerated conditions depends mainly on the initial microbiological load and its storage temperature [8–12]. A decisive factor limiting the storage of meat in refrigerated conditions is the minimum temperature of growth for microorganisms, non-specific and pathogenic inclusive [12]. When assessing the microbiological quality of refrigerated poultry meat, the total number of microorganisms and the number of bacteria of the genus *Pseudomonas* are most often taken into account [12–15]. *Pseudomonas* spp. bacteria are considered to be specific microorganisms that, through the degradation of meat proteins and the production of metabolites, cause physical and chemical changes in chicken meat, which, in turn, causes adverse sensory changes, including unpleasant smell, unacceptable colour changes, and the appearance of mucus [16–19]. Refrigerated storage of fresh meat requires a quick and effective reduction of the product temperature below 4 °C and maintaining a temperature not exceeding 4 °C throughout the entire period of storage and transport [20]. On the other hand, it is important to maintain the temperature at a constant level throughout the storage period, as temperature stability is also important for maintaining its microbiological quality and sensory characteristics [8,9,21,22]. The preservation of food products' quality during storage is a crucial step for reducing food waste and enhancing sustainability [11].

The development and improvement of research methods, including the MALDI-TOF MS (matrix-assisted laser desorption ionisation time-of-flight mass spectrometry) method, has made it possible not only to quantify the microbiological assessment of stored meat, but also to assess the profile of the meat microflora [23,24]. The matrix-assisted laser desorption/ionisation time-of-flight mass spectrometry (MALDI TOF MS) method is based on the analysis of the protein profile of the organism. The identification of microorganisms is based primarily on the detection of ribosomal proteins, but also mitochondrial proteins that can be isolated [23–25]. This method has found its special place in food microbiology, as a fast and inexpensive method, additionally characterised by high accuracy in identifying bacteria [25]. An important advantage of the method is also the small amount of material required for analysis, i.e., one bacterial colony [24]. Assessment of the microflora profile of broiler chicken breast meat may have both knowledge and application significance for improving the methods and conditions of its storage [25,26].

The quality of raw broiler chicken meat is determined by its positive sensory evaluation [27] as it is associated with consumer acceptance during purchase and during the preparation of meat for consumption [28]. Cooling and maintaining a stable refrigeration temperature throughout the storage period reduces the growth and multiplication of microorganisms and slows down the intensity of chemical transformations associated with the deterioration of the sensory quality of meat [29–31].

The aim of the study was to assess the bacterial flora of broiler chicken breast meat using the MALDI-TOF MS [matrix-assisted laser desorption ionisation time-of-flight] method, as well as the sensory evaluation of meat during storage under stable cold temperature (0.5 °C+/−0.5 °C).

2. Materials and Methods

2.1. Research Material

The raw material for the study was the breast meat of broiler chickens obtained under production conditions in a local poultry slaughterhouse located in the Podkarpacie region of Poland. All breast muscles were obtained from 40-day-old ROSS 308 broiler chickens reared in the same flock, with a stocking density of 33 kg/m^2 and delivered to the slaughterhouse in a specialised vehicle adapted for the transportation of broiler chickens. The slaughtering of the birds, including the butchering of the carcass and shearing of the pectoral muscles, was carried out in controlled production conditions. After chilling, 180 pieces of weight-balanced (250 ± 50 g) individual pectoral muscles not hampered with any defects were randomly collected. While maintaining sterility, they were randomly assigned to 9 study groups of 20 pieces each. Chicken breast fillets were placed in E2-type meat storage crates lined with foil bags. The foil bags and crates were approved for contact with food. The meat was stored in production conditions in a refrigerated warehouse at a stable temperature of 0.5 °C+/−0.5 °C. Temperature stability in the refrigerated warehouse was maintained using a cooling system operating in a continuous mode with temperature monitoring. Temperature monitoring in the refrigerated warehouse was carried out using specialised temperature sensors (SIMEX RS-485, SIMEX, Gdańsk, Poland) connected and configured with the SimCorder automatic temperature measurement software (SimCorder version 4, SIMEX, Gdańsk, Poland), with an accuracy of 0.1 °C and recording every 15 min. Microbiological and sensory quality was assessed on the 1st (24 h after slaughter) and 3rd, 5th, 7th, 8th, 9th, 10th, 11th, and 12th days of refrigerated storage. To achieve this objective, one randomly selected container containing 20 pieces of pectoral meat was transported at a temperature of 0.5 °C+/−0.5 °C to the Laboratory for the Evaluation of Poultry Products, University of Rzeszów on each day of the assessment. In the laboratory, while one research team immediately undertook the meat's microbiological assessment, the other performed the sensory evaluation of the meat.

2.2. Microbiological Analysis

The preparation of the breast muscle samples for storage is described in detail in Section 2.1. Research Material. For microbiological evaluation, all breast muscles were stored in E2-type meat storage containers lined with food-grade film. On each day of the study, 10 breast muscles were allocated for microbiological analysis, from which 10 replicates were randomly taken. Using sterile scalpels and forceps, 5 g of chicken meat was removed. It was then quickly transferred into a sterile Stomacher bag with 45 mL of 0.1% buffered peptone water (BPW, pH 7.0, Basingstoke, UK) and homogenised for 60 s in the Stomacher at room temperature. For each sample, appropriate serial decimal dilutions in 0.1% BPW solution were performed. The surface of the dry medium was covered with 0.1 mL of serial dilutions of the prepared samples. After two days of incubation at 30 °C, total viable counts (TVC) were counted on Tryptone Soya Agar (TSA, Oxoid, Basingstoke, UK), and after 24–48 h of incubation at 37 °C, total viable counts (TVC) were counted on Endo Agar (EA, Oxoid, Basingstoke, UK) the number of Enterobacteriaceae was counted, and after 48–72 h of incubation at 25 °C the number of *Pseudomonas* genera was counted on *Pseudomonas* agar supplemented with CFC/CN (PA, Oxoid, Basingstoke, UK).

After incubation, the bacterial biomass from the microbial medium was transferred to 300 μL of distilled water. Then, 900 μL of ethanol was added and mixed using Votrex (Vortex Classic, Velp Scientifica, Usmate, Italy). The mixture was centrifuged (MPW-150R, MPW MED. INSTRUMENTS, Warsaw, Poland) at 14,000 rpm (RCF = 22,570× *g*) for two minutes. The pellet was centrifuged once more after the supernatant was discarded. After pipetting off all remaining ethanol, the precipitate was allowed to dry at ambient temperature. The pellet was then combined with 30 μL of 70% formic acid using a paddle. Then, 30 μL of acetonitrile was added and mixed thoroughly. After centrifuging the solution for two minutes at maximum speed, 1.5 μL of supernatant was applied to a polished MALDI target plate (Bruker Daltonics, Bremen, Germany). Immediately after drying, 1.5 μL of matrix solution was added to each spot and allowed to air dry. Samples on a polished MALDI target plate were dried at room temperature (19 °C–21 °C), which took about 15–20 min. The metal plate with the overlaid samples was placed in the measuring chamber of the apparatus; the air was then removed to create a vacuum and the plate was exposed to a laser beam. The spectra were generated by MALDI-TOF and analysed with a Microflex LT (Bruker Daltonics, Germany) instrument using Flex Control 3.4 software and Biotyper Realtime Classification v3.1. An average of 40 laser shots taken in automatic mode using the lowest laser power required to ionise the samples was used to produce each spectrum. A real-time program, Classification v3.1 (Bruker Daltonics, Bremen, Germany) was used to evaluate the spectra and compare the measurement results to the database. The manufacturer specified that successful identification must be within a confidence score of ≥2.0 for the species level and ≥1.7 for the genus level [26].

2.3. Sensory Evaluation

The sensory attributes of raw breast muscles stored refrigerated were evaluated by a trained 10-person panel with proven sensory sensitivity (defined on the basis of the PN-EN ISO 8586 standard [32]. Sensory panellists were previously recruited from among researchers working at the Institute of Food Technology and Nutrition of the University of Rzeszów. The panel had previous experience in sensory evaluation of poultry meat and cold-stored products. Prior to the main study, the sensory panellists were calibrated in a pre-test. During the pre-test, the panellists were introduced to the identification of sensory attributes required to describe the aroma, colour, and texture of the overall acceptability of broiler chicken breast meat during storage. Prior to the evaluation, the panelists were briefed with the evaluation questionnaire. For purposes of evaluation, 10 cube-shaped samples measuring 2 cm × 2 cm × 2 cm were cut from each pectoral meat piece, which were subsequently coded. Following this initial preparation, the samples were presented to evaluators in white containers with a transparent lids, which were placed on white trays [27]. The procedure was the same for each sample. Each panellist received 10 samples from each study group on the day of the evaluation, which were given at random. Each of the panelists received 10 samples from each study group on the day of the evaluation. The test was carried out in a duly prepared room with appropriate lighting, room temperature, free from foreign smells, as well as any distracting factors, in accordance with the PN-EN ISO 8589 standard [33]. The attributes of the assessment were: smell, external colour, consistency, and general appearance. Sensory assessment was performed using a 5-point hedonic scale according to the specifications in Table 1.

Table 1. Specification of sensory characteristics adopted for raw meat from broiler chicken breasts stored refrigerated.

Score [CU]	Hedonic Scale	Sensory Descriptors			
		Smell	External Colour	Consistency	General Appearance
Desirability RANGE (3.51–5.0)	5	very conclusive, typical	balanced typical	elastic meat tissue, compact	faultless, moist surface, typical
	4	conclusive, typical	desired, less balanced, typical	rather elastic, when pressed, the deformation evens out	desirable, slightly dried surface
Adjusted desirability RANGE (2.51–3.5)	3	non-perceptible, slight change	rather desired, non-balanced, localised changes	permanent deformation on meat tissue when pressed	rather desirable, dried surface or slightly moist
Undesirable RANGE (1–2.5)	2	changed, light intensive, unacceptable	undesirable, localised changes, drips, yellow	meat tissue relaxed, flattens out after pressing	undesirable slimy, slightly sticky surface, locally changed colour
	1	changed, rotten	very undesirable, yellow or green in places	meat tissue relaxed after pressure, easily falls apart	very undesirable, surface sticky with mucus

2.4. Statistical Analysis

The results of the study are presented as means and standard deviations. The effect of the storage duration on the microbiological parameters of the broiler chicken meat was assessed using unidirectional analysis of variance (using post hoc and Tukey's HSD assays), while that of sensory characteristics was achieved using nonparametric Kruskal–Wallis assays. The differences were considered significant at $p < 0.05$. The calculation was performed using the Statistica 13.3 software package [34].

3. Results and Discussion

Cold conditions are common throughout the broiler chicken processing and distribution chain. While chill line controls used during butchering increase microbiota selection pressure, chilling remains one of the best ways to reduce microbial contamination of chicken carcasses during processing [4,5,35]. Besides cooling the carcass, another important factor in the process of obtaining meat from broiler chickens, which affects its final quality, is cooling the meat to a temperature not exceeding 4 °C [8,36]. Storage temperature has a significant impact on the growth rate of microorganisms, especially when it fluctuates [9]. The total number of microorganisms ranged from 2.34 on the first day to 6.79 log CFU/g in twelve days (Table 2). The TVC indicate several types of bacterial growth during storage, which varied between samples. This corroborates previous studies which showed that the range of total viable bacterial counts from chicken sampled after two-thirds of its storage time at 4 °C was 3 to 8 log CFU/g [37]. There was a significant association between storage temperatures and total viable counts.

Table 2. Results of the assessment of microbiological parameters of raw broiler chicken breast meat in refrigerated storage [Log CFU/g ± SD].

Time of Refrigerated Storage [d]	Total Count of Microorganisms	Number of Bacteria from:	
		Family Enterobacteriaceae	*Pseudomonas* Genera
1	2.34 ^{a} ± 0.19	<1.00 *	1.76 ^{a} ± 0.29
3	2.57 ^{a} ± 0.45	2.02 ^{a} ± 0.02	1.85 ^{a} ± 0.27
5	3.03 ^{b} ± 0.27	2.10 ^{a} ± 0.54	2.94 ^{b} ± 0.13
7	4.24 ^{c} ± 0.36	3.86 ^{b} ± 0.61	3.67 ^{c} ± 0.19
8	4.75 ^{d} ± 0.15	4.20 ^{bc} ± 0.13	4.70 ^{d} ± 0.59
9	5.06 ^{d} ± 0.22	4.45 ^{bc} ± 0.04	5.00 ^{d} ± 0.21
10	6.01 ^{e} ± 0.21	4.90 ^{c} ± 0.48	5.25 ^{d} ± 0.20
11	6.40 ^{e} ± 0.39	5.53 ^{d} ± 0.40	5.75 ^{e} ± 0.15
12	6.79 ^{f} ± 0.31	5.56 ^{d} ± 0.48	6.62 ^{f} ± 0.44
p Value	<0.0001	<0.0001	<0.0001

*—below the detection threshold [1 log CFU/g etc.]; $^{a, b, c}$ ...—values in columns marked with different letters differ at $p < 0.05$.

The number of Enterobacteriaceae genera ranged from <1.00 on the first day to 5.56 on the last day of the study (Table 2). In an investigation conducted by Balamatsia et al. [38], the *Pseudomonas* count in air-packed chicken samples reached approximately 7.0 log cfu/g after approximately 8 days of storage. *E. coli* is the most commonly used indicator of faecal contamination of food [39]. Its counts usually correlate more closely with those of Enterobacteriaceae, which are commonly associated with elevated counts on poultry carcasses due to mishandling, improper or unsanitary processing, and/or storage conditions [40]. Faecal contamination of beef and chicken meat with Enterobacteriaceae, including *Salmonella* spp., *E. coli*, *Proteus* spp. and *Klebsiella* spp., is a significant food hygiene problem [41,42]. Effective monitoring of the presence and accurate identification of zoonotic bacterial pathogens in food is critical to minimising the incidence of foodborne illness and reducing microbial contamination of food.

The current study revealed that *Pseudomonas* counts ranged from 1.76 on the first day to 6.62 log CFU/g (Table 2). In the study conducted, the deterioration in sensory characteristics, mainly meat odour during storage, was strongly correlated with an increase in the number of *Pseudomonas* ssp. bacteria, which were the predominant bacteria present in the meat evaluated (Table S1). The metabolic activity of *Pseudomonas* spp. bacteria during aerobic refrigerated storage results in the formation of metabolites, sensorially perceptible as unpleasant odour [2,21]. According to Franke et al. [43], *Pseudomonas* spp. are commonly responsible for the spoilage of chicken meat stored under aerobic packaging conditions. At the end of the meat's shelf life, *Pseudomonas* were not only found as dominating, but contributed to meat spoils through proteolytic, lipolytic, saccharyllytic, and biosurfactant processes [14]. The action of these microorganisms consists in the enzymatic acceleration of protein proteolysis as well as the oxidative and hydrolytic processes of tissue fats [16,44]. It is worthy of note that 7–8 log CFU/g for *Pseudomonas* spp. is a factor that determines the putrefaction of fresh meat according to Nychas et al. [45]. In an investigation conducted by Balamatsia et al. [38], *Pseudomonas* counts in air-packed chicken samples reached approximately 7.0 log CFU/g after approximately 8 days of storage. Another study found that at 2 °C, total Enterobacteriaceae and *Pseudomonas* spp. counts in meat increased over time and reached 4.64 log CFU/g, 4.16 log CFU/g, and 4.48 log CFU/g, respectively, on the eighth day of cold storage. The rate of bacterial growth in meat stored at 6 °C, in each of the evaluated periods, was higher than in meat stored at 2 °C [26].

In total, 48 isolates were identified from meat samples on the first day (Figure 1). On the first day, samples of 10 families, 10 genera, and 28 species were isolated. The most isolated species in day one were *Pseudomonas fluorescens* [8%], followed by *Microbacterium liquefaciens* and *Micrococcus luteus* [6%]. The species *Pseudomonas fragi*, *Pseudomonas lundensis*, and *Pseudomonas fluorescens* are the most common pseudomonads found in poultry meat [46,47]. The primary genera of Enterobacteriaceae isolated are *Serratia* (*Serratia fonticola*, *Serratia grimesii*, *Serratia liquefaciens*, *Serratia proteamaculans*, and *Serratia quinivorans*), *Hafnia* (*Hafnia alvei*, *Hafnia paralvei*), *Rahnella*, *Yersinia*, and *Buttiauxella* [48]. Several novel species of *Enterococcus* or *Lactobacillus*, including *Enterococcus viikkiensis*, *Enterococcus saigonensis*, and *Lactobacillus oligofermentans*, have also been identified in poultry meat products [49–51]. *Brochothrix thermosphacta* has also been frequently observed in poultry meat. Some of the numerous reports that can be found in the literature concentrated more explicitly on rotting bacteria, while others were more pathogen-oriented.

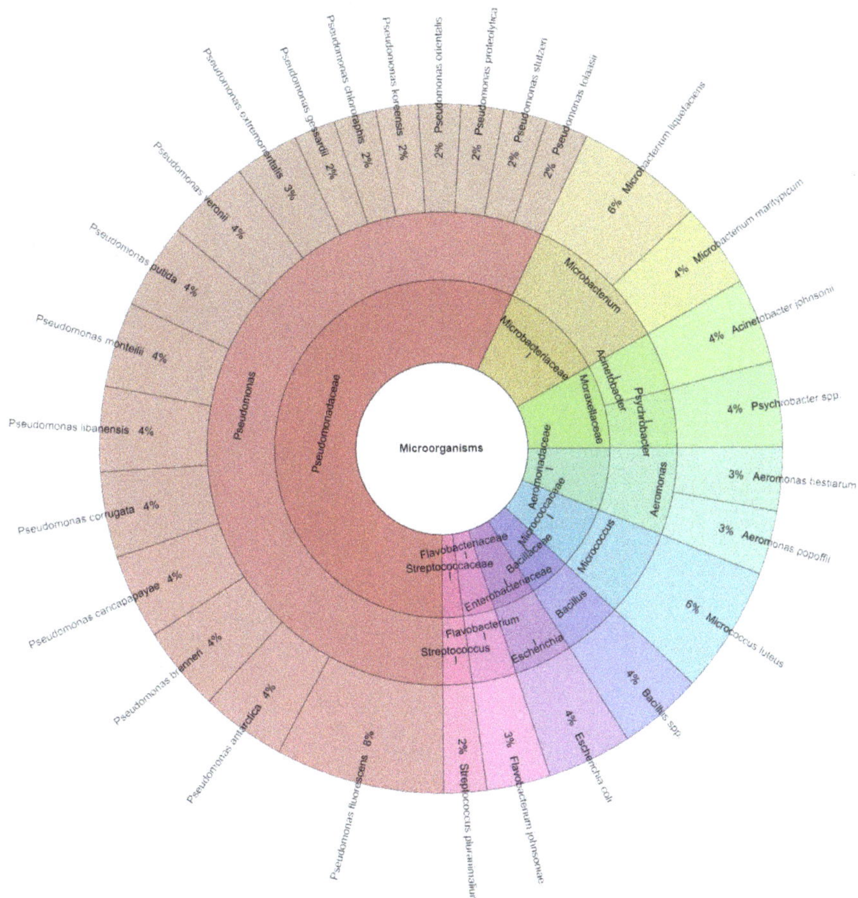

Figure 1. Krona chart: Isolated species from day 1.

On day 3, the number of isolates with identification at good level was 55 (Figure 2). On the third day of refrigerated storage, 10 families, 12 genera, and 24 species were identified in chicken meat. The most isolated species on day two were *Hafnia alvei*, *Pseudomonas fragi*, and *Micrococcus luteus* with 7%. *P. fragi* in cleaned non-contaminated samples and unknown bacteria in unwashed non-contaminated samples accounted for the majority of the factors. It has been demonstrated that *P. fragi*, a psychrotrophic bacterium, plays a major role in meat spoilage [52]. It is frequently discovered on fresh and aerobically damaged meat, even during storage at low temperatures [53,54].

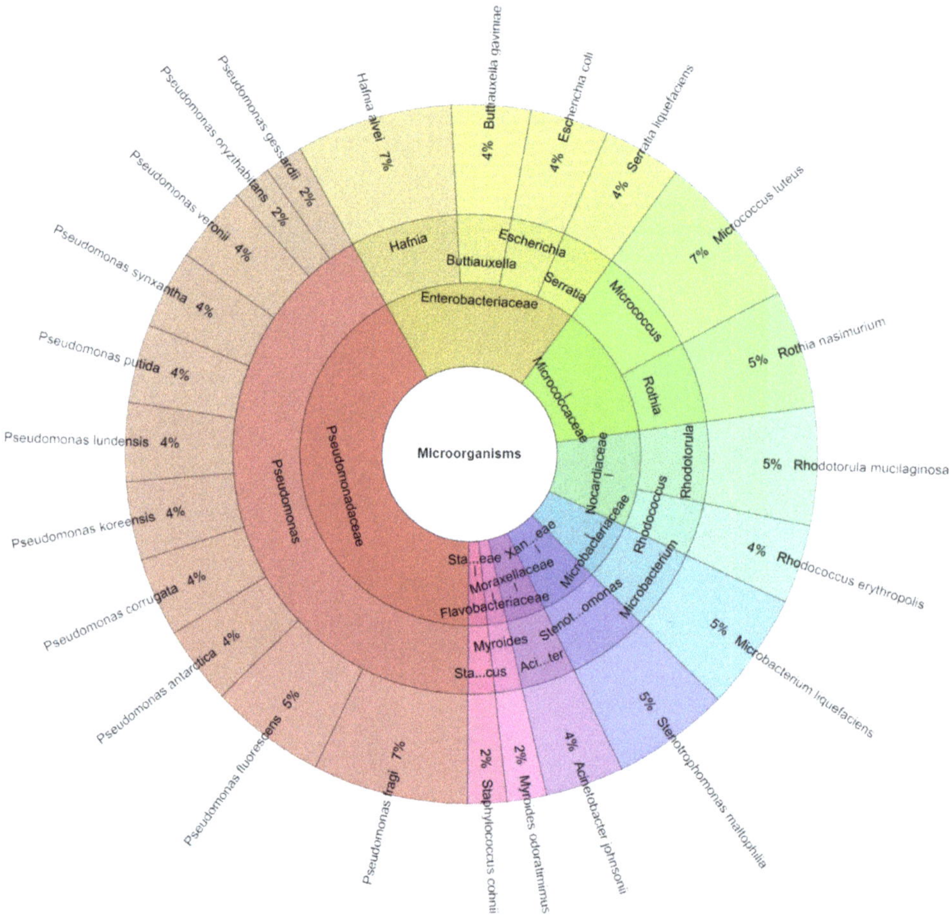

Figure 2. Krona chart: Isolated species from day 3.

A total of 63 isolates were identified on day 5 (Figure 3). Isolates from meat samples included 11 families, 15 genera, and 33 species with the most commonly isolated being *P. fragi* (16%), followed by *P. gessardi* (13%), and *P. lundensis* (11%). Since the psychrotrophic microorganism *Pseudomonas* causes surface change, it is a very important indicator of food spoiling in items that are stored at low temperatures [55].

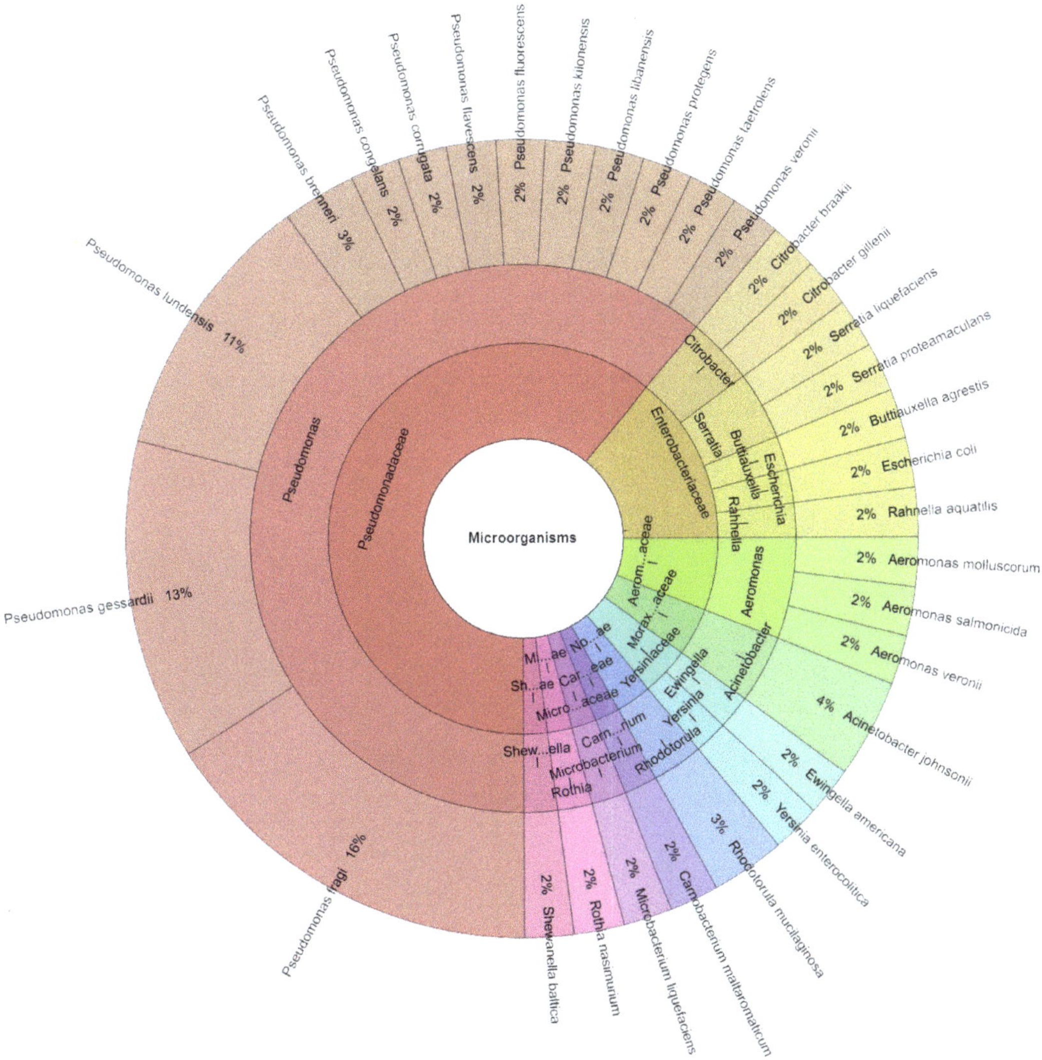

Figure 3. Krona chart: Isolated species from day 5.

The same tendency as in day 5 was found on day 7, where the most isolated species were *P. fragi* (16%), followed by *P. gessardi* (13%), and *P. lundensis* (11%). A total of 11 families, 15 genera, and 30 species were identified on the seventh day (Figure 4). The study identified 62 isolates with successful scores. In comparison to what was initially present, the microbiota diversity is reported to decline after storage while the bacterial load rises. [56,57]. The proliferation and metabolic activity of rotting bacteria lead to microbial deterioration. While the majority of research has concluded that the predominating bacteria in damaged food are the ones that cause spoiling, other studies have defined spoilage using the microbiological acceptability criterion (TVC of 7 log CFU/g) [57,58].

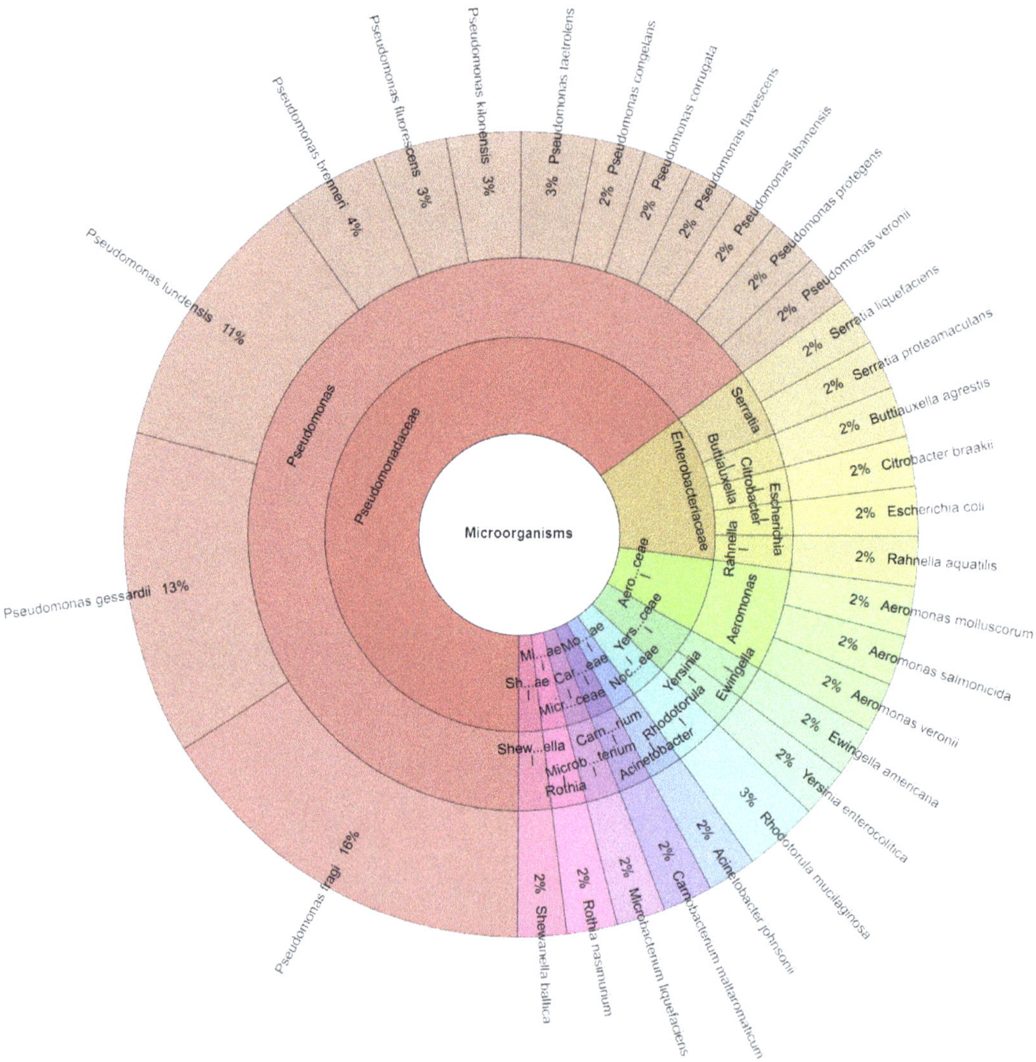

Figure 4. Krona chart: Isolated species from day 7.

Making use of the mass spectrometry with high score 52 isolates, 4 families, 6 genera, and 14 species were identified on day 8 (Figure 5). The most isolated species from chicken meat on day 8 was *P. fragi* with 56%, followed by *P. lundensis* with 13%. By using MALDI-TOF MS, spoilage microbes were identified. The primary organisms discovered after eight days at 4 °C and 10 °C were *Brochothrix thermosphacta*, *Carnobacterium* spp., and *Pseudomonas* spp. The primary spoilage microbiota was represented by the species *Hafnia alvei* at 10 °C and the genera *Carnobacterium* spp., *Serratia* spp., and *Yersinia* spp. at 4 °C [57].

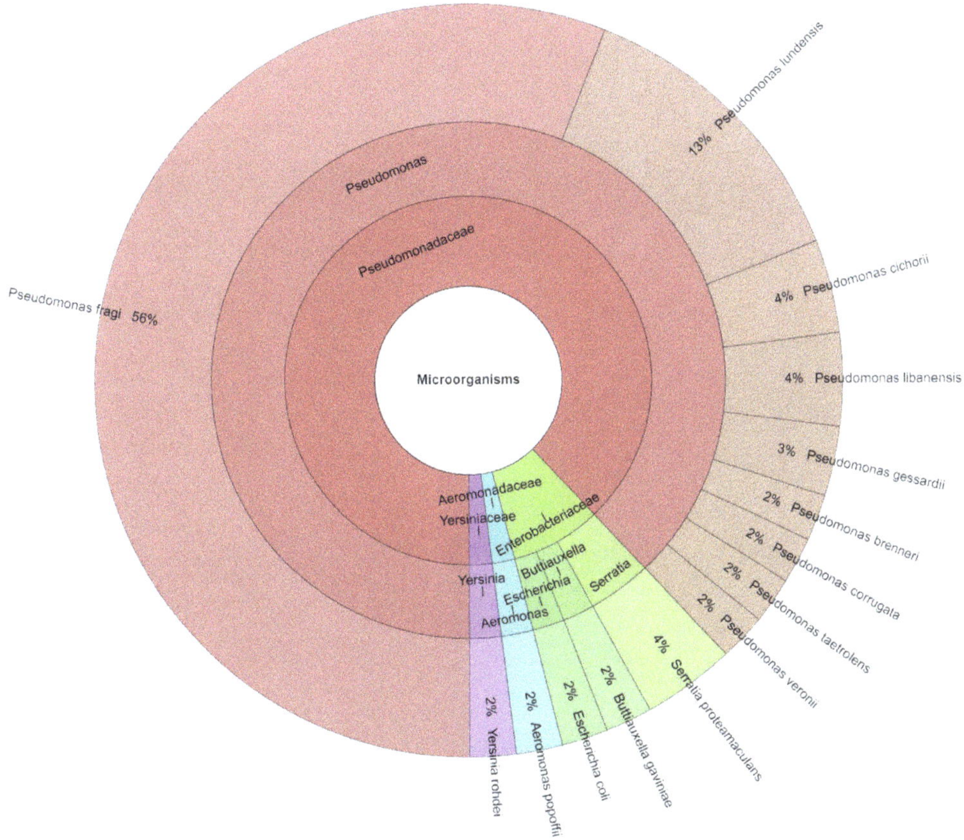

Figure 5. Krona chart: Isolated species from day 8.

From the meat samples on day 9, 55 isolates were identified (Figure 6). Overall, 4 families, 5 genera, and 13 species of bacteria were isolated. The most identified species on day 9 were *P. fragi* with 44%, *P. lundensis* with 14% and *P. fluorescens* with 11%. *Pseudomonas* spp. and other psychrophilic and psychrotrophic bacteria grow more readily when stored at low temperatures. Certain organisms are able to endure the entire process. An example of such species is *Shewanella putrefaciens*, which is commonly detected on carcasses during the slaughtering process and persists even after being stored in aerated places for 14 days [7].

On the seventh day, 44 isolates were found in meat samples (Figure 7). In a single day, 12 species, 4 genera, and 3 families were isolated from the samples. *P. fragi* (36%) was the most isolated species on the seventh day, followed by *P. gessardii* and *P. ludensis* (18%). It is commonly believed that spoiling is exclusively caused by a small number of representative species, also referred to as the autochthonous microbiota, which arise from the original microbial association [13].

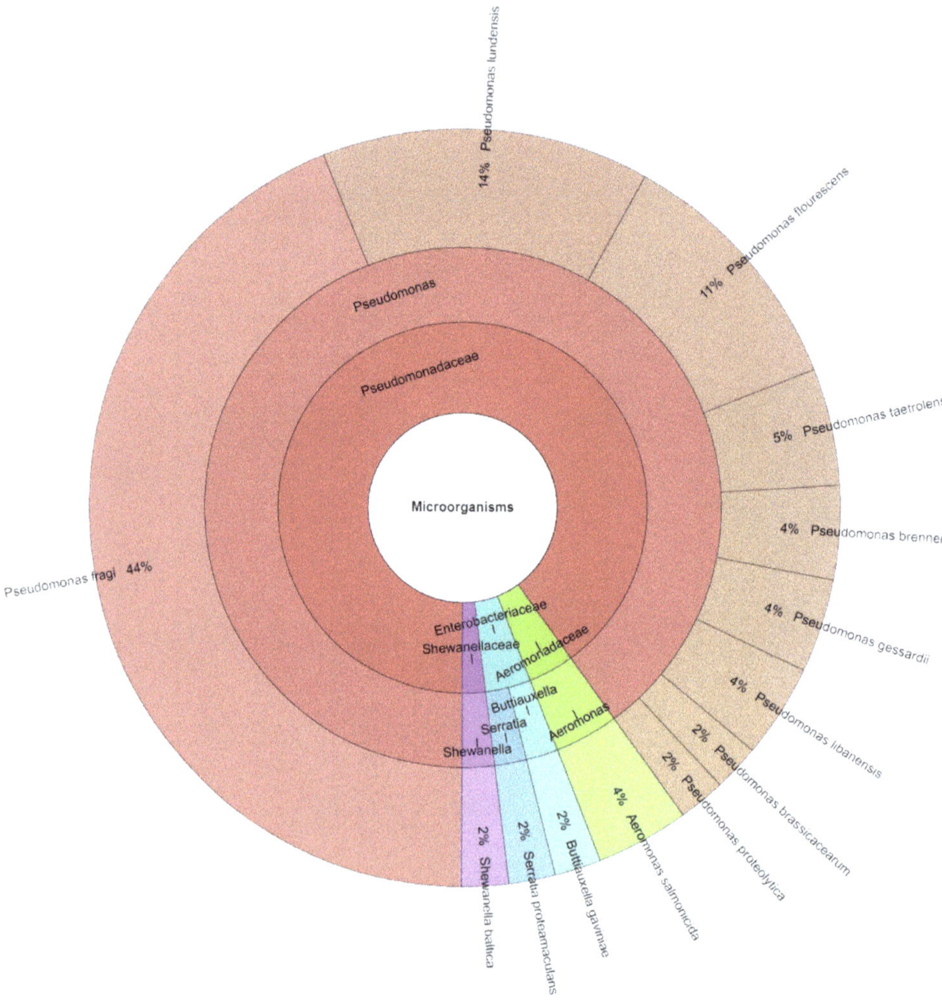

Figure 6. Krona chart: Isolated species from day 9.

Using meat samples, 42 isolates were found in all on the eleventh day. Three families, three genera, and twelfth species were separated from the samples on the eleventh day (Figure 8). *P. fragi* (48%) and *P. taetrolens* (12%) were the next most isolated species in the eleventh day, with *P. gessardii* and *P. lundensis* coming in at 10%. *Pseudomonas* spp. is one of the most common genera at aerobic storage, according to a number of studies. *P. fragi*, the dominant species in this group, is also found in meat that has been packaged in a modified atmosphere [13,59].

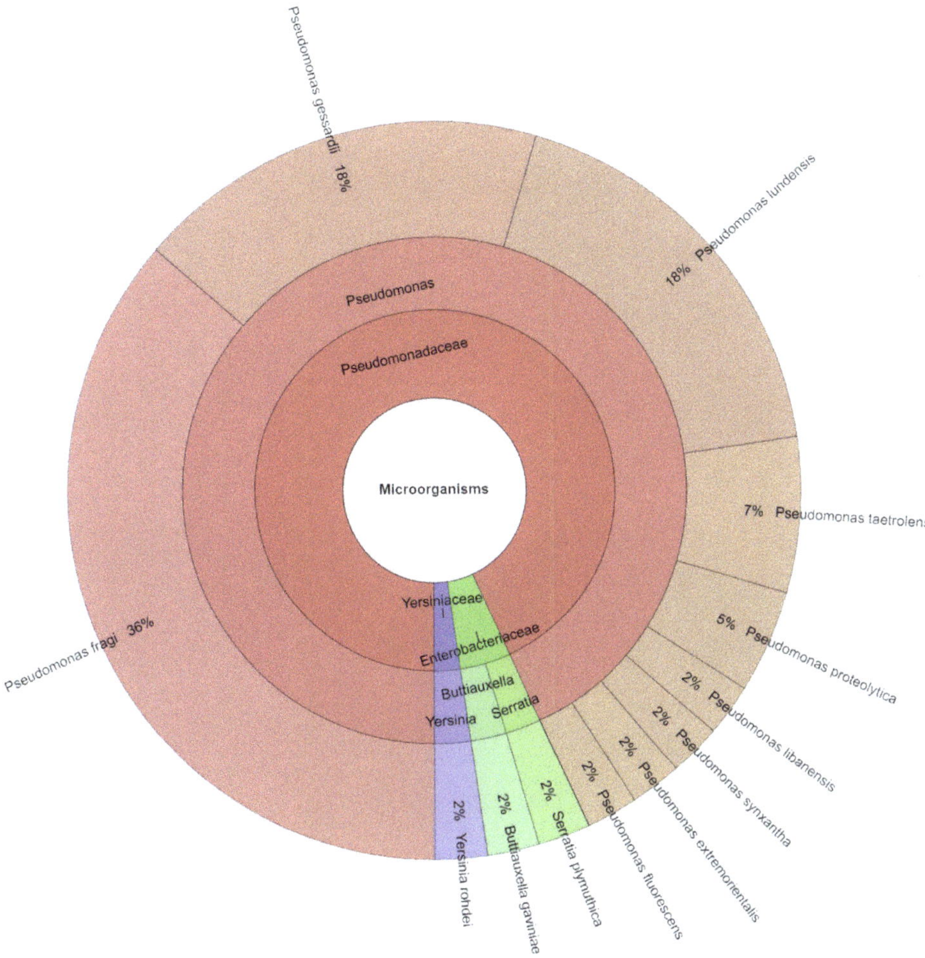

Figure 7. Krona chart: Isolated species from day 10.

On day twelve, forty-five isolates were found in the meat samples. Ten species, five genera, and five families of bacteria were identified (Figure 9). On the ninth day, *P. fragi* (40%), *P. lundensis* (33%), and *P. brenneri* (9%) were the most frequently recognised species. All the isolated species from all days are shown Table 3. The MALDI-TOF MS Biotyper was used to identify the meat and confirmed that culture tests revealed changes in the bacterial microflora's composition throughout the preservation period. Psychrophilic bacteria were the predominant microflora, irrespective of the temperature used, as they are characterised by their ability to flourish in chilled circumstances. Nevertheless, it was noted that the profile of the detected bacteria altered after storage at 6 °C, with the majority of unfavourable microflora emerging, suggestive of ongoing spoiling processes. *Aeromonas* spp., *Alcaligenes* spp., *Klebsiella* spp., and *Yersinia* spp. are a few possible examples of these bacteria [60,61].

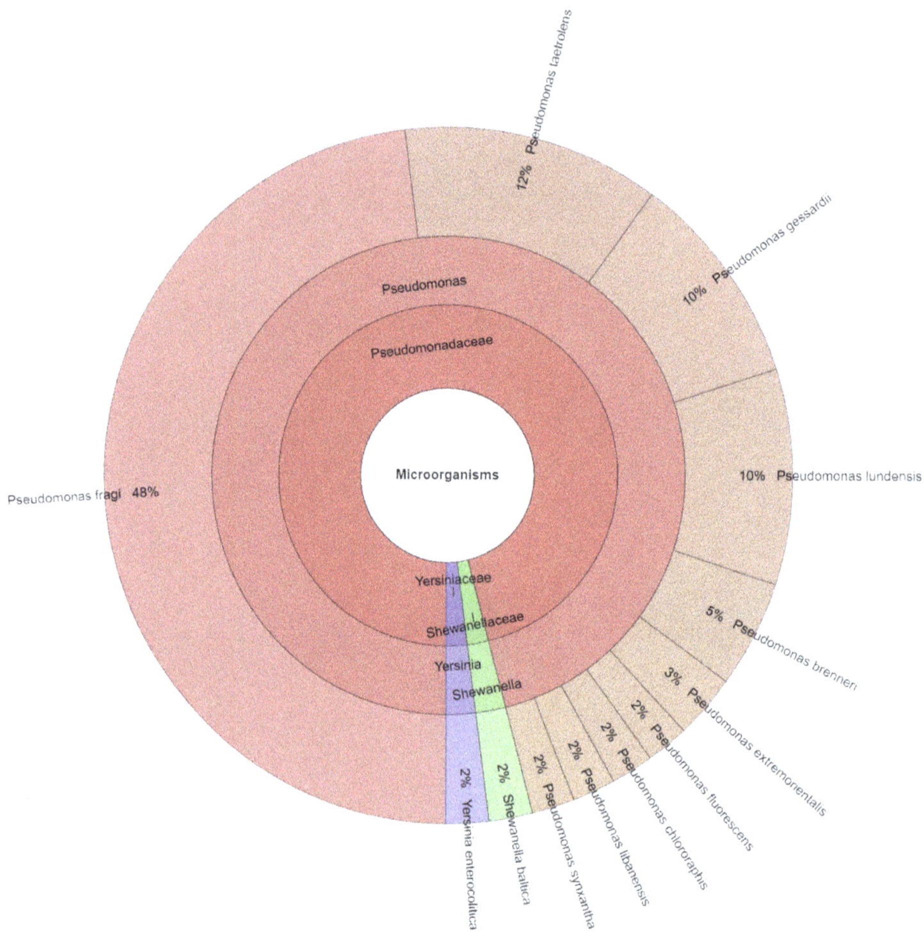

Figure 8. Krona chart: Isolated species from day 11.

The sensory characteristics of poultry products are crucial for determining their quality and have a major impact on consumer choices [27]. The metabolic activity of microbes results in the formation of metabolites that cause physical and chemical changes in poultry meat, perceived sensorially as unpleasant smell, discoloration, and mucus [17,18]. The effect of refrigerated storage conditions on the sensory characteristics of poultry meat was presented in the studies by Augustyńska-Prejsnar et al. [26], Yimenu et al. [62], Sujiwo et al. [21], Kondratowicz et al. [44], and Garavito et al. [63]. The research findings showed that all the tested sensory characteristics of meat deteriorated with the time of refrigerated storage (Table 4). It was indicated that meat stored at 0.5 °C+/−0.5 °C was of acceptable quality until the 8th day of storage (Table 4). Significant changes in smell took place on the 10th day of meat storage. Similarly, in Katiyo et al. [2], the smell of chicken meat deteriorated faster than the colour and general appearance and was correlated with an increase in the number of microorganisms. Also, in a study by Sujiwo et al. [21] on 12 days refrigerated storage (temp. 4 °C) of broiler chicken breasts packed on polystyrene trays wrapped in low-density polyethylene, odour and colour changed faster than texture and overall acceptability. The unpleasant odour is related to the oxidation of lipids in the meat. In addition, protein degradation products released by microorganisms also lead to an unpleasant odour, so meat spoilage can be determined by an unusual odour [21,64].

Our own study showed that on the 11th day of storage, the meat odour was found to be unacceptable (altered, putrid), the colour undesirable, altered in places, with external infiltration, the texture unacceptable, with muscle tissue loosening and flattening after pressure, and an overall undesirable appearance. Changes in the deterioration in sensory characteristics were observed more rapidly compared to the results of quantitative microbiological assessment. Different results during the cold storage of turkey meat for slaughter were shown by Kondratowicz et al. [44]. The evaluation of sensory smell and colour of raw broiler chicken meat is, at the consumer level, of great significance since it is the most visible quality trait and, therefore, linked to consumer acceptance [27,28]. As reported by Katiyo et al. [2] for the consumer, odour is a more reliable indicator of spoilage of raw broiler chicken meat at retail and home than changes in appearance. The detection of an uncharacteristically pungent, fishy ammonia smell can be a warning signal to the consumer. Franke et al. [43] also reported that an unpleasant odour is a signal of microbiological spoilage in chicken meat. Colour may be a more distinguishing factor for red meat, such as beef, due to its relatively high myoglobin content [65].

Figure 9. Krona chart: Isolated species from day 12.

Table 3. Isolated species from meat samples. "+"—occurrence of a bacterial isolate on the tested day.

Family	Genera	Species	Number of Isolates	Day								
				1	3	5	7	8	9	10	11	12
Moraxellaceae	*Acinetobacter*	*Acinetobacter johnsonii*	8	+	+	+	+					
Aeromonadaceae	*Aeromonas*	*Aeromonas bestiarum*	1	+								
Aeromonadaceae	*Aeromonas*	*Aeromonas eucrenophila*	1									+
Aeromonadaceae	*Aeromonas*	*Aeromonas molluscorum*	2			+	+					
Aeromonadaceae	*Aeromonas*	*Aeromonas popoffii*	2	+				+				
Aeromonadaceae	*Aeromonas*	*Aeromonas salmonicida*	6			+	+		+			
Aeromonadaceae	*Aeromonas*	*Aeromonas veronii*	2			+	+					
Bacillaceae	*Bacillus*	*Bacillus* spp.	2	+								
Enterobacteriaceae	*Buttiauxella*	*Buttiauxella agrestis*	2			+	+					
Enterobacteriaceae	*Buttiauxella*	*Buttiauxella gaviniae*	5		+			+	+	+		
Carnobacteriaceae	*Carnobacterium*	*Carnobacterium maltaromaticum*	2			+	+					
Enterobacteriaceae	*Citrobacter*	*Citrobacter braakii*	2			+	+					
Enterobacteriaceae	*Citrobacter*	*Citrobacter gillenii*	1			+						
Enterobacteriaceae	*Escherichia*	*Escherichia coli*	9	+	+	+	+	+				
Yersiniaceae	*Ewingella*	*Ewingella americana*	2			+	+					
Flavobacteriaceae	*Flavobacterium*	*Flavobacterium johnsoniae*	1	+								
Enterobacteriaceae	*Hafnia*	*Hafnia alvei*	6		+							+
Microbacteriaceae	*Microbacterium*	*Microbacterium liquefaciens*	8	+	+	+	+					
Microbacteriaceae	*Microbacterium*	*Microbacterium maritypicum*	2	+								
Micrococcaceae	*Micrococcus*	*Micrococcus luteus*	7	+	+							
Flavobacteriaceae	*Myroides*	*Myroides odoratimimus*	1		+							
Pseudomonadaceae	*Pseudomonas*	*Pseudomonas antarctica*	4	+	+							
Pseudomonadaceae	*Pseudomonas*	*Pseudomonas brassicacearum*	1						+			
Pseudomonadaceae	*Pseudomonas*	*Pseudomonas brenneri*	18	+		+	+	+	+		+	+

Table 3. *Cont.*

Family	Genera	Species	Number of Isolates	Day								
				1	3	5	7	8	9	10	11	12
Pseudomonadaceae	*Pseudomonas*	*Pseudomonas caricapapayae*	2	+								
Pseudomonadaceae	*Pseudomonas*	*Pseudomonas cichorii*	2					+				
Pseudomonadaceae	*Pseudomonas*	*Pseudomonas congelans*	2			+	+					
Pseudomonadaceae	*Pseudomonas*	*Pseudomonas corrugata*	8	+	+	+	+	+				+
Pseudomonadaceae	*Pseudomonas*	*Pseudomonas extremorientalis*	3	+						+	+	
Pseudomonadaceae	*Pseudomonas*	*Pseudomonas flavescens*	4			+	+					
Pseudomonadaceae	*Pseudomonas*	*Pseudomonas fluorescens*	19	+	+	+	+		+	+	+	
Pseudomonadaceae	*Pseudomonas*	*Pseudomonas fragi*	132		+	+	+	+	+	+	+	+
Pseudomonadaceae	*Pseudomonas*	*Pseudomonas gessardii*	36	+	+	+	+	+	+	+	+	+
Pseudomonadaceae	*Pseudomonas*	*Pseudomonas chlororaphis*	2	+							+	
Pseudomonadaceae	*Pseudomonas*	*Pseudomonas kilonensis*	4			+	+					
Pseudomonadaceae	*Pseudomonas*	*Pseudomonas koreensis*	3	+	+							
Pseudomonadaceae	*Pseudomonas*	*Pseudomonas libanensis*	10	+		+	+	+	+	+	+	
Pseudomonadaceae	*Pseudomonas*	*Pseudomonas lundensis*	58		+	+	+	+	+	+	+	+
Pseudomonadaceae	*Pseudomonas*	*Pseudomonas monteilii*	2	+								
Pseudomonadaceae	*Pseudomonas*	*Pseudomonas orientalis*	1	+								
Pseudomonadaceae	*Pseudomonas*	*Pseudomonas oryzihabitans*	1		+							
Pseudomonadaceae	*Pseudomonas*	*Pseudomonas protegens*	2			+	+					
Pseudomonadaceae	*Pseudomonas*	*Pseudomonas proteolytica*	4	+					+	+		
Pseudomonadaceae	*Pseudomonas*	*Pseudomonas putida*	4	+	+							
Pseudomonadaceae	*Pseudomonas*	*Pseudomonas stutzeri*	1	+								
Pseudomonadaceae	*Pseudomonas*	*Pseudomonas synxantha*	4		+					+	+	
Pseudomonadaceae	*Pseudomonas*	*Pseudomonas taetrolens*	17			+	+	+	+	+	+	+

Table 3. *Cont.*

Family	Genera	Species	Number of Isolates	Day								
				1	3	5	7	8	9	10	11	12
Pseudomonadaceae	*Pseudomonas*	*Pseudomonas tolaasii*	1	+								
Pseudomonadaceae	*Pseudomonas*	*Pseudomonas veronii*	7	+	+	+	+	+				
Moraxellaceae	*Psychrobacter*	*Psychrobacter spp.*	2	+								
Enterobacteriaceae	*Rahnella*	*Rahnella aquatilis*	2			+	+					
Nocardiaceae	*Rhodococcus*	*Rhodococcus erythropolis*	2		+							
Nocardiaceae	*Rhodotorula*	*Rhodotorula mucilaginosa*	7		+	+	+					
Micrococcaceae	*Rothia*	*Rothia nasimurium*	5		+	+	+					
Enterobacteriaceae	*Serratia*	*Serratia liquefaciens*	4		+	+	+					
Enterobacteriaceae	*Serratia*	*Serratia plymuthica*	1							+		
Enterobacteriaceae	*Serratia*	*Serratia proteamaculans*	6			+	+	+	+			+
Shewanellaceae	*Shewanella*	*Shewanella baltica*	4			+	+		+		+	
Staphylococcaceae	*Staphylococcus*	*Staphylococcus cohnii*	1		+							
Xanthomonadaceae	*Stenotrophomonas*	*Stenotrophomonas maltophilia*	3		+							
Streptococcaceae	*Streptococcus*	*Streptococcus pluranimalium*	1	+								
Yersiniaceae	*Yersinia*	*Yersinia enterocolitica*	4			+	+				+	+
Yersiniaceae	*Yersinia*	*Yersinia rohdei*	2					+		+		
Total isolates			**468**									

Explanation: "+"—occurrence of the bacterial isolate on the tested day.

Table 4. Results of the evaluation of sensory characteristics of raw broiler chicken meat in refrigerated storage (points).

Day of Refrigerated Storage [d]	Studied Traits			
	Smell	Colour	Consistency	General Appearance
1	4.96 ^{a} ± 0.20	4.98 ^{a} ± 0.14	5.00 ^{a} ± 0.00	5.00 ^{a} ± 0.00
3	4.68 ^{a} ± 0.49	4.94 ^{a} ± 0.24	4.96 ^{a} ± 0.20	4.95 ^{a} ± 0.22
5	4.46 ^{ab} ± 0.60	4.84 ^{ab} ± 0.39	4.65 ^{ab} ± 0.52	4.80 ^{ab} ± 0.45
7	4.01 ^{b} ± 0.22	4.10 ^{ab} ± 0.36	4.07 ^{ab} ± 0.33	4.09 ^{bc} ± 0.40
8	3.82 ^{bc} ± 0.44	3.92 ^{bc} ± 0.44	4.12 ^{bc} ± 0.41	3.82 ^{c} ± 0.54
9	3.15 ^{c} ± 0.48	3.48 ^{c} ± 0.50	3.50 ^{c} ± 0.50	3.35 ^{c} ± 0.48
10	2.48 ^{d} ± 0.17	3.14 ^{cd} ± 0.35	3.10 ^{cd} ± 0.30	3.12 ^{cd} ± 0.33
11	2.36 ^{de} ± 0.22	2.40 ^{de} ± 0.14	2.46 ^{de} ± 0.20	2.38 ^{de} ± 0.14
12	1.80 ^{e} ± 0.42	2.35 ^{e} ± 0.48	2.42 ^{e} ± 0.50	2.20 ^{e} ± 0.49
p Value	0.0001	0.0130	0.0217	0.0002

$^{a, b, c}$...—Values in columns marked with different letters differ at $p < 0.05$.

4. Conclusions

To sum up, the study has indicated that by using the MALDI-TOF MS method, it is possible to assess the bacterial profile of poultry meat stored in conditions of stable, low refrigeration temperature maintained at 0.5 °C+/−0.5 °C. A detailed analysis of the dynamics of the development of individual bacterial species performed in this paper using the MALDI-TOF MS method is capable of contributing to expanding knowledge on the participation of individual bacterial species in the spoilage process of poultry meat stored in refrigerated conditions at a narrow, but stable level temperature range. The research and its findings should be treated as preliminary, paving the way to further research into optimising the chilling of poultry meat to extend its shelf life. However, further extensive research is necessary before relevant conclusions can be drawn regarding its application in the meat industry.

Supplementary Materials: The following supporting information can be downloaded at: https://www.mdpi.com/article/10.3390/foods13244063/s1, Table S1. Correlation matrix on sensory analysis parameters and microbiological analysis results in broiler chicken breast meat during refrigerated storage.

Author Contributions: Conceptualisation, Z.S. and M.S.; methodology, M.K., A.A.-P. and P.H.; software, A.A.-P. and M.K.; validation, M.K. and A.A.-P. formal analysis, M.K. and Z.S.; investigation, M.K., A.A.-P. and P.H.; resources, Z.S. and M.K.; data curation, A.A.-P. and P.H.; writing—original draft preparation, M.K., A.A.-P. and Z.S.; writing—review and editing, A.A.-P. and M.K.; visualisation, M.K.; supervision, Z.S., M.K. and M.S.; project administration, A.A.-P. All authors have read and agreed to the published version of the manuscript.

Funding: This research received no external funding.

Institutional Review Board Statement: Not applicable.

Informed Consent Statement: Not applicable.

Data Availability Statement: The data presented in this study are available on request from the corresponding author. The data are not publicly available due to privacy restrictions.

Conflicts of Interest: The authors declare no conflicts of interest.

References

1. Fernández-Pan, I.; Carrión-Granda, X.; Maté, J.I. Antimicrobial Efficiency of Edible Coatings on the Preservation of Chicken Breast Fillets. *Food Control* **2014**, *36*, 69–75. [CrossRef]
2. Katiyo, W.; De Kock, H.L.; Coorey, R.; Buys, E.M. Sensory Implications of Chicken Meat Spoilage in Relation to Microbial and Physicochemical Characteristics during Refrigerated Storage. *LWT* **2020**, *128*, 109468. [CrossRef]
3. McMillin, K.W. Advancements in Meat Packaging. *Meat Sci.* **2017**, *132*, 153–162. [CrossRef] [PubMed]

4. Hutchison, M.L.; Taylor, M.J.; Tchòrzewska, M.A.; Ford, G.; Madden, R.H.; Knowles, T.G. Modelling-Based Identification of Factors Influencing Campylobacters in Chicken Broiler Houses and on Carcasses Sampled after Processing and Chilling. *J. Appl. Microbiol.* **2017**, *122*, 1389–1401. [CrossRef] [PubMed]
5. Marmion, M.; Ferone, M.T.; Whyte, P.; Scannell, A.G.M. The Changing Microbiome of Poultry Meat; from Farm to Fridge. *Food Microbiol.* **2021**, *99*, 103823. [CrossRef] [PubMed]
6. Bełkot, Z.; Pełczyńska, E. Influence of the Chilling System on Bacterial Contamination and the Quality of Slaughter Chicken Carcasses. *Med. Weter.* **2008**, *64*, 1225–1231. (In Polish)
7. Hinton, A.; Cason, J.A.; Ingram, K.D. Tracking Spoilage Bacteria in Commercial Poultry Processing and Refrigerated Storage of Poultry Carcasses. *Int. J. Food Microbiol.* **2004**, *91*, 155–165. [CrossRef]
8. Silva, N.B.D.; Longhi, D.A.; Martins, W.F.; Aragão, G.M.F.D.; Carciofi, B.A.M. Mathematical Modeling of Lactobacillus Viridescens Growth in Vacuum Packed Sliced Ham under Non Isothermal Conditions. *Procedia Food Sci.* **2016**, *7*, 33–36. [CrossRef]
9. Necidová, L.; Zouharová, A.; Haruštiaková, D.; Bursová, Š.; Bartáková, K.; Golian, J. The Effect of Cold Chain Disruption on the Microbiological Profile of Chilled Chicken Meat. *Poult. Sci.* **2024**, *103*, 104290. [CrossRef]
10. Kaewthong, P.; Pomponio, L.; Carrascal, J.R.; Knøchel, S.; Wattanachant, S.; Karlsson, A.H. Changes in the Quality of Chicken Breast Meat Due to Superchilling and Temperature Fluctuations during Storage. *J. Poult. Sci.* **2019**, *56*, 308–317. [CrossRef] [PubMed]
11. Mahato, P.; Rajagopal, R.; Goyette, B.; Adhikary, S. Low-Temperature Anaerobic Digestion of Chicken Manure at High Organic and Nitrogen Loads—Strategies for Controlling Short Chain Fatty Acids. *Bioresour. Technol.* **2022**, *351*, 127049. [CrossRef]
12. Wickramasinghe, N.N.; Ravensdale, J.; Coorey, R.; Chandry, S.P.; Dykes, G.A. The Predominance of Psychrotrophic Pseudomonads on Aerobically Stored Chilled Red Meat. *Compr. Rev. Food Sci. Food Saf.* **2019**, *18*, 1622–1635. [CrossRef]
13. Casaburi, A.; Piombino, P.; Nychas, G.-J.; Villani, F.; Ercolini, D. Bacterial Populations and the Volatilome Associated to Meat Spoilage. *Food Microbiol.* **2015**, *45*, 83–102. [CrossRef] [PubMed]
14. Morales, P.A.; Aguirre, J.S.; Troncoso, M.R.; Figueroa, G.O. Phenotypic and Genotypic Characterization of *Pseudomonas* Spp. Present in Spoiled Poultry Fillets Sold in Retail Settings. *LWT* **2016**, *73*, 609–614. [CrossRef]
15. Ntzimani, A.; Kalamaras, A.; Tsironi, T.; Taoukis, P. Shelf Life Extension of Chicken Cuts Packed under Modified Atmospheres and Edible Antimicrobial Coatings. *Appl. Sci.* **2023**, *13*, 4025. [CrossRef]
16. Odeyemi, O.A.; Alegbeleye, O.O.; Strateva, M.; Stratev, D. Understanding Spoilage Microbial Community and Spoilage Mechanisms in Foods of Animal Origin. *Compr. Rev. Food Sci. Food Saf.* **2020**, *19*, 311–331. [CrossRef] [PubMed]
17. Andreani, N.A.; Fasolato, L. Pseudomonas and Related Genera. In *The Microbiological Quality of Food*; Elsevier: Amsterdam, The Netherlands, 2017; pp. 25–59, ISBN 978-0-08-100502-6.
18. Hinton, A., Jr. Formulating Poultry Processing Sanitizers from Alkaline Salts of Fatty Acids. *J. Food Microbiol. Saf. Hyg.* **2017**, *1*, 116–117. [CrossRef]
19. Rouger, A.; Tresse, O.; Zagorec, M. Bacterial Contaminants of Poultry Meat: Sources, Species, and Dynamics. *Microorganisms* **2017**, *5*, 50. [CrossRef]
20. European Union. Regulation (EU) 2017/1981 of 31 October 2017 Amending Annex III to Regulation (EC) No 853/2004 of the European Parliament and of the Council as Regards Temperature Conditions During Transport of Meat. Official Journal of the European Union, L 285. 31 October 2017, pp. 1–38. Available online: https://eur-lex.europa.eu/legal-content/EN/TXT/?uri=OJ:L:2017:285:TOC (accessed on 16 September 2024).
21. Sujiwo, J.; Kim, D.; Jang, A. Relation among Quality Traits of Chicken Breast Meat during Cold Storage: Correlations between Freshness Traits and Torrymeter Values. *Poult. Sci.* **2018**, *97*, 2887–2894. [CrossRef] [PubMed]
22. Dourou, D.; Spyrelli, E.D.; Doulgeraki, A.I.; Argyri, A.A.; Grounta, A.; Nychas, G.-J.E.; Chorianopoulos, N.G.; Tassou, C.C. Microbiota of Chicken Breast and Thigh Fillets Stored under Different Refrigeration Temperatures Assessed by Next-Generation Sequencing. *Foods* **2021**, *10*, 765. [CrossRef]
23. Singhal, N.; Kumar, M.; Kanaujia, P.K.; Virdi, J.S. MALDI-TOF Mass Spectrometry: An Emerging Technology for Microbial Identification and Diagnosis. *Front. Microbiol.* **2015**, *6*, 791. [CrossRef] [PubMed]
24. Kačániová, M.; Garzoli, S.; Ben Hsouna, A.; Bianchi, A.; Kluz, M.I.; Elizondo-Luevano, J.H.; Ban, Z.; Ben Saad, R.; Mnif, W.; Haščík, P. The Potential of *Thymus serpyllum* Essential Oil as an Antibacterial Agent against *Pseudomonas aeruginosa* in the Preservation of Sous Vide Red Deer Meat. *Foods* **2024**, *13*, 3107. [CrossRef]
25. Kačániová, M.; Mellen, M.; Vukovic, N.L.; Kluz, M.; Puchalski, C.; Haščík, P.; Kunová, S. Combined Effect of Vacuum Packaging, Fennel and Savory Essential Oil Treatment on the Quality of Chicken Thighs. *Microorganisms* **2019**, *7*, 134. [CrossRef] [PubMed]
26. Augustyńska-Prejsnar, A.; Hanus, P.; Ormian, M.; Kačániová, M.; Sokołowicz, Z.; Topczewska, J. The Effect of Temperature and Storage Duration on the Quality and Attributes of the Breast Meat of Hens after Their Laying Periods. *Foods* **2023**, *12*, 4340. [CrossRef] [PubMed]
27. Tura, M.; Gagliano, M.A.; Valli, E.; Petracci, M.; Gallina Toschi, T.A. Methodological Review in Sensory Analyses of Chicken Meat. *Poult. Sci.* **2024**, *103*, 104083. [CrossRef] [PubMed]
28. Troy, D.J.; Kerry, J.P. Consumer Perception and the Role of Science in the Meat Industry. *Meat Sci.* **2010**, *86*, 214–226. [CrossRef] [PubMed]
29. Al-Jasser, M.S. Effect of Cooling and Freezing Temperatures on Microbial and Chemical Properties of Chicken Meat during Storage. *J. Food Agric. Environ.* **2012**, *10*, 113–116.
30. Stonehouse, G.G.; Evans, J.A. The Use of Supercooling for Fresh Foods: A Review. *J. Food Eng.* **2015**, *148*, 74–79. [CrossRef]

31. Sivarajan, M.; Lalithapriya, U.; Mariajenita, P.; Vajiha, B.A.; Harini, K.; Madhushalini, D.; Sukumar, M. Synergistic Effect of Spice Extracts and Modified Atmospheric Packaging towards Non-Thermal Preservation of Chicken Meat under Refrigerated Storage. *Poult. Sci.* **2017**, *96*, 2839–2844. [CrossRef]
32. *ISO 8586:2023*; Sensory Analysis—Selection and Training of Sensory Assessors. International Organization for Standarization: Geneva, Switzerland, 2023.
33. *ISO 8589:2007/Amd 1:2014*; General Guidelines for the Design of a Sensory Analysis Laboratory. iTeh Standards, International Organization for Standarization: Newark, NJ, USA, 2014.
34. *Data Analysis Software System*, version 13.3; StatSoft Electronic Statistics Textbook; StatSoft, Inc.: Kraków, Poland, 2018.
35. Gonçalves-Tenório, A.; Silva, B.; Rodrigues, V.; Cadavez, V.; Gonzales-Barron, U. Prevalence of Pathogens in Poultry Meat: A Meta-Analysis of European Published Surveys. *Foods* **2018**, *7*, 69. [CrossRef]
36. Tuncer, B.; Sireli, U.T. Microbial Growth on Broiler Carcasses Stored at Different Temperatures After Air- or Water-Chilling. *Poult. Sci.* **2008**, *87*, 793–799. [CrossRef] [PubMed]
37. Rouger, A.; Remenant, B.; Prévost, H.; Zagorec, M. A Method to Isolate Bacterial Communities and Characterize Ecosystems from Food Products: Validation and Utilization in as a Reproducible Chicken Meat Model. *Int. J. Food Microbiol.* **2017**, *247*, 38–47. [CrossRef] [PubMed]
38. Balamatsia, C.C.; Paleologos, E.K.; Kontominas, M.G.; Savvaidis, I.N. Correlation between Microbial Flora, Sensory Changes and Biogenic Amines Formation in Fresh Chicken Meat Stored Aerobically or under Modified Atmosphere Packaging at 4 °C: Possible Role of Biogenic Amines as Spoilage Indicators. *Antonie Van Leeuwenhoek* **2006**, *89*, 9–17. [CrossRef] [PubMed]
39. Erkmen, O. Isolation and Counting of Indicator and Pathogenic Microorganisms. In *Microbiological Analysis of Foods and Food Processing Environments*; Elsevier: Amsterdam, The Netherlands, 2022; pp. 103–104, ISBN 978-0-323-91651-6.
40. Whyte, P.; McGill, K.; Collins, J.D. An Assessment of Steam Pasteurization and Hot Water Immersion Treatments for the Microbiological Decontamination of Broiler Carcasses. *Food Microbiol.* **2003**, *20*, 111–117. [CrossRef]
41. Zhao, C.; Ge, B.; De Villena, J.; Sudler, R.; Yeh, E.; Zhao, S.; White, D.G.; Wagner, D.; Meng, J. Prevalence of *Campylobacter* Spp., *Escherichia coli*, and *Salmonella* Serovars in Retail Chicken, Turkey, Pork, and Beef from the Greater Washington, D.C., Area. *Appl. Environ. Microbiol.* **2001**, *67*, 5431–5436. [CrossRef]
42. Paterson, D.L. Resistance in Gram-Negative Bacteria: Enterobacteriaceae. *Am. J. Med.* **2006**, *119*, S20–S28. [CrossRef] [PubMed]
43. Franke, C.; Höll, L.; Langowski, H.-C.; Petermeier, H.; Vogel, R.F. Sensory Evaluation of Chicken Breast Packed in Two Different Modified Atmospheres. *Food Packag. Shelf Life* **2017**, *13*, 66–75. [CrossRef]
44. Kondratowicz, J.; Chwastowska-Siwiecka, I.; Burczyk, E.; Piekarska, J.; Kułdo, Ż. Sensory & Microbiological Assessment of Turkey Hens Breast Muscles Depending on Method and Time of Cold Storage. *Zywnosc Nauka Technol. Jakosc* **2011**, *18*, 143–152. [CrossRef]
45. Nychas, G.-J.E.; Skandamis, P.N.; Tassou, C.C.; Koutsoumanis, K.P. Meat Spoilage during Distribution. *Meat Sci.* **2008**, *78*, 77–89. [CrossRef] [PubMed]
46. Arnaut-Rollier, I. Identities of the *Pseudomonas* Spp. in Flora from Chilled Chicken. *Int. J. Food Microbiol.* **1999**, *48*, 87–96. [CrossRef] [PubMed]
47. Arnaut-Rollier, I.; Vauterin, L.; De Vos, P.; Massart, D.L.; Devriese, L.A.; De Zutter, L.; Van Hoof, J. A Numerical Taxonomic Study of the *Pseudomonas* Flora Isolated from Poultry Meat. *J. Appl. Microbiol.* **1999**, *87*, 15–28. [CrossRef]
48. Säde, E.; Murros, A.; Björkroth, J. Predominant Enterobacteria on Modified-Atmosphere Packaged Meat and Poultry. *Food Microbiol.* **2013**, *34*, 252–258. [CrossRef] [PubMed]
49. Harada, T.; Dang, V.C.; Nguyen, D.P.; Nguyen, T.A.D.; Sakamoto, M.; Ohkuma, M.; Motooka, D.; Nakamura, S.; Uchida, K.; Jinnai, M.; et al. Enterococcus Saigonensis Sp. Nov., Isolated from Retail Chicken Meat and Liver. *Int. J. Syst. Evol. Microbiol.* **2016**, *66*, 3779–3785. [CrossRef]
50. Rahkila, R.; Johansson, P.; Säde, E.; Björkroth, J. Identification of Enterococci from Broiler Products and a Broiler Processing Plant and Description of *Enterococcus viikkiensis* Sp. Nov. *Appl. Environ. Microbiol.* **2011**, *77*, 1196–1203. [CrossRef] [PubMed]
51. Koort, J.; Murros, A.; Coenye, T.; Eerola, S.; Vandamme, P.; Sukura, A.; Björkroth, J. *Lactobacillus oligofermentans* sp. Nov., Associated with Spoilage of Modified-Atmosphere-Packaged Poultry Products. *Appl. Environ. Microbiol.* **2005**, *71*, 4400–4406. [CrossRef] [PubMed]
52. Gill, C.O. Active Packaging in Practice: Meat. In *Novel Food Packaging Techniques*; Elsevier: Amsterdam, The Netherlands, 2003; pp. 365–383, ISBN 978-1-85573-675-7.
53. Doulgeraki, A.I.; Paramithiotis, S.; Nychas, G.-J.E. Characterization of the Enterobacteriaceae Community That Developed during Storage of Minced Beef under Aerobic or Modified Atmosphere Packaging Conditions. *Int. J. Food Microbiol.* **2011**, *145*, 77–83. [CrossRef]
54. Ercolini, D.; Ferrocino, I.; Nasi, A.; Ndagijimana, M.; Vernocchi, P.; La Storia, A.; Laghi, L.; Mauriello, G.; Guerzoni, M.E.; Villani, F. Monitoring of Microbial Metabolites and Bacterial Diversity in Beef Stored under Different Packaging Conditions. *Appl. Environ. Microbiol.* **2011**, *77*, 7372–7381. [CrossRef]
55. González-Miret, M.L.; Escudero-Gilete, M.L.; Heredia, F.J. The Establishment of Critical Control Points at the Washing and Air Chilling Stages in Poultry Meat Production Using Multivariate Statistics. *Food Control* **2006**, *17*, 935–941. [CrossRef]

56. Chaillou, S.; Chaulot-Talmon, A.; Caekebeke, H.; Cardinal, M.; Christieans, S.; Denis, C.; Hélène Desmonts, M.; Dousset, X.; Feurer, C.; Hamon, E.; et al. Origin and Ecological Selection of Core and Food-Specific Bacterial Communities Associated with Meat and Seafood Spoilage. *ISME J.* **2015**, *9*, 1105–1118. [CrossRef]
57. Höll, L.; Behr, J.; Vogel, R.F. Identification and Growth Dynamics of Meat Spoilage Microorganisms in Modified Atmosphere Packaged Poultry Meat by MALDI-TOF MS. *Food Microbiol.* **2016**, *60*, 84–91. [CrossRef] [PubMed]
58. Zhang, Q.Q.; Han, Y.Q.; Cao, J.X.; Xu, X.L.; Zhou, G.H.; Zhang, W.Y. The Spoilage of Air-Packaged Broiler Meat during Storage at Normal and Fluctuating Storage Temperatures. *Poult. Sci.* **2012**, *91*, 208–214. [CrossRef] [PubMed]
59. Ercolini, D.; Russo, F.; Blaiotta, G.; Pepe, O.; Mauriello, G.; Villani, F. Simultaneous Detection of *Pseudomonas fragi*, *P. lundensis*, and *P. putida* from Meat by Use of a Multiplex PCR Assay Targeting the *carA* Gene. *Appl. Environ. Microbiol.* **2007**, *73*, 2354–2359. [CrossRef] [PubMed]
60. Vergara-Figueroa, J.; Cerda-Leal, F.; Alejandro-Martín, S.; Gacitúa, W. Evaluation of the PLA-nZH-Cu Nanocomposite Film on the Micro-Biological, Organoleptic and Physicochemical Qualities of Packed Chicken Meat. *Foods* **2022**, *11*, 546. [CrossRef]
61. Doulgeraki, A.I.; Ercolini, D.; Villani, F.; Nychas, G.-J.E. Spoilage Microbiota Associated to the Storage of Raw Meat in Different Conditions. *Int. J. Food Microbiol.* **2012**, *157*, 130–141. [CrossRef]
62. Yimenu, S.M.; Koo, J.; Kim, B.S.; Kim, J.H.; Kim, J.Y. Freshness-Based Real-Time Shelf-Life Estimation of Packaged Chicken Meat under Dynamic Storage Conditions. *Poult. Sci.* **2019**, *98*, 6921–6930. [CrossRef] [PubMed]
63. Garavito, J.; Moncayo-Martínez, D.; Castellanos, D.A. Evaluation of Antimicrobial Coatings on Preservation and Shelf Life of Fresh Chicken Breast Fillets Under Cold Storage. *Foods* **2020**, *9*, 1203. [CrossRef]
64. Duan, D.; Wang, H.; Xue, S.; Li, M.; Xu, X. Application of Disinfectant Sprays after Chilling to Reduce the Initial Microbial Load and Extend the Shelf-life of Chilled Chicken Carcasses. *Food Control* **2017**, *75*, 70–77. [CrossRef]
65. Esmer, O.K.; Irkin, R.; Degirmencioglu, N.; Degirmencioglu, A. The Effects of Modified Atmosphere Gas Composition on Microbiological Criteria, Color and Oxidation Values of Minced Beef Meat. *Meat Sci.* **2011**, *88*, 221–226. [CrossRef] [PubMed]

foods

MDPI

Article

Evaluation of Antioxidant and Antibacterial Activity of Gelatin Nanoparticles with Bitter Orange Peel Extract for Food Applications

Adamaris García-Juárez [1,2], Alba Mery Garzón-García [3,4], José Rogelio Ramos-Enríquez [2], José Agustín Tapia-Hernández [1], Saúl Ruiz-Cruz [1,*], Dalila Fernanda Canizales-Rodríguez [2], Carmen Lizette Del-Toro-Sánchez [1], Francisco Rodríguez-Félix [1], Víctor Manuel Ocaño-Higuera [2] and José de Jesús Ornelas-Paz [5]

1 Departamento de Investigación y Posgrado en Alimentos, Universidad de Sonora, Encinas y Rosales s/n, Hermosillo 83000, Sonora, Mexico; a218203582@unison.mx (A.G.-J.); joseagustin.tapia@unison.mx (J.A.T.-H.); carmen.deltoro@unison.mx (C.L.D.-T.-S.); francisco.rodriguezfelix@unison.mx (F.R.-F.)

2 Departamento de Ciencias Químico Biológicas, Universidad de Sonora, Encinas y Rosales s/n, Hermosillo 83000, Sonora, Mexico; rogelio.ramos@unison.mx (J.R.R.-E.); dalila.canizales@unison.mx (D.F.C.-R.); victor.ocano@unison.mx (V.M.O.-H.)

3 Tecnología Agroambiental, Universidad del Valle Sede Regional Caicedonia, Carrera 14 No 4-48, Caicedonia 76250, Valle del Cauca, Colombia; garzon.alba@correounivalle.edu.co or almgarzonga@unal.edu.co

4 Facultad de Ingeniería y Administración, Universidad Nacional de Colombia—Sede Palmira, Carrera 32 # 12-00, Palmira 763533, Valle del Cauca, Colombia

5 Centro de Investigación en Alimentación y Desarrollo, Av. Río Conchos S/N Parque Industrial, Cuauhtémoc 31570, Chihuahua, Mexico; jornelas@ciad.mx

* Correspondence: saul.ruizcruz@unison.mx; Tel.: +52-644-1250653

Abstract: Bitter orange is a citrus fruit rich in bioactive compounds, but its waste is currently underutilized. One potential solution is to encapsulate these bioactive compounds. This research aims to synthesize gelatin nanoparticles loaded with an ethanolic extract of bitter orange peel and to evaluate their in vitro antioxidant and antibacterial activities. Coaxial electrospray was used to encapsulate the ethanolic extract of bitter orange with bovine gelatin as wall material, considering a voltage of 15 kV, a wall solution flow rate of 0.1 mL/h, and a core solution flow rate of 0.08 mL/h. Characterization of the nanoparticles was performed using scanning electron microscopy (SEM) and Fourier transform infrared spectroscopy (FT-IR). Antioxidant activity was evaluated by the total phenolic content, flavonoids, and antioxidant capacity by the DPPH$^{\bullet}$, ABTS$^{\bullet+}$, and FRAP assays. Antibacterial activity was assessed by the well diffusion technique on Mueller–Hinton agar against *Listeria monocytogenes* and *Escherichia coli* O157:H7 bacteria. SEM images confirmed that the nanoparticles were spherical in shape, while FT-IR analysis indicated that the incorporation of the extract did not alter the amide bonds of the gelatin protein. The nanoparticles containing the extract exhibited higher antioxidant activity and heightened inhibition against *E. coli* O157:H7, indicating their potential food applications.

Keywords: *Citrus aurantium*; nanotechnology; coaxial electrospray; bioactive compounds; foodborne pathogens

1. Introduction

The bitter orange (*Citrus aurantium*) is a fruit that grows on a tree belonging to the *Rutaceae* family. Native to East Africa, Arabia, and Syria, it is actively cultivated in regions such as Spain, Italy, and North America [1]. In Sonora, Mexico, *Citrus aurantium* is primarily used as an ornamental tree, and its fruits are typically discarded [2]. Bitter orange is a globose fruit with a slightly textured surface and an intense orange color when ripe (Figure 1). Due to its rich content of diverse bioactive compounds such as phenolics, flavonoids, essential oils, and vitamins, bitter orange is commonly employed for medicinal purposes (antibacterial, anti-inflammatory, anticancer, antioxidant, and cardiovascular properties) and in the food industry (juices and essential oils) [3]. Bitter orange biowaste currently produces atmospheric and aquatic pollution contamination since it accumulates in the soil when it is not consumed or it is not fully used [4]. The predominant residue of the bitter orange is its peel, comprising approximately 40% of its fresh weight. In the peel, the documented phenolic compounds primarily consist of phenolic acids (74%) and flavonoids (23%). Among these, the most prevalent phenolic compounds identified are p-coumaric acid and ferulic acid [5].

Figure 1. Tree (**a**) and fruit (bitter orange) (**b**) of the *Citrus aurantium* species.

The previously mentioned compounds can help prevent food spoilage and cross-contamination, which pose significant health risks to consumers and can lead to temporary digestive issues such as gastroenteritis, diarrhea, fever, or even infections in other systems. Bacteria such as *Escherichia coli* and *Listeria monocytogenes* are often found in raw or poorly processed foods, including meat, dairy products, seafood, and fruits, making them two of the primary etiological agents of foodborne diseases [6]. There is currently a growing interest in obtaining bioactive compounds from biowaste due to their antioxidant and antimicrobial properties. These compounds also modulate the intestinal microbiota and promote the immune system [2]. Biowaste is rich in phenolic molecules, which contain aromatic rings with hydroxyl groups. These molecules are known to inhibit microorganisms by interacting with their cytoplasmic membranes, cell walls, and nucleic acids, thereby disrupting cellular functions, slowing growth, and causing cell death [7,8].

Nanoencapsulation is a process in which an active ingredient, an active molecule of a food (antioxidants, essential fatty acids, vitamins), or living cells (probiotics) are trapped within a wall material (polymers of carbohydrates, proteins, and lipids). Encapsulation serves not only to enhance the stability, bioavailability, and controlled release characteristics of biomolecules but also aids in concealing undesirable odor and taste [9]. Different nanoencapsulation techniques exist, notable among them are nanoprecipitation, emulsion–diffusion, double-emulsification, emulsion–coacervation, polymer coating, and electrospray techniques. However, electrospray is relatively simple, low cost, requires a low amount of solvents and the particles are obtained in a single step [10]. The principle of electrospray lies in the ability of an electric field to deform the droplet interface to produce nanometer-range droplets. The Rayleigh method serves to determine the rupture of the drop. This limit occurs when the surface tension of the drop is elevated by the electrostatic force [11]. The electrospray control variables, such as solution flow rate, electric potential, collector distance, environmental humidity, and polymer solution viscosity, conductivity, surface tension, and concentration, must be considered [12]. On the other hand, coaxial electrospray modifies the uniaxial electrospray process by using a coaxial capillary needle and syringes for two solutions [13]. It is an electrohydrodynamic process in which multilayer nanoparticles are formed by introducing coaxial electrified jets. Coaxial electrospray offers several potential benefits, including high efficiency, a simple and cost-effective process, the ability to preserve biological activity, and the production of particles with a uniform size distribution [12]. Therefore, the objective of this research is to produce by coaxial electrospray gelatin nanoparticles containing an ethanolic extract of bitter orange peel (*Citrus aurantium*) and to analyze their morphological, structural, antioxidant, and antibacterial properties as potential food applications.

2. Materials and Methods

2.1. Plant Material

Bitter orange fruits were harvested in the first quarter of 2022 at the University of Sonora, Hermosillo, Sonora, Mexico. A total of 900 g of fruit was washed with a neutral detergent, rinsed with potable water, and the peel was removed. The orange peel was cut into 3 cm by 3 cm squares and dried for 24 h at 60 °C in a convection oven (Blue M, New Columbia, PA, USA) until it reached a moisture content of 17%. Following drying, the material underwent grinding and sieving (size 60 sieve).

2.2. Chemical and Reagents

Absolute ethanol (≥99.5%) and hydrochloric acid (HCl, 37.5%) were from Meyer Reagents (Ciudad de México, MEX); gallic acid, 2 N Folin solution, quercetin, 2,2-diphenyl-1-picrylhydrazyl ($DPPH^{\bullet}$) radical, 6-hydroxy-2,5,7,8-tetramethylchroman-2-carboxylic acid (Trolox), tripyridyl-s-triazin (TPTZ), 2,2′-azino-bis(3-ethylbenzothiazoline-6-sulfonic acid) ($ABTS^{\bullet+}$) radical, glacial acetic acid, and bovine gelatin type B were obtained from Sigma-Aldrich (St. Louis, MO, USA); potassium persulfate ($K_2S_2O_8$, ≥99.0%) was from Jalmek (San Nicolás de los Garza, NL, MEX); anhydrous sodium carbonate (Na_2CO_3, ≥99.5%), anhydrous sodium acetate ($C_2H_3NaO_2$, ≥99.0%), and iron trichloride ($FeCl_3$) were procured from CTR SCIENTIFIC (Monterrey, NL, MEX); Tryptic Soy Broth and Oxford Agar were purchased from MCD LAB (Tlaneplanta de Baz, EM, MEX); MacConkey Sorbitol Agar and Mueller–Hinton Agar were from Difco (Sparks, MD, USA); Ultrapure water was obtained using a purification system (Merck, Darmstadt, HE, GER).

2.3. Preparation of the Extract of Bitter Orange Peel

The initial solutions were prepared with 1.5 g of orange peel and 10 mL of absolute ethanol. These solutions were then homogenized, subjected to sonication for 20 min, and subsequently centrifuged at 1500 rpm for 15 min (Compact II Centrifuge, BD, East Rutherford, NJ, USA). The supernatants were collected, and the residues underwent a second extraction [14]. Thus, the supernatants were combined and concentrated with the rotary evaporator R-100 V (Flawil, SG, CH) at 200 rpm, 40 °C, and 90 kP. The concentrated samples were suspended in ethanol until a known concentration was obtained. This process was conducted in two batches.

2.4. Preliminary Test for Nanoencapsulation

Uniaxial electrospray was used to test the production of gelatin nanoparticles [15,16]. For the feed solution, a concentration of 10% w/v gelatin was considered. This solution was prepared with 20% aqueous acetic acid, then sealed with parafilm, and stirred under ambient conditions until complete dissolution was achieved, approximately 1 h. In the electrospray process, a flow rate in the pump (KD Scientific, Holliston, MA, USA) of 0.1 mL/h was maintained for the polymer solution. The voltage applied using a CZE 1000R high-voltage power supply (Spellman, Hauppauge, NY, USA) was 15 kV, and the distance to the collector varied from 5, 10, and 15 cm. The distance to produce the gelatin nanoparticles was selected based on the lowest mean diameter (MD) and polydispersity index (PI). MD was measured using ImageJ 1.53 k (Wayne Rasband, Bethesda, MD, USA), while PI was calculated according to Estrella-Osuna et al. [17] and a frequency analysis.

$$PI = \sigma/\overline{X}, \quad (1)$$

where σ corresponds to standard deviation $\overline{X}$ refers to the mean diameter of nanoparticles.

2.5. Preparation of Gelatin Nanoparticles Containing an Extract of Bitter Orange Peel

The coaxial electrospray process was conducted considering the optimal distance for gelatin nanoparticle production, a voltage of 15 kV, a flow rate of 0.1 mL/h for the polymer solution (outer layer material), and a flow rate of 0.08 mL/h for the extract of bitter orange peel, which was the inner material.

2.6. Morphological and Structural Characterization of Nanoparticles

The morphological characterization of the nanoparticles was conducted by scanning electron microscopy (SEM) using a JEOL 5410 LV equipment (Arkishima, Tokyo, Japan). An acceleration voltage of up to 15 kV and samples were positioned onto double-sided conductive carbon tape affixed to the aluminum pin stub of the SEM [18].

The infrared spectrum was obtained using an FT-IR spectrophotometer (Frontier, Perkin Elmer, Waltham, MA, USA). The spectra were recorded using the attenuated total reflectance (ATR) technique in transmittance mode. A spectrum scan from 4000 to 400 cm^{-1} was considered [19]. The determinations were performed three times.

2.7. Preparation of Nanoparticle Suspension

The empty nanoparticles and those that included the ethanolic extract of bitter orange peel were suspended in absolute ethanol for approximately 12 h. Subsequently, these solutions were centrifuged at 1500 rpm for 15 min (COMPACT II CENTRIFUGE, BD, NJ, USA). The supernatants at a concentration of 2000 μg/mL were used to determine the antioxidant and antibacterial activity.

2.8. Determination of Total Phenolic and Flavonoids Contents

The total phenolic content was determined in accordance with Garzón-García et al. [20]. A total of 10 μL of samples (extract of bitter orange peel, solutions of empty gelatin nanoparticles, and nanoparticles containing the extract) and 25 μL of 1 N Folin solution were added to the microplate and refrigerated for 5 min. After refrigeration, 25 μL of 20% Na_2CO_3 and 140 μL of distilled water were added to the wells. The microplate was put to rest for 30 min, and the absorbance was measured at 760 nm (Multiskan GO, Thermo Fisher Scientific, Waltham, MA, USA). A calibration curve was plotted with the gallic acid standard, considering concentrations from 0.06 to 1 mg/mL (R^2 = 0.998). The results were reported as mg of gallic acid equivalents per gram of sample (mg GAE/g DS).

The determination of flavonoid content was performed according to Del-Toro-Sánchez et al. [21]. A total of 80 μL of each sample and 80 μL of an ethanolic solution of $AlCl_3$ (20 g/L) were added to wells of the microplate, shaken for 30 s, and left in the dark at room temperature for 1 h. Subsequently, it was shaken again for 30 s, and the absorbance was measured at 415 nm. A calibration curve was plotted with the quercetin standard at concentrations from 0 to 0.3 mg/mL (R^2 = 0.998). The results were reported as mg of quercetin equivalents per gram of sample (mg EQ/g DS).

2.9. Antioxidant Capacity Essays

For the measurement of antioxidant capacity by scavenging 2,2-diphenyl-1-picrylhydrazyl radical (DPPH•), 1.5 mg of DPPH• radical was dissolved in 50 mL of methanol. The absorbance of the solution was adjusted to 0.7 ± 0.01 at a wavelength of 515 nm. A total of 200 μL of the solution and 20 μL of extract and sample solutions were added to the microplate wells, left to rest in the dark for 30 min, and the absorbance was measured at a wavelength of 515 nm [22]. A calibration curve was plotted with the Trolox standard, considering concentrations between 0 and 0.08 mg/mL (R^2 = 0.993). The results were expressed as μM Trolox equivalents per gram of sample (μM TE/g DS).

For the determination of antioxidant capacity by 2,2′-azino-bis (3-ethylbenzothiazoline-6-sulfonic acid) ($ABTS^{\bullet+}$) assay, 19.3 mg of ABTS were dissolved in 5 mL of distilled water. Separately, 0.0378 g of potassium persulfate and 1 mL of water were mixed. Subsequently, 88 μL of the potassium persulfate solution was added to the ABTS solution and left to rest in the dark for 12 h under refrigeration. A total of 1 mL of this last solution was added to 88 mL of ethanol, and the absorbance was adjusted to 0.7 ± 0.01 at a wavelength of 734 nm. For measurements, 270 μL of the adjusted solution and 20 μL of each sample were added to the microplate wells. The absorbance was measured at 734 nm in a microplate reader after 30 min of rest [23]. A calibration curve was plotted with the Trolox standard at concentrations from 0 to 0.1 mg/mL (R^2 = 0.992). The results were expressed in μTrolox equivalents per gram of sample (μM TE/g DS).

To measure the ferric-reducing antioxidant power, three stock solutions were prepared: a sodium acetate buffer solution (300 mM/L at a pH of 3.6), a 20 mM $FeCl_3 \times 6H_2O$, solution, and a 10 mM TPTZ in 40 mM HCl solution. The working solution was obtained by adding the three stock solutions in a 10:1:1 ratio (buffer/$FeCl_3 \times 6H_2O$/TPTZ-HCl). For measurements, 20 μL of samples (extract of bitter orange peel, solutions of empty gelatin nanoparticles, and nanoparticles containing the extract) and 280 μL of the working solution were added to the microplate wells and the absorbance was measured at 638 nm after 30 min of rest. A calibration curve was plotted with the Trolox standard, considering concentrations between 0 and 1 mg/mL (R^2 = 0.995). The results were expressed as μM Trolox equivalents per gram of sample [22].

2.10. Antimicrobial Activity

Escherichia coli O157:H7 (ATCC 43890) and *Listeria monocytogenes* (ATCC 7664) strains were used in this study. An aliquot of each strain was added to 3 mL of tryptic soy broth and left to incubate at 37 °C for 18 h. To obtain pure cultures, a loop of *E. coli* O157:H7 was streaked on Sorbitol MacConkey Agar and *L. monocytogenes* on Oxford Agar. The plates were left incubated at 37 °C for 24 h. For each bacterium, a colony from the plates was inoculated in 9 mL of trypticase soy broth [24]. The population of the suspension was adjusted to 1.5×10^8 cells/mL using the McFarland standard. Serial decimal dilutions were performed to obtain inoculums with a population of 10^3 CFU/mL. For the agar well diffusion method, 100 μL of each inoculum were streaked on plates with Mueller–Hinton Agar. Subsequently, three 6 mm wells were made per plate, into which 50 μL of each sample was added [25]. The plates were incubated for 18 h at 37 °C. The wells with 50 μL of ethanol were considered as negative control. After incubation, the inhibition zones were measured with a vernier caliper.

2.11. Statistical Analysis

The results of the antioxidant and antibacterial activity determinations were performed in triplicate and were presented as mean ± standard deviation. A one-way analysis of variance (ANOVA) and a Tukey test were carried out to determine whether there was a significant difference in the antioxidant and antibacterial activity of the extract of bitter orange peel, the empty gelatin nanoparticles, and the nanoparticles containing the extract ($p < 0.05$) [26]. The statistical analyses were performed using Minitab 17 (State College, PA, USA).

3. Results and Discussion

3.1. Preparation of Bitter Orange Peel Extract

After the ultrasound-assisted extraction process and the concentration of the extract by evaporation under reduced pressure, an average yield of 1.076% of ethanolic extract of bitter orange peel was estimated considering the two batches. A concentration of 70 mg/mL was chosen for electrospray nanoencapsulation and subsequent analyses.

3.2. Obtaining of Nanoparticles

The determination of the best conditions of nanoencapsulation was based on the values of mean diameter (MD) and polydispersity index (PI) for the empty particles produced by uniaxial electrospray. This involved preparing a feed solution consisting of 10% *w*/*v* gelatin in 20% acetic acid, positioned at distances of 5, 10, and 15 cm from the collector. SEM micrographs indicate that the distance between the collecting plate and the needle significantly impacted the uniformity of the particles (Figure 2). Consequently, this led to the generation of more scattered particles with varying sizes, contributing to an increase in the PI. Considering the above, the most favorable condition for producing gelatin nanoparticles was achieved when the collector was positioned at 5 cm. On the contrary, Torkamani et al. [16] investigated different combinations of voltage (15, 20, and 25 kV) and flow rate (0.25, 0.5, and 0.75 mL/h) while maintaining a polymer concentration of 5% *w*/*w*. The authors found that the optimal conditions for producing spherical gelatin beads with an average diameter of 297 ± 70 nm were 20 kV, a flow rate of 0.5 mL/h, and 10 cm of collector distance. Therefore, the formation of spherical gelatin beads in the present study and the reported studies could be differentiated based on the experimental conditions. Torkamani et al. [16] utilized a higher voltage, a higher flow rate, a larger collector distance, and a lower polymer concentration than in our study.

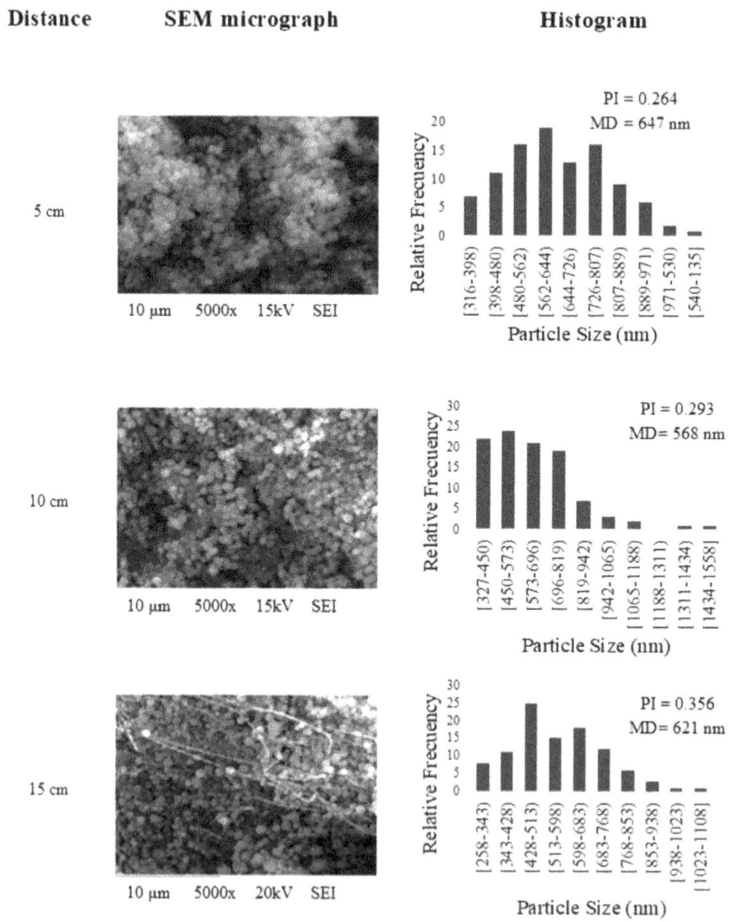

Figure 2. Preliminary tests for the preparation of gelatin nanoparticles using electrospray: SEM micrographs (5000× magnification) and histograms at collector distances of 5, 10, and 15 cm.

3.3. Preparation of Gelatin Nanoparticles Containing an Extract of Bitter Orange Peel

The ethanolic extract of bitter orange peel was encapsulated with bovine type B gelatin by coaxial electrospray, considering the previously established conditions (Figure 3). These materials were chosen because proteins, such as gelatin, are biocompatible and readily available as raw materials. It should be noted that there is currently no available data in the literature regarding the nanoencapsulation of an ethanolic extract of bitter orange peel using protein as an encapsulating agent. However, some authors synthesized buriti oil nanoparticles using water-in-oil emulsification with porcine gelatin alone and in combination with alginate. The particles formed with only gelatin were spherical, had a smooth surface, showed a homogeneous size distribution, exhibited low agglomeration, and were less than 100 nm in size [27].

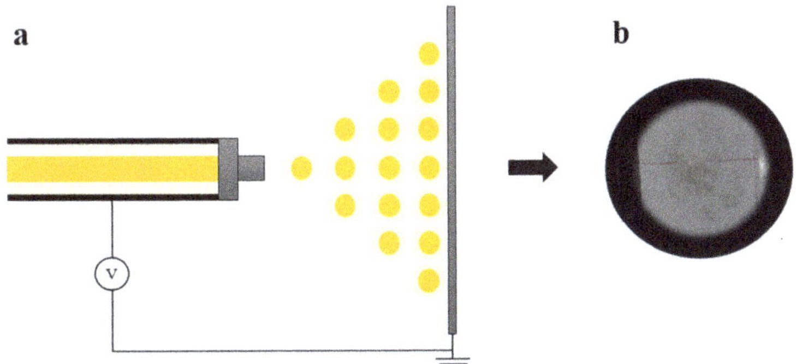

Figure 3. Coaxial electrospray process (**a**) and powder of gelatin nanoparticles containing bitter orange peel extract (**b**).

The coaxial electrospray method employed a flow rate of 0.1 mL/h for the gelatin solution and 0.08 mL/h for the extract. The collection distance was set at 5 cm, operating under 15 kV, and the collection process lasted approximately 1 h. The resultant gelatin nanoparticles, incorporating bitter orange peel extract, exhibited a spherical shape and a uniform size, as depicted in Figure 4. This was confirmed by assessing the polydispersity index (PI) and mean diameter (MD), whose values were 0.435 and 0.675 μm, respectively. In a separate study, Figueroa et al. [26] utilized a 10% w/v gelatin solution to encapsulate betalains extracted from pitaya at concentrations of 1%, 3%, and 5% w/v. Interestingly, the mean diameter and polydispersity index of the nanoparticles did not vary significantly across these different betalains concentrations. Specifically, the mean diameter values were reported as 846, 832, and 839 nm, while the polydispersity index values were 0.12, 0.13, and 0.13, respectively. Discrepancies in PI values between studies can be attributed to differences in environmental conditions. In the present study, nanoparticle production occurred under humidity conditions ranging from 48% to 55% throughout the day, as efforts were made to identify optimal conditions for nanoparticle preparation. Furthermore, it was discovered that higher humidity levels corresponded to larger nanoparticle sizes [28].

Figure 5 shows the infrared spectrum of the gelatin, gelatin nanoparticles containing bitter orange peel extract, and ethanolic extract of bitter orange peel. A total of 14 different bonds were identified, four of them characteristic of gelatin and typical of proteins: 3258 cm^{-1} (Amide A); 1625 cm^{-1} (Amide I), characteristic of the carbonyl bond (C=O); 1431 cm^{-1} (Amide II) associated with N-H bending; and 1224 cm^{-1} (Amide III), correlated with the vibrations of C-N and N-H bonds [29]. Four bonds were also observed for the gelatin nanoparticles containing orange peel extract: 3260 cm^{-1} (Amide A); 1625 cm^{-1} (Amide I), attributed to the carbonyl bond (C=O); 1427 cm^{-1} (Amide II), resulting from N-H bending; and 1231 cm^{-1} (Amide III), corresponding to the vibrations of C-N and N-H bonds. Additionally, other bonds were detected at specific frequencies: 3326 cm^{-1}, representing O-H bond stretching and associated with phenols and carboxylic acids found in pectin and lignin; 2966 cm^{-1}, indicating C-H bond stretching originated from methyl and methoxy groups; 1663 cm^{-1}, indicating C=C bond stretching possibly due to the presence of benzenes or aromatic rings in carotenoids; 1380 cm^{-1}, reflecting CH_3 bending attributed to compounds such as hesperidin and carotenoids; 1041 cm^{-1}, corresponding to the C-O group of alcohols and carboxylic acids found in bioactive compounds (ascorbic acid and naringin). Finally, a bond at 880 cm^{-1}, corresponding to the C-H bond of aromatic compounds such as beta-carotene and phenols [30]. Encapsulating the bitter orange peel extract in the gelatin core did not significantly affect any of the characteristic amide bonds. This indicates that the inclusion of the extract did not alter the chemical structure of the gelatin and was a physical process [31].

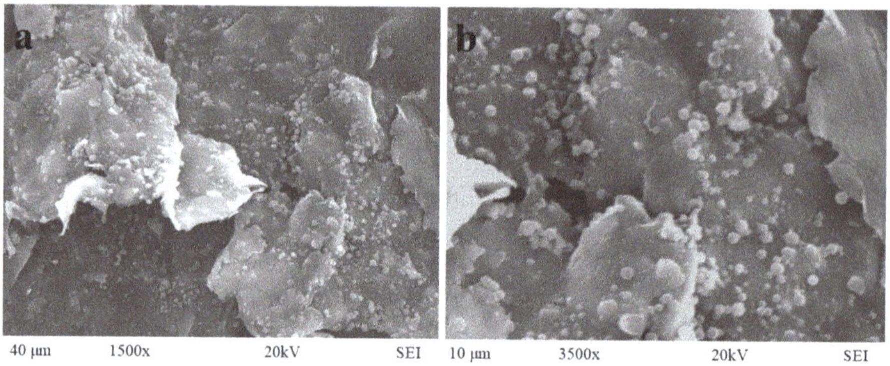

Figure 4. SEM micrograph of gelatin nanoparticles containing bitter orange peel extract. Magnification (**a**) 1500× and (**b**) 3500×.

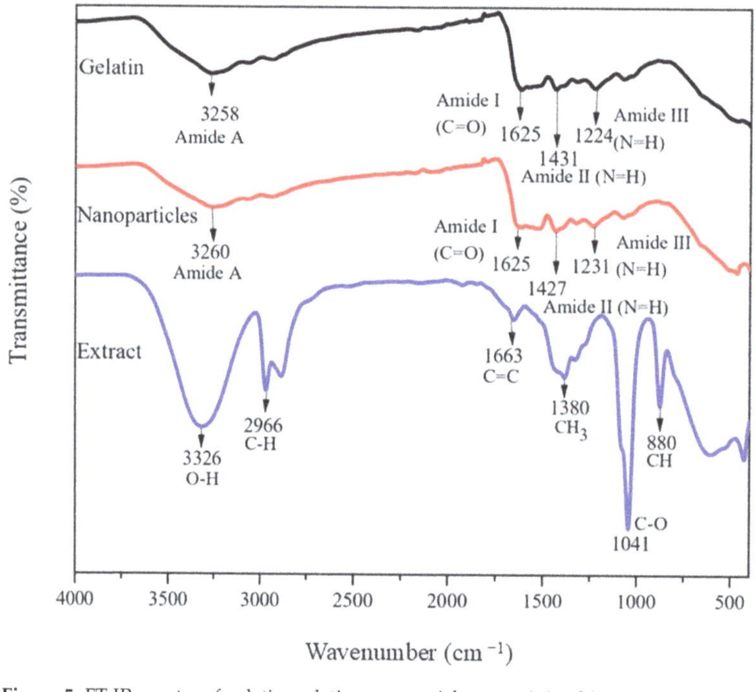

Figure 5. FT-IR spectra of gelatin, gelatin nanoparticles containing bitter orange peel extract, and ethanolic extract of bitter orange.

3.4. Antioxidant Activity

As per Table 1, an assessment of total phenols, flavonoids, and antioxidant capacity was conducted for the ethanolic extract of bitter orange peel, gelatin nanoparticles, and gelatin nanoparticles containing ethanolic extract of bitter orange peel. The nanoparticles containing the extract showed the highest total phenol content (30.7 ± 3.02 mg EAG/g DS) compared to the extract (7.3 ± 0.59 mg EAG/g DS). This may occur because the technique used is sensitive to various compounds and bonds that can interact in the reactive system through electron transfer [32]. Conversely, the value obtained for the extract is closely similar to those found in the literature. Divya et al. [33] reported variations in the extraction of total phenols from lyophilized *Citrus aurantium* peel using different solvents: acetone (10.0 ± 0.70 ETA mg/g), hexane (5.0 ± 0.50 mg ETA/g), methanol (22.5 ± 0.80 mg ETA/g), ethyl acetate (5.0 ± 0.21 mg ETA/g), and water (45.0 ± 0.80 mg ETA/g). Therefore, the biological activity of the extract could depend on the extraction method and the solvent used.

Table 1. Total phenolic and flavonoid content and antioxidant capacity of the ethanolic extract of bitter orange peel, the empty gelatin nanoparticles, and the nanoparticles containing the extract.

Determination	Ethanolic Extract of Bitter Orange Peel	Empty Gelatin Nanoparticles	Nanoparticles Containing the Extract of Bitter Orange Peel
Total phenols [mg EAG/g DS]	7.337 ± 0.586 [b]	-	30.656 ± 3.015 [a]
Flavonoids [mg EQ/g DS]	4.179 ± 0.012 [a]	0.295 ± 0.042 [b]	0.321 ± 0.085 [b]
DPPH• [μM ET/g DS]	1.246 ± 0.003 [b]	1.885 ± 0.183 [b]	3.589 ± 0.534 [a]
ABTS•+ [μM ET/g DS]	17.171 ± 0.534 [b]	13.238 ± 1.820 [b]	65.671 ± 5.359 [a]
FRAP [μM ET/ g DS]	180.607 ± 2.382 [a]	62.916 ± 1.226 [c]	112.431 ± 2.941 [b]

Note: Distinct lowercase letters in the same row indicate a statistical difference ($p < 0.05$).

The extract showed the highest flavonoid content (4.179 ± 0.012 mg EQ/g DS), presumably because when the flavonoids interact with gelatin, there may be alterations in the arrangement of hydroxyl groups. The substitution of hydroxyl groups by glycosylation could decrease the antioxidant activity [34]. On the other hand, it was noted that the addition of the ethanolic extract of bitter orange peel to the gelatin nanoparticles increased its activity of neutralizing radicals by DPPH• and ABTS•+ essays (3.589 ± 0.534 and 65.671 ± 5.359 μM ET/g DS, respectively). There are few reports about the impact of the gelatin coating on antioxidant activity. However, bioactive compounds are protected by this gelatin protein layer. Moreover, the presence of a high concentration of antioxidant compounds acts through mechanisms involving hydrogen atom transfer to neutralize free radicals. For FRAP, the highest value was observed for the extract (180.607 ± 2.382 μM ET/g DS). This occurred because the antioxidant compounds responsible for neutralizing free radicals through electron transfer were less available in the gelatin nanoparticles, making it difficult for them to traverse the protein barrier and reduce the iron complex. Based on the findings, it can be deduced that, overall, gelatin nanoparticles serve as effective encapsulants, offering protection to the bioactive compounds found in the bitter orange peel.

3.5. Antimicrobial Activity

According to Table 2, greater inhibition of the Gram-negative bacteria (*E. coli*) was obtained with the gelatin nanoparticles containing the extract of bitter orange peel (1.967 ± 0.058 cm), possibly due to the phenolic compounds it presented. Electrostatic attractions, hydrophobic interactions, Van der Waals forces, and receptor–ligand interactions facilitate the adhesion between nanomaterials and microbial cells, resulting in the breakdown of the microorganism's cell wall. Similarly, the destruction of the cell membrane is encouraged by the production of free radicals and reactive oxygen species (ROS), which compromise the antioxidant defense system and cause mechanical damage to the cell membrane. Following this, nanomaterials interact with key cellular organelles such as DNA, enzymes, ribosomes, and lysosomes. This interaction leads to various issues, including protein deactivation, oxidative stress, altered membrane permeability, heterogeneous changes, electrolyte imbalance, and modifications in gene expression [35].

Table 2. Antimicrobial activity (zone of inhibition in centimeters) of the ethanolic extract of bitter orange peel, the empty gelatin nanoparticles, and the nanoparticles containing the extract.

Microorganism	Ethanolic Extract of Bitter Orange Peel	Empty Gelatin Nanoparticles	Nanoparticles Containing the Extract of Bitter Orange Peel
E. coli O157:H7	1.533 ± 0.321^{a}	0.667 ± 0.058^{b}	1.967 ± 0.058^{a}
L. monocytogenes	1.833 ± 0.289^{a}	0.020 ± 0.00^{b}	0.700 ± 0.300^{b}

Note: Distinct lowercase letters in the same row indicate a statistical difference ($p < 0.05$).

The extract exhibited superior inhibition against Gram-positive bacteria (*L. monocytogenes*), with a diameter of inhibition measured at 1.833 ± 0.289 cm, owing to its elevated flavonoid content. Studies indicate that aglycone-type citrus flavonoids possess higher activity compared to glycosylated counterparts [36]. Additionally, the gelatin's effect on the extract involves glycosylation, potentially leading to a reduction in both its antibacterial and antioxidant efficacy by replacing hydroxyl groups. In another research, gelatin nanoparticles containing buriti oil exhibited superior inhibition against Gram-negative bacteria *Pseudomonas aeruginosa* and *Klebsiella pneumoniae* compared to the Gram-positive microorganism *Staphylococcus aureus* [27].

4. Conclusions

Gelatin nanoparticles were prepared using coaxial electrospraying under specific conditions: a 5 cm distance between the needle and the collector, a pump flow rate of 0.1 mL/h for the polymer solution, and 0.08 mL/h for the ethanolic extract of bitter orange peel, with a voltage of 15 kV. The characterization of the nanoparticles containing an extract of bitter orange peel by SEM confirmed that they were monodisperse spheres of uniform size, while FT-IR analysis indicated that the incorporation of the extract into the gelatin nanoparticles consisted of a physical process. Additionally, it was found that encapsulation enhanced the antioxidant capacity by the $DPPH^{\bullet}$ and $ABTS^{\bullet+}$ assays. Both the extract and the nanoparticles containing extract showed effectiveness in inhibiting the Gram-negative microorganism *E. coli* O157:H7. Therefore, encapsulating the extract with gelatin provides significant benefits due to the protection of bioactive compounds for food applications.

Author Contributions: Conceptualization, A.M.G.-G., J.R.R.-E. and D.F.C.-R.; methodology, A.G.-J.; software, A.M.G.-G.; validation, A.M.G.-G., J.R.R.-E. and S.R.-C.; formal analysis, J.R.R.-E., V.M.O.-H. and J.d.J.O.-P.; investigation, A.G.-J. and A.M.G.-G.; resources, S.R.-C., C.L.D.-T.-S. and F.R.-F.; data curation, A.M.G.-G. and J.R.R.-E.; writing—original draft preparation, A.G.-J. and A.M.G.-G.; writing—review and editing, J.A.T.-H. and S.R.-C.; visualization, J.A.T.-H.; supervision, V.M.O.-H. and J.d.J.O.-P.; project administration, J.R.R.-E. and S.R.-C.; funding acquisition, J.R.R.-E. and S.R.-C. All authors have read and agreed to the published version of the manuscript.

Funding: This research received no external funding.

Institutional Review Board Statement: Not applicable.

Informed Consent Statement: Not applicable.

Data Availability Statement: The original contributions presented in the study are included in the article; further inquiries can be directed to the corresponding authors.

Acknowledgments: Adamaris García-Juárez expresses gratitude to the Department of Chemical-Biological Sciences at the University of Sonora for supporting this investigation and its scientific dissemination.

Conflicts of Interest: The authors declare no conflicts of interest.

References

1. Mannucci, C.; Calapai, F.; Cardia, L.; Inferrera, G.; D'Arena, G.; Di Pietro, M.; Navarra, M.; Gangemi, S.; Ventura-Spagnolo, E.; Calapai, G. Clinical Pharmacology of *Citrus aurantium* and *Citrus sinensis* for the Treatment of Anxiety. *Evid. Based Complement. Alternat. Med.* **2018**, *2018*, 3624094. [CrossRef] [PubMed]
2. Mazorra-Manzano, M.A.; Moreno-Hernández, J.M.; Ramírez-Suarez, J.C.; de Jesús Torres-Llanez, M.; González-Córdova, A.F.; Vallejo-Córdoba, B. Sour orange *Citrus aurantium* L. flowers: A new vegetable source of milk-clotting proteases. *LWT-Food Sci. Technol.* **2013**, *54*, 325–330. [CrossRef]
3. Sarrou, E.; Chatzopoulou, P.; Dimassi-Theriou, K.; Therios, I. Volatile constituents and antioxidant activity of peel, flowers and leaf oils of *Citrus aurantium* L. growing in Greece. *Molecules* **2013**, *18*, 10639–10647. [CrossRef] [PubMed]
4. Farag, M.A.; Abib, B.; Ayad, L.; Khattab, A.R. Sweet and bitter oranges: An updated comparative review of their bioactives, nutrition, food quality, therapeutic merits and biowaste valorization practices. *Food Chem.* **2020**, *331*, 127306. [CrossRef]
5. Karimi, E.; Oskoueian, E.; Hendra, R.; Oskoueian, A.; Jaafar, H.Z. Phenolic compounds characterization and biological activities of *Citrus aurantium* bloom. *Molecules* **2012**, *17*, 1203–1218. [CrossRef]
6. Pereira, J.A.; Berenguer, C.V.; Câmara, J.S. Delving into Agri-Food Waste Composition for Antibacterial Phytochemicals. *Metabolites* **2023**, *13*, 634. [CrossRef]
7. Gonçalves, L.A.; Lorenzo, J.M.; Trindade, M.A. Fruit and agro-industrial waste extracts as potential antimicrobials in meat products: A brief review. *Foods* **2021**, *10*, 1469. [CrossRef] [PubMed]
8. Sha, S.P.; Modak, D.; Sarkar, S.; Roy, S.K.; Sah, S.P.; Ghatani, K.; Bhattacharjee, S. Fruit waste: A current perspective for the sustainable production of pharmacological, nutraceutical, and bioactive resources. *Front. Microbiol.* **2023**, *14*, 1260071. [CrossRef]
9. Fathi, M.; Mozafari, M.R.; Mohebbi, M. Nanoencapsulation of food ingredients using lipid based delivery systems. *Trends Food Sci. Technol.* **2012**, *23*, 13–27. [CrossRef]
10. Tapia-Hernández, J.A.; Torres-Chávez, P.I.; Ramírez-Wong, B.; Rascón-Chu, A.; Plascencia-Jatomea, M.; Barreras-Urbina, C.G.; Rangel-Vázquez, N.A.; Rodríguez-Félix, F. Micro-and nanoparticles by electrospray: Advances and applications in foods. *J. Agric. Food Chem.* **2015**, *63*, 4699–4707. [CrossRef]
11. Hassan, N.A.; Darwesh, O.M.; Smuda, S.S.; Altemimi, A.B.; Hu, A.; Cacciola, F.; Haoujar, I.; Abedelmaksoud, T.G. Recent trends in the preparation of nano-starch particles. *Molecules* **2022**, *27*, 5497. [CrossRef] [PubMed]
12. Tanhaei, A.; Mohammadi, M.; Hamishehkar, H.; Hamblin, M.R. Electrospraying as a novel method of particle engineering for drug delivery vehicles. *J. Control. Release* **2021**, *330*, 851–865. [CrossRef] [PubMed]
13. Zhang, L.; Huang, J.; Si, T.; Xu, R.X. Coaxial electrospray of microparticles and nanoparticles for biomedical applications. *Expert Rev. Med. Devices* **2012**, *9*, 595–612. [CrossRef] [PubMed]
14. Borrás-Enríquez, A.J.; Reyes-Ventura, E.; Villanueva-Rodríguez, S.J.; Moreno-Vilet, L. Effect of ultrasound-assisted extraction parameters on total polyphenols and its antioxidant activity from mango residues (*Mangifera indica* L. var. Manililla). *Separations* **2021**, *8*, 94. [CrossRef]
15. Torkamani, A.E.; Syahariza, Z.A.; Norziah, M.H.; Mahmood, W.A.K.; Juliano, P. Production and characterization of gelatin spherical particles formed via electrospraying and encapsulated with polyphenolic antioxidants from *Momordica charantia*. *Food Bioprocess Technol.* **2018**, *11*, 1943–1954. [CrossRef]
16. Tapia-Hernández, J.A.; Del-Toro-Sánchez, C.L.; Cinco-Moroyoqui, F.J.; Ruiz-Cruz, S.; Juárez, J.; Castro-Enríquez, D.D.; Barreras-Urbina, C.G.; López-Ahumada, G.A.; Rodríguez-Félix, F. Gallic acid-loaded zein nanoparticles by electrospraying process. *J. Food Sci.* **2019**, *84*, 818–831. [CrossRef]
17. Estrella-Osuna, D.E.; Tapia-Hernández, J.A.; Ruíz-Cruz, S.; Márquez-Ríos, E.; Ornelas-Paz, J.D.J.; Del-Toro-Sánchez, C.L.; Ocaño-Higura, V.M.; Rodríguez-Féliz, F.; Estrada-Alvarado, M.I.; Cira-Chávez, L.A. Nanoencapsulation of Eggplant (*Solanum melongena* L.) Peel Extract in Electrospun Gelatin Nanofiber: Preparation, Characterization, and In Vitro Release. *Nanomaterials* **2022**, *12*, 2303. [CrossRef] [PubMed]
18. Tapia-Hernández, J.A.; Rodríguez-Félix, D.E.; Plascencia-Jatomea, M.; Rascón-Chu, A.; López-Ahumada, G.A.; Ruiz-Cruz, S.; Barreras-Urbina, C.G.; Rodríguez-Félix, F. Porous wheat gluten microparticles obtained by electrospray: Preparation and characterization. *Adv. Polym. Technol.* **2018**, *37*, 2314–2324. [CrossRef]
19. Rodríguez-Félix, F.; Del-Toro-Sánchez, C.L.; Tapia-Hernández, J.A. A new design for obtaining of white zein micro-and nanoparticles powder: Antisolvent-dialysis method. *Food Sci. Biotechnol.* **2020**, *29*, 619–629. [CrossRef] [PubMed]
20. Garzón-García, A.M.; Ruiz-Cruz, S.; Dussán-Sarria, S.; Hleap-Zapata, J.I.; Márquez-Ríos, E.; Del-Toro-Sánchez, C.L.; Tapia-Hernández, J.A.; Canizales-Rodríguez, D.F.; Ocaño-Higuera, V.M. Effect of UV-C Postharvest Disinfection on the Quality of Fresh-Cut Tommy Atkins' Mango. *Pol. J. Food Nutr. Sci.* **2023**, *73*, 39–49. [CrossRef]
21. Del-Toro-Sánchez, C.L.; Rodríguez-Félix, F.; Cinco-Moroyoqui, F.J.; Juárez, J.; Ruiz-Cruz, S.; Wong-Corral, F.J.; Borboa-Flores, J.; Castro-Enríquez, D.D.; Barreras-Urbina, C.G.; Tapia-Hernández, J.A. Recovery of phytochemical from three safflower (*Carthamus tinctorius* L.) by-products: Antioxidant properties, protective effect of human erythrocytes and profile by UPLC-DAD-MS. *J. Food Process. Preserv.* **2021**, *45*, e15765. [CrossRef]
22. González-Vega, R.I.; Cárdenas-López, J.L.; López-Elías, J.A.; Ruiz-Cruz, S.; Reyes-Díaz, A.; Perez-Perez, L.M.; Cinco-Moroyoqui, F.J.; Robles-Zepeda, R.E.; Borboa-Flores, J.; Del-Toro-Sánchez, C.L. Optimization of growing conditions for pigments production from microalga *Navicula incerta* using response surface methodology and its antioxidant capacity. *Saudi J. Biol. Sci.* **2021**, *8*, 1401–1416. [CrossRef] [PubMed]

23. Perez-Perez, L.M.; Huerta-Ocampo, J.Á.; Ruiz-Cruz, S.; Cinco-Moroyoqui, F.J.; Wong-Corral, F.J.; Rascón-Valenzuela, L.A.; Robles-García, M.A.; González-Vega, R.I.; Rosas-Burgos, E.C.; Corella-Madueño, M.A.G.; et al. Evaluation of quality, antioxidant capacity, and digestibility of chickpea (*Cicer arietinum* L. cv Blanoro) stored under N_2 and CO_2 atmospheres. *Molecules* **2021**, *26*, 2773. [CrossRef] [PubMed]
24. Ruiz-Cruz, S.; Acedo-Félix, E.; Díaz-Cinco, M.; Islas-Osuna, M.A.; González-Aguilar, G.A. Efficacy of sanitizers in reducing *Escherichia coli* O157: H7, *Salmonella* spp. and *Listeria monocytogenes* populations on fresh-cut carrots. *Food Control* **2007**, *18*, 1383–1390. [CrossRef]
25. Balouiri, M.; Sadiki, M.; Ibnsouda, S.K. Methods for *in vitro* evaluating antimicrobial activity: A review. *J. Pharm. Anal.* **2016**, *6*, 71–79. [CrossRef]
26. Figueroa-Enriquez, C.E.; Rodríguez-Félix, F.; Plascencia-Jatomea, M.; Sánchez-Escalante, A.; Vargas-López, J.M.; Tapia-Hernández, J.A.; Canizales-Rodríguez, D.F.; Castro-Enriquez, D.D.; Ruiz-Cruz, S.; Santos-Sauceda, I.; et al. Nanoparticles of Betalain–Gelatin with Antioxidant Properties by Coaxial Electrospraying: Preparation and Characterization. *ACS Omega* **2023**, *8*, 41156–41168. [CrossRef]
27. Castro, G.M.M.A.; Passos, T.S.; Nascimento, S.S.D.C.; Medeiros, I.; Araújo, N.K.; Maciel, B.L.L.; Padilha, C.E.; Ralmalho, A.M.Z.; Júnior, F.C.S.; de Assis, C.F. Gelatin nanoparticles enable water dispersibility and potentialize the antimicrobial activity of Buriti (*Mauritia flexuosa*) oil. *BMC Biotechnol.* **2020**, *20*, 55. [CrossRef]
28. Abedi Ostad, M.; Arezuman, R.; Oroojalian, F.; Hanafi, A.; Amani, A. Introducing humidity and temperature as important parameters determining the size of chitosan nanoparticles prepared by electrospray. *Nanomed. Res. J.* **2021**, *6*, 385–395.
29. Pandia-Estrada, S.; Romero-Santivañez, R.; Céspedes-Chombo, R.; Solari-Godiño, A. Edible films gelatin-based obtained from mahi-mahi skin (*Coryphaena hippurus*) and oregano extract: Physicochemical, antimicrobial, structural and surface characteristics. *Sci. Agropecu.* **2021**, *12*, 229–237. [CrossRef]
30. Liang, S.; Guo, X.; Feng, N.; Tian, Q. Isotherms, kinetics and thermodynamic studies of adsorption of Cu^{2+} from aqueous solutions by Mg^{2+}/K^{+} type orange peel adsorbents. *J. Hazard. Mater.* **2010**, *174*, 756–762. [CrossRef]
31. Kwak, H.W.; Kim, J.E.; Lee, K.H. Green fabrication of antibacterial gelatin fiber for biomedical application. *React. Funct. Polym.* **2019**, *136*, 86–94. [CrossRef]
32. Bibi Sadeer, N.; Montesano, D.; Albrizio, S.; Zengin, G.; Mahomoodally, M.F. The versatility of antioxidant assays in food science and safety—Chemistry, applications, strengths, and limitations. *Antioxidants* **2020**, *9*, 709. [CrossRef]
33. Divya, P.J.; Jamuna, P.; Jyothi, L.A. Antioxidant properties of fresh and processed *Citrus aurantium* fruit. *Cogent Food Agric.* **2016**, *2*, 1184119. [CrossRef]
34. Dias, M.C.; Pinto, D.C.; Silva, A.M. Plant flavonoids: Chemical characteristics and biological activity. *Molecules* **2021**, *26*, 5377. [CrossRef] [PubMed]
35. Basavegowda, N.; Baek, K.H. Combination strategies of different antimicrobials: An efficient and alternative tool for pathogen inactivation. *Biomedicines* **2022**, *10*, 2219. [CrossRef] [PubMed]
36. Miklasińska-Majdanik, M.; Kępa, M.; Wojtyczka, R.D.; Idzik, D.; Wąsik, T.J. Phenolic compounds diminish antibiotic resistance of *Staphylococcus aureus* clinical strains. *Int. J. Environ. Res. Public Health* **2018**, *15*, 2321. [CrossRef]

foods

MDPI

Article

Dry-Cured Sausages "Salchichón" Manufactured with a Valorized Ingredient from Red Grape Pomace (Var. Tempranillo)

Matilde D'Arrigo [1], María Jesús Petrón [2,*], Jonathan Delgado-Adámez [1], Jesús Javier García-Parra [1], María Jesús Martín-Mateos [1] and María Rosario Ramírez-Bernabé [1,*]

[1] Centro de Investigaciones Científicas y Tecnológicas de Extremadura (CICYTEX), Instituto Tecnológico Agroalimentario (INTAEX), 06007 Badajoz, Spain; matilde.darrigo@juntaex.es (M.D.); jonathan.delgado@juntaex.es (J.D.-A.); jesusjavier.garcia@juntaex.es (J.J.G.-P.); mariajesus.martinmat@juntaex.es (M.J.M.-M.)

[2] Escuela Ingenierías Agrarias, Universidad de Extremadura, 06006 Badajoz, Spain

* Correspondence: mjpetron@unex.es (M.J.P.); mariarosario.ramirez@juntaex.es (M.R.R.-B.)

Abstract: The inclusion of an ingredient made from red grape pomace (RGP) var. Tempranillo was evaluated for the preservation of a traditional dry-cured sausages (salchichón). The pomace was valorized through thermal blanching (103 °C for 1 min) and hydrostatic high-pressure treatment (600 MPa/5 min) before the addition to salchichón. Four formulations of salchichón were evaluated, including a negative control (NC—without red grape pomace or synthetic additives), positive control (PC—with ascorbic acid and nitrites), low level (LL—0.5%), and high level (HL—1%) of RGP. Physicochemical, microbiological, and sensorial effects were analyzed. RGP reduced the final pH of salchichón and favored the growth of lactic acid bacteria at similar levels as PC. The addition of ascorbic acid and nitrites resulted in a final product with a redder and less yellow color than the other formulations. This cured color was not reached with the addition of RGP. However, its inclusion slightly reduced lipid and protein oxidation in salchichón. PC showed high levels of sulfur and terpene levels in a volatile profile, although at a sensory level, only differences in spicy taste were not noticed by panelists. The incorporation of the ingredient could enable the substitution of nitrites with valorized red grape pomace in sausages, although the desirable color achieved with nitrifying salts was not fully attained.

Keywords: dry-cured sausages; red grape pomace; phenolic compounds; antioxidant; nitrifying salts

1. Introduction

The "salchichón" is a fermented dry-cured sausage which has a prolonged process of drying and ripening before consumption. It is generally made from a mixture of chopped meat (pork and/or beef) and lard, is seasoned with spices such as salt, black pepper, additives (nitrate, nitrite, and antioxidants), and optionally has starter cultures. This process originates physico-chemical [1] and microbial changes, including fermentation, dehydration [2], color changes [3], lipolysis [4], and proteolysis [5]. Oxidation reactions in meat and meat products during processing and storage under conventional settings are common. Maturation favors the formation of the typical aroma compounds of dry-cured meat products, which can be provided by the seasoning added and by the main pathways of lipid-oxidation reactions [6]. However, an excess of lipid oxidation can lead to the formation of hydroperoxides, promoting the generation of some negative volatile compounds, such as some aldehydes, ketones, acids, and alcohols [7]. Additionally, the oxidation process of proteins and amino acids may decrease their bioavailability, digestibility, solubility, and proteolytic activity [5,8]. Oxidation reactions can also adversely change the product's appearance by oxidizing myoglobin to oxymyoglobin and metmyoglobin and producing brown pigments [9]. Sensory and nutritional attributes, as well as consumer satisfaction, could be negatively altered by these factors.

The addition of nitrate and nitrite salts has a key role in the inhibition of foodborne pathogens, like the Gram-positive spore-forming anaerobic bacteria *Clostridium botulinum* and the most resistant Gram-negative aerobic/facultative anaerobic bacteria (*Escherichia coli* and *Salmonella*), which are inhibited by nitrite salts [10]. On the other hand, nitrifying salts contribute to the development of the typical flavor and red color, also retarding lipid oxidation in cured meat products [11–13]. Despite the significant technological and safety goals of nitrites, they can react with secondary amines in the gastrointestinal tract to form carcinogenic N-nitrosamines, which are linked to the incidence of some types of cancer [14].

Consequently, given the potential health risks related to the residual nitrite in meat products, consumers are demanding nitrite-free products that maintain sensory and microbiological quality [13] or technological strategies that involve the applications of natural antioxidants directly to meat and meat products [15]. Researchers and meat industries search for clean labels on the possibility of reducing [16] and substituting [17,18] sodium nitrite. A substantial source of bioactive and natural additives, or their extracts, can be obtained from agro-industrial by-products, thereby creating a circular and sustainable economy.

This interest of consumers in sustainable industrial practice, along with the benefits of reducing industrial waste discharge, increases the economy of reuse systems or valorization strategies. Grape pomace is the main by-product of winemaking, accounting for 62% of organic waste, and mainly consists of skin, pulp, stalk residuals, and seeds. The most common destinations, in Europe, for solid grape residues are land spreading, distillation, incineration, serving as a source of tartaric acid, oil recovery, or animal feeding [19]. Nevertheless, this material contains several compounds (fiber, proteins, fat, and phenolic compounds) with health benefits that may add value to products in the food, cosmetics, or pharmaceutical industries [20]. In addition, this residue may be used for other purposes, such as natural dyes, preservatives, and/or antioxidants in meat products [18,21].

Red grape pomace (RGP) is recognized for its rich content of polyphenols, mainly catechins, epicatechins, gallic acid, and procyanidins, with a multitude of biological properties, such as antioxidant, antimicrobial, or anticancer activities [22,23]. However, its bioactive compounds content is variable, and it is influenced by environmental factors and grape varieties, among others [24]. For better recovery of valuable compounds from grape pomace, many extraction techniques have been studied, such as Soxhlet extraction, maceration, supercritical fluid extraction, subcritical water extraction, and ultrasound-assisted extraction [22,25], but today, these are being replaced by the so-called non-conventional methods. Traditional methods often require lengthy procedures and a variety of specialized solvents. In contrast, new techniques prioritize sustainability and efficient preservation of these valuable components. In this sense, HHP (hydrostatic high-pressure) is considered to be energy-efficient and environmentally friendly in food processing [26,27]. In previous studies, HHP has been described as a suitable technology for the valorization of the red/white wine pomace to produce a possible ingredient where antioxidant and antimicrobial properties—phenolic compounds—could be maintained after processing [28,29]. HHP allowed for a sustainable valorization of grape pomace since no solvents are required, and the whole by-product is reused, thus generating no residues. However, HHP does not always reduce the activity of the polyphenol oxidase, necessitating other hurdles (such as thermal blanching) to stabilize grape pomace [29–31].

Several studies have examined the impact of grape pomace as an alternative ingredient to reduce the quantity of nitrite in different meat and meat products. Regarding dry fermented sausages, there are only a few studies that have used grape-seed extract in dry-cured sausage [12], in Spanish "Chorizo" [9,32], in "Cinta Senese" or Italian dry fermented sausage [17,33], and with grape-seed flour in Turkish dry fermented sausage or "Sucuk" [34]. These extracts were found to be effective inhibitors of oxidative reactions, with TBARS (Thiobarbituric acid-reactive substance) values under 1 mg MDA (malondialdehyde) kg^{-1}, which is considered as the threshold of sensory perception for lipid oxidation, thus not producing "off-odours" [9]. Moreover, extracts of grape pomace helped to avoid the growth of food pathogens (*Listeria monocytogenes*, *Salmonella* spp., *Staphylococcus aureus*, coliforms, and *E. coli*). Phenolic compounds exhibit antibacterial properties by depriving microbes of essential nutrients like iron and disrupting key microbial processes [15]. They specifically target the microbial cell membrane, causing structural damage and increasing permeability. This allows the phenolic compounds to penetrate the bacterial cell, leading to cell lysis, the release of intracellular ATP, and the loss of vital cellular contents. The antimicrobial and antioxidant properties of phenolic compounds make them viable alternatives to nitrates and nitrites in meat products. A study [17] showed that adding dry grape pomace extract to boiled sausages reduced the nitrite levels while maintaining both microbiological stability and sensory quality. This suggests that grape pomace can be used to lower the amount of curing salts with a toxic potential without affecting the taste or quality of the meat.

Therefore, most of the reported studies deal with the use of polyphenol extracts from grape pomace, seed, or skin to promote antioxidant and antimicrobial activity in meat products. Nevertheless, the integral use of RGP provides the advantage of incorporating dietary fibers together with phenolic compounds. This is particularly beneficial since dietary fibers were identified as predominant compounds in red grape pomace [23]. To the best of our knowledge, no previous studies have investigated the impact of incorporating integral red grape pomace (RGP) stabilized with the HHP technique for preservation in dry-cured sausages. Therefore, the main objective of this study is to evaluate the incorporation of valorized RGP from Tempranillo var. into dry-cured sausages to improve their preservation and replace the use of synthetic additives like nitrifying salts.

2. Materials and Methods

2.1. Manufacture of the Ingredient from Red Grape Pomace (RGP)

In September 2022, a wine manufacturing company located in Santa Marta de los Barros (Badajoz, Spain) provided red grape pomace (cv *Tempranillo*). About 5 kg of red grape pomace were collected ("initial" pomace). The process includes thermal blanching, grounding, vacuum packaging, and finally, HHP. Before high-pressure processing (600 MPa/5 min), RGP was thermally blanched (TB) to inactivate the PPO enzyme by an exhausting unit (Chaconsa, Murcia, Spain) applying steam at 103 °C for 1 min of residence. These conditions were chosen based on our previous experiments, where scalding was applied for durations ranging from 1 to 5 min, followed by an evaluation of polyphenol oxidase enzyme activity and total phenol content. The enzyme activity of the PPO was decreased completely at 1 min of blanching with a stabilized phenol content [31].

The thermally blanched RGP was vacuum packaged and frozen at −18 °C for 24 h and then crushed using an Ultra Centrifugal Mill (RETSCH ZM200, Haan, Germany) until achieving a fine powder. This powder of RGP was vacuum packaged into Eurobag plastic bags with the following characteristics: polyamide/polyethylene 20/100, oxygen permeability of 50 cm^3 m^{-2} 24 h^{-1}, 0% relative humidity, and 120 µm of thickness. Immediately, the packaged ground pomace was processed in semi-industrial equipment (6000/55, Hiperbaric, S.A., Burgos, Spain), with a container capacity of 55 L at 600 MPa for 5 min at 16 °C of initial temperature of the water (TB + HHP pomace). The ingredient from red grape pomace was stored at −80 °C until being added in different proportions into dry-cured sausages.

To perform the analysis of the initial pomace and the TB pomace samples, they were finely crushed after being well-frozen in a domestic blender Thermomix TM-6 (Vorwerk, Vorwerk, Germany) at maximum speed for 1–2 min. In the same way, it was carried out for the TB + HHP pomace.

2.2. Dry-Cured Sausages Preparation

Meat (12 kg) for the manufacture of the dry-cured sausages was acquired in a local market. Minced meat (60% minced pork, 40% minced fresh bacon) was mixed with the following ingredients: 12 g kg^{-1} fresh minced garlic, 20 g kg^{-1} sodium chloride, 1 g kg^{-1} of nutmeg, 1 g kg^{-1} white pepper and 1 g kg^{-1} black pepper. Masses of 3 kg were prepared consecutively for each formulation: negative control (NC), positive control (PC), low-level pomace (LL), and high-level pomace (HL). First, the NC batch (3 kg) was prepared with the previous formulation. The PC batch (3 kg) was prepared with the previous mixture and by adding synthetic additives (0.5 g kg^{-1} of acid L-ascorbic (Laffort, Valladolid, Spain) and 0.1 g kg^{-1} of sodium nitrite (Panreac, Barcelona, Spain)). The LL batch was prepared with the initial recipe and the RGP ingredient (0.5% *w*/*w*), and the HL batch was made with the initial recipe and the RGP ingredient (1% *w*/*w*). Levels of pomace were chosen according to previous studies, taking the highest levels that were not rejected at the sensory level [31]. All the masses were mixed using an automatic vacuum mixer (Talleres Cato, Barcelona, Spain) for 5 min and stored in refrigeration for 1–2 h, and then, these were stuffed into a 45–50 mm diameter natural dried pork casing.

Once the sausages were formed, they were dried in the industrial dryer of our pilot plant at temperatures between 12 °C and 14 °C and 85% RH for 21 days. The weight losses at the end of the maturation process were between 43.3–54.1% (Figure 1). Seven sausages per batch were prepared, although 5 per batch were analyzed. Therefore, a total of 20 sausages were evaluated. At the end of the maturation, each sausage reached a weight of around 190 g.

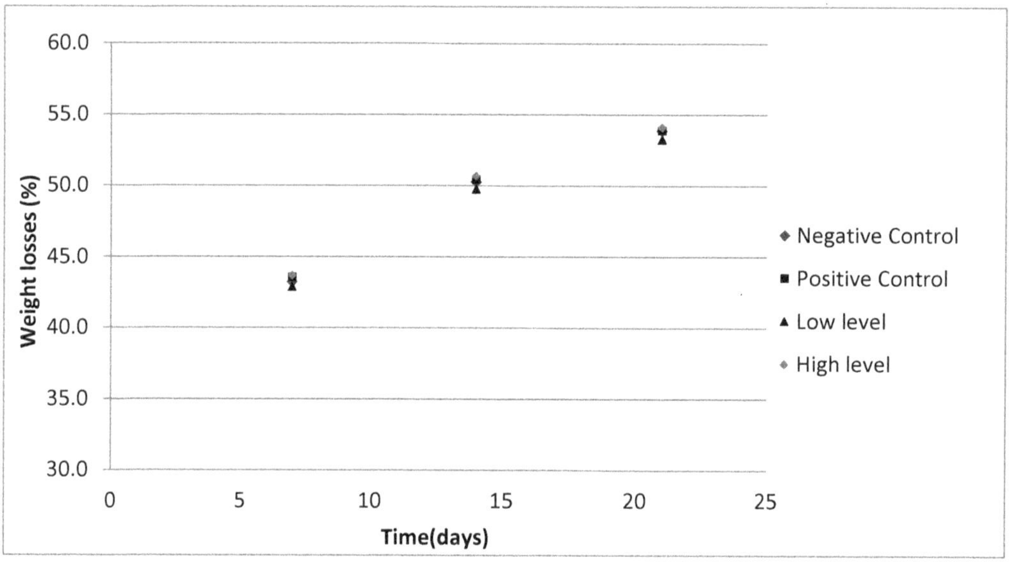

Figure 1. Weight loss (%) of dry-cured sausages during maturation.

2.3. Analysis of the Ingredient from RGP

The phenolic compound content was determined by the Folin–Ciocalteu reagent-based colorimetric assay. The absorbance was measured at 765 nm using a Thermo-Evolution 201 spectrophotometer (Fisher Scientific SL, Madrid, Spain). A calibration curve using Gallic acid as the reference standard was generated. The total phenolic content was expressed as the Gallic acid equivalents per sample (wet base) (mg GAE 100 g^{-1}).

Polyphenol oxidase (PPO) activity was measured at 420 nm and 25 °C for 3 min in a Thermo Scientific Evolution UV Vis spectrophotometer (Fisher Scientific SL, 187 Madrid, Spain), in a kinetic model. The results were expressed as a percentage of activity with respect to the control samples.

The composition of the initial pomace was analyzed in 3 independent vacuum-packaged bags. For pH and water activity measurements of the pomace, a pHmeter Crison pH 25 + (Crison, Barcelona, Spain) and a Novasina (Labmaster, Lachen, Switzerland) were used. Moisture and protein analyses were conducted according to the AOAC methodology [35]. The fat content was assessed by the Folch method [36] and fiber according to the modified Southgate method [37], all of them in wet base (WB). Fatty acid methyl esters (FAMEs) from the initial pomace were analyzed using an Agilent 6890 gas chromatograph (Agilent Technologies, Santa Clara, CA, USA), equipped with a flame ionization detector (FID) and a fused silica column (60 mm length, 0.25 mm inner diameter, and 0.25 m film thickness). The injector and detector temperatures were 260 °C and 280 °C, respectively. The column oven temperature was raised to 220 °C on a ramp temperature, and helium was used as a carrier gas, with a constant flow of 1.2 mL min^{-1} and make-up of 25 mL min^{-1}. The injection mode was used with a split ratio of 1:100. Individual FAME identification was carried out on the basis of Sigma standards (Supelco 37 component FAME mix standard, Sigma Aldrich, St. Louis, MO, USA) compared with the retention times obtained. The results are expressed as a percentage of total fatty acid methyl esters.

For microbiological analysis, 10 g of RPG was aseptically weighted in a sterile plastic bag and homogenized with 90 mL of a sterile solution peptone water (Merck, Darmstadt, Germany) in a masticator blender (Stomacher 400 Circulator, Seward, West Sussex, UK) at a 1/10 dilution (*w*/*v*). Mesophilic, molds and yeasts and *Enterobacteriaceae* counts of RGP were determined. All microbial counts were expressed as the log of colony-forming units (CFU) per g of sample weight (log CFU g^{-1}).

2.4. Analysis of Dry-Cured Sausage

The analysis of pH, proximate composition, water activity, and FAMEs followed the procedure previously described.

For the microbiological analysis, 10 g of sausage were taken and homogenized with 90 mL of sterile peptone water. Serial decimal dilutions were prepared in sterile peptone water, and 1 mL of each sample was spread on suitable culture media. Mesophilic aerobic counts were analyzed by a standard Plate Count Agar (Merck, 1.07881), and the plates were incubated at 30 °C for 72 h. Psychrophilic counts were determined on Plate Count Agar (Merck, 1.07881) after incubation at 7 °C for 10 days. Lactic Acid Bacteria (LAB) were incubated on Man Rogosa Sharpe Agar (MRSA, Scharlau, Barcelona, Spain) at 37 °C for 72 h; anaerobic sulfite-reducing *Clostridium* spp. were incubated on Tryptose Sulfite Cycloserine Agar (Merck, 1.10235) at 37 °C for 24 h. Staphylococcus aureus was determined on Baird Parker Agar (Merck, 1.05406) after incubation at 37 °C for 24–48 h. The total coliforms and *E. coli* and were incubated on Chromocult Agar (Merck, 1.10426) at 37 °C for 24–48 h. The molds and yeasts were incubated at 25 °C for 4–5 days. Finally, *Salmonella* and *L. monocytogenes* were determined according to ISO 6579, 1993 and ISO 11290-1, 1996, respectively. Microbial counts were expressed as the log of colony-forming units (CFU) per g of sample weight (log CFU g^{-1}) and the absence of *L. monocytogenes* or *Salmonella* spp. in 25 g of sample.

For the instrumental color, the color coordinates lightness (L*), redness (a* red–green axis), and yellowness (b* yellow–blue axis) in the CIE Lab color space were analyzed. In addition, the Hue angle was calculated ($h^\circ = \tan^{-1}(b^*/a^*)$), as well as the saturation index or Chroma (C*) ($C = (a^{*2} + b^{*2})^{0.5}$). One slice of one-centimeter thickness of the sample was prepared. Two readings were recorded (one on each side of the slice), and a value was determined as the mean of the readings.

For oxidative stability, lipid oxidation was assessed by thiobarbituric acid reactive substances (TBA-RS) according to Sørensen and Jørgensen (1996) [38]. TBARS values were calculated from the standard (1,1,3,3-Tetraethoxypropane, TEP) curve, and the results were expressed as mg of malondialdehyde per kg of sample (mg MDA kg^{-1}). Protein oxidation was obtained by measuring the carbonyl groups formed during incubation with 2,4-Dinitrophenylhydrazine (DNPH) in 2 N HCl following the method described by Oliver,

Ahn, Moerman, Goldstein, and Stadtman (1987) [39]. Protein oxidation was expressed as nmol carbonyls mg protein^{-1}.

For sensory evaluation, a trained sensory panel was formed by eight judges with specific training in the sensory analysis of dry-cured products. Two slices from each dry-cured sausage were presented to each panelist. At each session, one dry-cured sausage from each batch was analyzed, so a total of five sessions were required. The following descriptors were examined: lean color (uncured–cured) and fat color (white–yellow), intensity "salchichón" odor, unpleasant odor, hardness, juiciness, unpleasant texture, salty, spicy, aromatic intensity, and unpleasant taste/flavor. The intensity of each parameter was calculated from 0 = low intensity to 10 points = high intensity. Panelists assessed the different parameters by a quantitative–descriptive analysis with a structured scale (0–10). Data were collected using the FIZZ software, 2.45 A version (sensory analysis and computer test management) (Biosystemes, Couternon, France). All sessions were conducted at room temperature in a sensory room equipped with white, fluorescent lighting. About 100 mL of water at room temperature were provided to the panelists between samples.

For the volatile compounds, two grams of ground sausages were placed into a 20 mL vial (screw-capped with a Teflon-silicone septum). A 1 cm 50/30 µm DVB/CAR/PDMS SPME fiber (Supelco, Bellefonte, PA, USA) was utilized for the analysis of volatile compounds in the headspace of the previously prepared vial. The fiber was exposed to the headspace at 37 °C for 30 min.

For the separation of volatile compounds, a Varian CP-3800 gas chromatograph with a CombiPAL autosampler (CTC Analytics, Zwingen, Switzerland) and with an HP-5 capillary column (30 m × 0.32 mm × 0.25 µm; Agilent Technology, Santa Clara, CA, USA) was used. The injection port was at 270 °C, and the oven temperature was held at 35 °C for 10 min, increased to 250 °C (7 °C min^{-1}), and held for 5 min, with a total running time of 45 min. The temperature of the transfer line, trap, and manifold were 280 °C, 200 °C, and 60 °C, respectively. Identification was carried out in a Varian Saturn 2200 MS mass spectrometer (Varian Inc., Palo Alto, CA, USA), and mass spectra were obtained by electronic impact at 70 eV, with one scan s^{-1} over the 40–300 m/z range.

The volatiles were identified by comparing their mass spectra and linear retention indexes (LRI) with commercial standards (Sigma-Aldrich, St. Louis, MO, USA) or by mass spectra identification using the NIST library (Agilent MSD Chemstation E.02.01.1177 software). The concentration of compounds was estimated by using the internal standard (4-methyl-1-pentanol), which was added to each sample (final concentration: 1.22 mg kg^{-1}) and expressed as µg kg^{-1}.

2.5. Statistical Analysis

The analyses of the RGP were performed in triplicate (three bags per batch), and the mean values and their standard deviations (SD) were calculated. In the assay of dry-cured sausages, five samples per batch were evaluated. In order to evaluate the changes during processing in grape pomace (Table 1), a Student's t-test was applied between initial pomace vs. TB pomace (p-value initial-TB); another Student's t-test was applied between TB vs. HHP pomace (p-value TB-HHP). Finally, an ANOVA was applied to evaluate differences among the three groups (p-value treatments; and then a Tukey's HSD test was applied when differences were significant). One-way ANOVA was employed to find differences among treatments using the SPSS 21.0 statistical program (SPSS Inc., Chicago, IL, USA). If the ANOVA detected significant differences between mean values, these were compared using Tukey's test ($p < 0.05$). Principal component analysis (PCA) was carried out with instrumental color and oxidative parameters to evaluate the relationships among the samples of the four groups.

Table 1. Changes in the total phenolic compounds content (PPC, mg 100 g^{-1}) and polyphenoloxidase (PPO, % percentage respect to the initial) enzyme during the process of manufacture of the ingredient of red grape pomace.

	Initial (1)	TB (2)	TB + HHP (3)	*p*-Value (1–2)	*p*-Value (2–3)	*p*-Value (1–2–3)
PPC (mg 100 g^{-1})	467.6 ± 3.3	882.0 ± 41.5	379.5 ± 43.7	<0.001	<0.001	<0.001
PPO (%)	100.0 ± 4.9	0.0 ± 0.0	0.0 ± 0.0	<0.001	0.341	<0.001

TB: thermal blanching; HHP hydrostatic high pressure; *p*-value (1–2) (Student's *t*-test between initial purée and TB purée); *p*-value (2–3) (Student's *t*-test between TB purée and HHP purée); *p*-value (1–2–3) (Tukey's HSD test between initial, TB, and HHP). Different letters in the same row indicate significant differences in the Tukey test ($p < 0.05$).

3. Results and Discussion

3.1. Valorization Process and Composition of the Red Grape Pomace

The valorization process consisted of a thermal blanching, crushing, and a hydrostatic high pressure (HHP) treatment. Previous studies were carried out to know the effect of HHP on RGP and demonstrated that HHP treatment does not reduce PPO activity, resulting in a reduction of bioactive compounds in the valorized products during storage [29,30]. For that reason, the grape pomace was thermally blanched (TB) with the purpose of inactivating the PPO enzyme before HHP treatment. The one-way analysis of the variance (Table 1) showed a complete inactivation of PPO after TB (1 min, 103 °C), and it remained inactivated until the end of the valorization process.

Table 1 shows the total phenolic compound content (PPC) during the valorization process, indicating that TB significantly increased PPC. This initial rise in the PPC of the RGP ingredient aligns with previous research, demonstrating that blanching vegetables increases their phenol content [40]. Some studies have attributed the reported PPC increase to the reduction and inactivation of polyphenol oxidase, along with the release of bound phenolic acids resulting from the breakdown of cellular constituents in the plant cell walls of the leafy vegetable [14]. On the other hand, a significant PPC reduction is observed on the RGP ingredient (TB + HHP) after HHP treatment, which is an unexpected result. Previous studies reported that HHP maintained or even increased the extraction of PCC in grape pomace [41,42]. This increase is induced by HHP through structural changes in the cell matrices, leading to the extraction of phenolic compounds [41]. In the same vein, a parallel study in our research group reported slight non-significant reductions in PCC in "Tempranillo" red grape pomace after treatments at 600 MPa for 5 min [30]. Similarly, the valorization process by HHP of white grape pomace has been recently published [31], and an important increase in PCC was observed after TB. However, the levels were well-preserved after HHP. Compared to those previous studies, in our study, the equipment used to crush the pomace was different, as it allowed the obtaining of a fine powder. The grapes were scalded and then frozen before crushing, resulting in an intense reduction of particle size before HHP treatment. This reduction in particle size was more intense than that obtained with white grape pomace, potentially facilitating the degradation of phenolic compounds by exposition to the air. At this point, an improvement in crushing methodology would be required to avoid the loss of phenolic compounds before HHP treatment.

The proximate composition of the percentage wet base (%WB), pH, and aw of the RGP ingredients are shown in Table 2.

Table 2. Physical–chemical composition and microbial counts of the valorized red grape pomace (RGP).

Valorized Red Grape Pomace			
pH	4.2	±	0.1
Aw	0.963	±	0.001
Proximate composition (g 100 g^{-1})			
Moisture	39.6	±	3.5
Protein	3.4	±	0.1
Fat	5.1	±	0.9
Fiber	50.3	±	1.5
Fatty acids profile (%)			
C12:0	0.0	±	0.1
C14:0	0.0	±	0.1
C16:0	9.8	±	0.2
C16:1	0.5	±	0.1
C17:1	0.1	±	0.0
C17:1	0.1	±	0.0
C18:0	4.1	±	0.1
C18:1	18.1	±	0.1
C18:2	66.0	±	0.5
C18:3	1.2	±	0.1
C20:0	0.0	±	0.0
C20:1	0.0	±	0.1
Microbial counts (log CFU g^{-1})			
Mesophilic	1.5	±	1.0
Molds and Yeasts	<1	±	0.0
Enterobacteriaceae	<1	±	0.0

The low pH and aw could provide long-term stability for the valorized pomace, as indicated by previous studies, which suggest a shelf life of at least 9 months [30]. This fact is particularly important for a seasonal by-product like this. Additionally, the low pH (<4.5) in the RGP ingredient also contributes to the great stability of anthocyanins [43]. The fiber was the major component in the RPG ingredient (50.3 ± 1.5%), even above moisture (39.6 ± 3.5%), followed by lipids (5.1 ± 0.9%) and proteins (3.4 ± 0.1%). All described values are within the range of those reported by Antonic et al. (2020) [23]. These data are difficult to compare, since the results of the proximate composition are often given on a dry basis or derived from flour of grape pomace [19]. And sometimes, only skins are used without seeds or stalk remains [24].

The major fatty acids found in our study were linoleic acid (C18:2 ω6, 66.0%), oleic acid (C18:1 ω9, 18.1%), palmitic acid (C16:0, 9.8%), stearic acid (C18:0, 4.1%), and linolenic acid (C18:3, 1.2%) (Table 2). The amounts of these major fatty acids were in the intervals of values indicated for grape-seed oil in previous studies. The lipid fraction, derived from the seeds, is rich in unsaturated fatty acids and powerful antioxidants, such as vitamin E [23].

From the point of view of microbial counts, the application of HHP treatment (600 MPa, 5 min) effectively reduced microbial counts in the RGP ingredient (Table 2). Microbial counts are within acceptable ranges. Previous research about the effect of HHP treatments on red and white grape pomace found that processing conditions of 600 MPa for 5 min reduced the microbial population and allowed a long shelf-life for the processed pomace [29].

3.2. *Effect of the Incorporation of the Valorized Pomace Ingredient in Dry-Cured Sausages*

Table 3 shows the proximate composition and fatty acids profile of dry-cured sausage. In general, dry-cured sausages showed a lower content of fat and moisture and a higher protein content than other similar dry-cured products [44]. Fatty acid composition presented high levels of oleic and polyunsaturated fatty acids, such as linoleic acid (C18:2) and linolenic acid (C18:3).

Table 3. Proximate composition (g 100 g^{-1}) and fatty acids profile (%) of dry-cured sausage ("salchichón") (negative control batch).

	Mean		SD
Moisture	28.1	±	1.5
Protein	40.5	±	1.9
Fat	13.5	±	2.1
C12:0	0.1	±	0.0
C14:0	1.4	±	0.0
C16:0	23.2	±	0.0
C16:1	3.1	±	0.0
C17:0	0.4	±	0.0
C17:1	0.4	±	0.0
C18:0	10.8	±	0.0
C18:1	45.6	±	0.1
C18:2	13.6	±	0.1
C18:3	0.6	±	0.0
C20:0	0.0	±	0.0
C20:1	0.9	±	0.0

The percentages shown in Table 3 are higher than those reported in other studies on dry fermented sausages [1,2,4], but our results are like those found by Moretti et al. (2004) [45] in a typical Sicilian salami under ripening-room conditions. Differences in fatty acid profile could be attributed to the different meat cuts utilized [7], the way of manufacture of the "salchichón" [1,2], or the differences in the feeding composition [44]. The fatty acid composition could affect the stability of dry-cured meat products, since unsaturated fatty acids are easily oxidized. Sausages with an RGP ingredient could not be considered fiber-enriched foods because the calculated level of fiber added from RGP is lower than 6 g per 100 g. This is the limit required by European legislation (CE N°1924/2006). However, in the present study, the objective would not be to increase the fiber levels but to obtain a substitute for nitrites through the incorporation of bioactive compounds from the pomace into the sausage.

The moisture and aw of dry-cured sausages (Table 4) were comparable to that reported in similar products [1–3,9,12,17,33,34]. Aw values of <0.89 contribute to controlling pathogenic organism growth [46]. NC sausages presented higher pH values than the rest of the groups, with PC being the batch with the lowest pH value. The pH values were consistent with those usually reported for dry-cured sausages without a natural antioxidant and with nitrite salts after fermentation and after about 3–5 weeks of dry ripening [1–3,7,44]. After fermentation, the pH values decrease due to the production of organic acids by lactic acid bacteria, and during the drying and ripening period, the pH increases again due to the release of peptides, amino acids, and ammonium from proteolytic reactions [9,12,17,32–34]. Dry-cured salchichón generally has a long shelf-life, which is determined by the low water activity and an acidic pH. The development of LAB during ripening produces lactic acid, which is responsible for the low pH of these products, which provides a microbiologically safe product. The combination of an acid pH, low moisture content, and water activity avoids the growth of other spoilage or pathogen microorganisms, so an adequate development of the fermentation and drying processes are critical points in these traditional products. In our study, the addition of the RGP ingredient in salchichón maintained the pH at similar levels as typical commercial dry-cured meat products, which is important to reach an adequate fermentation.

Table 4. Physicochemical characteristics and microbiological counts of dry-cured sausages "salchichón".

	Negative Control	Positive Control	Low Level	High Level	*p*-Value
Moisture	28.1 ± 1.5	28.3 ± 1.1	26.9 ± 1.1	27.7 ± 1.0	0.302
Aw	0.829 ± 0.011	0.833 ± 0.013	0.824 ± 0.009	0.832 ± 0.010	0.599
pH	5.8 a ± 0.1	5.5 c ± 0.0	5.7 b ± 0.0	5.7 b ± 0.0	0.000
		Microbial counts (log CFU g^{-1})			
Mesophilic	8.1 ± 0.2	7.0 ± 3.9	8.4 ± 0.1	8.7 ± 0.2	0.556
Psychrophilic	8.1 ± 0.1	8.2 ± 0.2	8.0 ± 0.2	8.1 ± 0.2	0.323
Lactic acid bacteria	7.9 b ± 0.1	8.7 a ± 0.2	8.6 a ± 0.4	8.6 a ± 0.2	0.000
Cl. perfringens	<1	0.9 ± 0.1	<1	<1	0.261
S. aureus	<2	2.0 ± 0.2	<2	1.9 ± 0.1	0.391
Total coliforms	<1	<1	<1	<1	-
E. coli	<1	<1	<1	<1	-
Molds and yeasts	4.7 ± 1.1	5.2 ± 0.7	4.5 ± 0.2	3.9 ± 0.4	0.069

Negative control (NC: sausages manufactured without nitrites/RGP); positive control (PC: sausages manufactured with nitrites); low level (LL: sausages manufactured with 0.5% (*w*/*w*) RGP); high level (sausages manufactured with 1% (*w*/*w*) RGP). Different letters in the same row indicate significant differences in the Tukey test ($p < 0.05$).

Microbiological counts of the dry-cured sausages are presented in Table 4. The initial counts (log CFU g−1) in masses before drying were mesophilic counts (7.0 ± 0.1), LAB (4.9 ± 0.1), *Clostridium perfringens* (0.9 ± 0.1), *Staphylococcus aureus* (1.9 ± 0.1), total coliforms (4.7 ± 0.2), *Escherichia coli* (2.3 ± 0.5), and molds and yeast counts (>4). The raw mixtures were in microbiological safety conditions, and these counts are considered in the range reported in comparable studies [2,7,12,32]. In the dry-cured sausages (Table 4), the counts (log CFU g^{-1}) of mesophilic and psychrophilic presented similar levels in all formulations of dry-cured sausages. However, the counts of LAB were lower in the NC group compared to the other groups, which agrees with the higher pH observed in these dry-cured sausages after maturation. The final pH of a fermented meat product is a consequence of the development of LAB during the maturation process, which is typical of the microbiota of fermented sausages [45]. Lactic aerobic bacteria produce lactic acid during the maturation process, with decreases in pH and the inhibition of microbial growth of other species and pathogenic bacteria, especially *Staphylococcus aureus* [7] or *Enterobacteriaceae* [2]. Therefore, high growth of LAB is linked with a low final pH in fermented meat products, and this low pH helps stabilize the fermented meat product microbiologically by inhibiting the growth of other pathogens [12].

The counts of *C. perfringens*, *S. aureus*, total coliforms, *E. coli*, and molds and yeasts were similar in all groups ($p < 0.05$), and they were not affected by the formulation. The counts of dry-cured sausages were in the normal range for fermented meat products, and they were in line with those reported in Sicilian Salami (Moretti et al., 2004), traditional dry-cured sausages [7,45], the dry-cured sausage "salchichón", or "chorizo" with grape pomace or its extracts [9,12,32]. Finally, *Salmonella* and *Listeria* were not detected in any group (absence in 25 g), as *S. aureus*, *E. coli*, total coliforms, and C. perfringens were also absent or inferior to the limit established by regulation (CE) nº2073/2005.

According to a previous in vitro study, the antimicrobial activity of extracts from grape seeds (*Vitis vinifera* L.) was estimated against *E. coli* and *Listeria innocua* [47], and this activity could remain unchanged after HHP [30]. This technology was utilized for the stabilization of the RGP. In fresh burgers, the ingredient from white pomace presented some antimicrobial effect, since its addition to the burger reduced the counts of molds and yeasts and total coliform [31]. In the case of dry-cured sausages, the antimicrobial effect was not evident, since most counts were similar in all groups. However, the inclusion of RGP favored the acidification of the product and the levels of LAB were comparable to the found in the PC, which was manufactured with nitrifying salts and ascorbic acid.

Nitrites control the proliferation of bacteria that causes meat spoilage [48], and they also contribute to the control of other pathogenic microorganisms, such as *E. coli* O157:H7,

S. aureus, and *Salmonella* [49]. The protective effect against pathogens by the addition of ascorbic acid and nitrites was not perceived in PC sausages. It should be considered that other types of studies, such as challenge tests, would be required to further investigate this effect. At the industrial level, ascorbic acid is generally added together with nitrifying salts to increase the stability of nitrites. Additionally, this acid influences reducing pH, as well as exhibiting antioxidant and antimicrobial activities [10].

Significant changes in color were detected among dry-cured sausages, and only the lightness was similar in all groups (Table 5). The PC group presented significantly higher a* and lower b* than the NC group. In general, the addition of the RGP ingredient to dry-cured sausages had no significant effect on color variation compared to the NC. However, a significantly lower chroma was found in the HL compared to the NC group. In addition, the redness values of dry-cured sausage with RGP (LL, HL) and NC were also similar, since the levels of pomace added were low enough to avoid color modifications in the meat product. In this line, previous studies have also reported that the L* values of dry-cured sausage were not affected by the incorporation of grape-seed extracts [17,32,33]. The redness value was like that reported by Kurt (2016) [34] in Turkish dry-cured sausage with grape-seed flour. This similarity might be justified by the color pigments present in grapes.

Table 5. Instrumental color parameters and oxidative parameters (TBA-RS and carbonyls) of dry-cured sausages "salchichón".

	Negative Control			Positive Control			Low Level			High Level			p-Value
						Instrumental color							
L*	45.0	±	2.6	44.5	±	1.7	44.4	±	1.1	42.4	±	2.1	0.197
a*	5.0 b	±	1.1	7.7 a	±	1.5	4.4 b	±	0.6	4.0 b	±	0.6	0.000
b*	12.1 a	±	1.7	9.0 b	±	1.7	10.3 ab	±	1.0	9.8 ab	±	1.1	0.019
Chroma	13.2 a	±	1.4	12.0 ab	±	1.0	11.2 ab	±	1.1	10.6 b	±	1.2	0.017
Hue	67.0 a	±	6.2	49.0 b	±	9.8	67.0 a	±	2.3	68.0 a	±	1.7	0.000
						Oxidative parameters							
TBA-RS	0.5 a	±	0.2	0.2 b	±	0.0	0.3 ab	±	0.0	0.4 ab	±	0.1	0.013
Carbonyls	3.1 a	±	1.0	2.3 ab	±	0.4	1.8 b	±	0.3	2.1 b	±	0.3	0.016

Negative control (NC: sausages manufactured without nitrites/RGP); positive control (PC: sausages manufactured with nitrites); low level (LL: sausages manufactured with 0.5% (*w*/*w*) RGP); high level (sausages manufactured with 1% (*w*/*w*) RGP). Oxidative parameters: lipid oxidation or TBA-RS (mg MDA kg^{-1}), protein oxidation (nmols carbonyls mg $protein^{-1}$). Different letters in the same row indicate significant differences in the Tukey test ($p < 0.05$).

The effect of the nitrite-enhancing a* value in PC was somewhat expected, considering the role that nitrites play in the formation of the red curing pigment nitrosomyoglobin [13]. Moreover, the PC batch exhibited a significant difference compared to the NC in terms of the b* value, which was the highest. These variations in instrumental color coordinates also resulted in changes in hue and chrome values. In line with our results, Higuero et al. (2020) [11] reported increasing values of CIE a* and reduced values of CIE b* in dry-cured loin due to the addition of nitrifying salts. Therefore, nitrifying salts are not only responsible for the higher redness but also for the lower yellowness of the dry-cured meat products.

The mean values of TBA-RS were low (<0.5 mg MDA kg^{-1}) for all groups, indicating low lipid oxidation during the drying process (Table 5). The PC batch showed a significantly lower TBA value compared to the NC batch. Both levels of the RGP ingredient (LL and HL batches) presented intermediate values, which could indicate a slight tendency to decrease lipid oxidation. The mechanism by which nitrite inhibits lipid oxidation is clearly defined, including the formation of nitrosyl–myoglobin (MbNO) and nitric oxide ferrous complexes, the synthesis of S-nitrosocysteine, and the inhibition of the Fenton reaction due to the neutralization of the release of Fe^{2+} from myoglobin [13]. Previous studies have reported an intense antioxidant effect from the addition of grape-seed extracts [9,17] and grape-seed flour [34]. A possible explanation for these differences regarding the antioxidant effect of RGP could be that the mentioned studies have added extracts from the grape seed, in which the concentration of polyphenols would be higher than in our valorized by-products (skin, seeds, and stalks). However, in a previous study, the addition of white grape pomace to burgers at different levels of inclusion (0.5, 1, or 3%) significantly decreased lipid oxidation compared to a negative control group without pomace, even compared to a positive control group including sulfites [31].

The addition of the RGP ingredient had a significant impact on protein oxidation, as measured by the carbonyl content (Table 5). Samples from the LL batch (0.5%) and HL batch (1%) presented significantly lower carbonyl contents than the samples from the NC group. However, the addition of nitrites (PC) did not lead to a decrease in protein oxidation in the dry-cured sausages. The effect of nitrite on protein oxidation is unclear, and there are contradictory results regarding its impact on protein carbonylation [5]. Depending on the concentration, sodium nitrite has both anti- and pro-oxidant outcomes on protein oxidation in meat products [50]. Regarding the observed effect of the RGP ingredient in our samples, this is consistent with the study by Yu et al. (2013) [51] who incorporated grape-seed phenolic compounds into Chinese-style sausage. In this study, the inhibition of protein oxidation was attributed to the prevention of metmyoglobin formation by the competitive chelation of iron from myoglobin. However, there are also studies that do not describe this effect and even report an increase in carbonyl groups compared to sausages including nitrites or phenolic compounds [9,12,17,32,34], despite the influence of phenolic compounds on protein oxidation due to the chemical structure of the phenolic compounds and their interactions with the myofibrillar proteins [5]. In our study, the effect of the addition of the RGP ingredient on decreasing protein oxidation was somewhat unexpected, since it presented more effectivity than nitrifying salts. A previous study on model systems reported that protein oxidation is a complex group of reactions, and nitrite can act as both an antioxidant and a pro-oxidant on proteins [8]. In line with our results, sliced dry-cured loin manufactured with nitrites presented lower values of carbonyls with respect to that free of nitrites [11].

Figure 2 displays the results of the sensory analysis of dry-cured sausages from different batches. There were no statistically significant differences in the attributes assessed by the panel for the addition of the RPG ingredient, except for the lean and fat color. Appearance descriptors showed higher scores regarding the intensity of lean color in the PC group compared to the NC group, while RGP (LL and HL) showed the highest lean color intensity. In addition, PC presented significant differences in fat color, since the yellowness of fat was lower than in the other groups. These results are consistent with the instrumental color differences (Table 5). On the other hand, the spicy taste was lower in PC than in other groups, although no explanation has been found for these results. Nitrites act as antioxidants and modulate the aroma and taste formation in dry-cured products [14]. Some panelists noticed an excess of spicy taste in the product, which could be modulated by the inclusion of nitrites. Therefore, the use of nitrites produced a positive effect at the sensory level which was not achieved by the addition of pomace.

It is important to highlight that the use of the RPG ingredient did not negatively affect the texture and aroma characteristics in the sensory analysis of the sausages. The tasters did not find any unpleasant texture, odor, or flavor/taste in any of the groups, regardless of the recipe used. This is particularly important in the case of the groups prepared with RPG ingredients (LL and HL), which provide a significant content of polyphenols (Table 1). A parallel study with white grape pomace [31], produced by the same methodology, which was added into burgers, found that high levels of the ingredient reduced the acceptability of burgers due to problems with the texture, since the pomace was perceived during chewing. In the current study, pomace was more intensely ground, and the mentioned texture problems have been solved (p-value for texture >0.05). But, this has led to a significant reduction of phenolic compounds in the ingredient (Table 1). Furthermore, this fact may have reduced the bioactivity of the ingredient, since as mentioned, the antioxidant effect on salchichón was not intense.

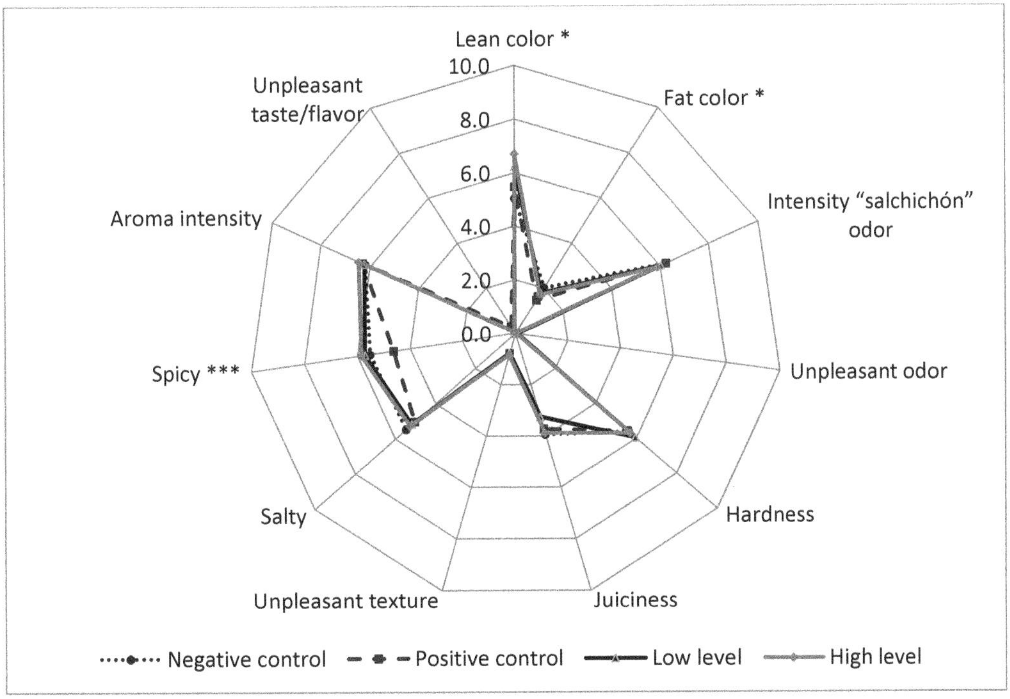

Figure 2. Radar chart for sensorial profile of dry-cured sausages. Negative control (sausages manufactured without nitrites); positive control (sausages manufactured with nitrites); low level (sausages manufactured with 0.5% (*w*/*w*) red wine pomace); high level (sausages manufactured with 1% (*w*/*w*) red wine pomace). ns. (non-significant differences). * $p < 0.05$; *** $p < 0.001$.

Dry-cured sausages from different groups presented significant differences in 15 of the 49 volatile compounds isolated (Table 6). Volatile compounds were grouped into sulfur compounds (9 compounds), terpenes (19 compounds), alcohols (8 compounds), aldehydes (4 compounds), ketones (3 compounds), lineal hydrocarbons (3 compounds), and aromatic hydrocarbons (3 compounds).

Terpenes and sulfur compounds were the most abundant groups of the volatile compounds isolated. Terpenes represented between 43–51%, while sulfur compounds represented between 32–36% with respect to the total compounds. The highest percentages were found in the PC group (51 and 36%), while the NC and the groups with pomace presented similar levels. On the other hand, the rest of the chemical groups represented minor compounds, such as alcohols (4–10%), ketones (4–5%), aldehydes (2–3%), lineal hydrocarbons (2–11%), or aromatic hydrocarbons (1%).

Despite the difficulties in establishing the origin of some volatile compounds, they could be grouped according to their most probable origin. In dry-cured sausages, most volatile compounds, like terpenes and sulfur, could be derived from the seasoning (in this case, garlic, pepper, and nutmeg) added during the manufacturing process. The second group would be the lipid-derived compounds, like some aldehydes, ketones, and alcohols. In addition, some volatile compounds would be originated from the Maillard reaction, such as Strecker aldehydes and hydroxyketones. In fermented dry-cured meat products, microorganisms also participate in the fermentation and in the degradation of amino acids, like some sulfur compounds. Amino acid degradation can also lead to aromatic aldehydes, such as benzaldehyde or benzene acetaldehyde [6].

Table 6. Effect of different formulations on the headspace volatile compounds (μg kg^{-1}) of dry-cured sausage ("salchichón").

	LRI	Negative Control			Positive Control			Low Level			High Level			p-Value	Descriptors [a]
Sulfur compounds															
Allyl methyl sulfide	678.61	573.5 ab	±	370.1	879.4 a	±	376.4	212.9 b	±	113.7	527.5 ab	±	310.3	0.030	Garlic
Dimethyldisulfide	720.40	123.9 ab	±	71.6	163.4 a	±	71.2	29.1 c	±	16.7	61.6 bc	±	32.1	0.005	Garlic
Allylsulfide	856.13	1021.5	±	782.6	1318.9	±	556.1	470.8	±	167.8	1144.0	±	645.2	0.160	Cabbage
Methyl 2-propenyl disulfide (methyl allyl disulfide)	911.37	2595.8 ab	±	1748.9	4226.2 a	±	2006.8	1130.8 b	±	441.1	2598.8 ab	±	1706.5	0.049	
Diallyl disulfide	1084.33	11,691.2	±	5637.8	10,850.7	±	4436.5	3971.4	±	1505.3	8689.8	±	5081.0	0.061	Garlic
Allyl disulfide	1105.74	865.5	±	532.1	1369.1	±	978.8	356.0	±	151.7	693.6	±	494.1	0.110	
1,3,5-Trithiane	1160.24	12.1	±	10.0	17.6	±	5.1	7.2	±	3.5	12.1	±	7.1	0.165	
Allyl trisulfide	1315.72	98.1	±	51.8	98.9	±	38.9	28.5	±	13.7	65.2	±	46.4	0.043	Pungent sulfur, garlic
Diallyl tetrasulfide	1566.49	63.0	±	34.7	57.2	±	17.7	19.4	±	9.3	43.8	±	28.6	0.057	
Terpenes															
1R-α-Pinene *	933.06	804.4	±	1665.1	328.6	±	135.1	825.8	±	294.0	495.0	±	705.3	0.790	Pine, turpentine
L-β-Pinene *	977.07	3424.8 ab	±	1763.5	4360.0 a	±	2100.3	1230.1 b	±	415.2	2759.0 ab	±	1845.4	0.049	pine, resin, turpentine
β-Myrcene	994.66	177.5	±	387.7	0.0	±	0.0	303.3	±	81.1	673.3	±	495.7	0.208	balsamic, must, spice
α-Phellandrene	1004.95	1157.7	±	747.6	1425.6	±	897.4	337.8	±	117.6	709.8	±	403.7	0.061	dill
α- Terpinene *	1019.19	9485.1	±	4989.5	10,656.7	±	5530.4	3215.3	±	822.7	6977.9	±	458.6	0.075	lemon
Limonene *	1033.14	6443.7	±	3893.3	6682.2	±	3533.9	1994.7	±	548.6	4374.3	±	3067.5	0.092	lemon, orange
γ- Terpinene	1064.63	626.8 a	±	269.9	685.1 a	±	226.3	192.2 b	±	111.2	492.4 ab	±	279.1	0.018	
Terpinolene	1093.76	226.9	±	130.1	226.1	±	210.9	113.1	±	46.2	160.8	±	83.2	0.479	Pine, wood, mint
β-Terpineol	1186.98	1033.6 a	±	529.4	968.3 a	±	272.2	315.9 b	±	93.9	713.0 ab	±	366.0	0.022	must
α-Terpineol	1200.39	94.5 a	±	53.1	83.4 ab	±	22.3	31.0 b	±	9.6	63.4 ab	±	32.0	0.040	oil, anise, mint
δ-Elemene	1355.56	176.0	±	123.3	155.2	±	69.0	45.3	±	14.8	104.1	±	63.1	0.072	wood
α-Copaene	1394.98	240.1	±	163.0	206.4	±	80.7	54.8	±	35.4	150.9	±	88.9	0.056	wood, spice
L-Caryophyllene	1429.80	46.3	±	36.4	38.0	±	16.2	11.9	±	4.7	27.8	±	17.5	0.117	wood, spice
β-Caryophyllene	1443.58	87.3	±	58.9	72.2	±	24.7	22.3	±	8.8	54.4	±	33.2	0.064	wood, spice
trans-α-Bergamotene	1456.64	12.6 a	±	8.0	10.1 ab	±	3.7	3.3 b	±	1.2	8.1 ab	±	4.4	0.052	wood, warm, tea
α-Caryophyllene	1478.68	1112.6	±	834.1	955.9	±	418.4	300.1	±	107.5	688.7	±	421.4	0.107	earth
β-Bisabolene	1530.86	7.3 b	±	2.7	12.7 a	±	3.6	0 c			5.0 b	±	3.6	0.000	balsamic
δ-Cadinene	1548.32	23.3	±	22.0	19.4	±	13.4	3.6	±	5.0	17.6	±	10.4	0.178	thyme, medicine, wood
Caryophyllene oxide	1613.59	15.2	±	11.9	13.1	±	4.3	5.0	±	1.8	12.5	±	7.3	0.189	herb, sweet, spice
Alcohols															
Ethanol	-	897.7 a	±	494.2	386.7 ab	±	290.3	47.7 b	±	65.7	398.2 ab	±	218.2	0.005	sweet
3-Methyl-1-butanol *	714.29	84.3	±	55.0	114.6	±	37.7	56.1	±	11.1	113.3	±	75.5	0.253	whiskey, malt, burnt
2,3-Butanediol *	767.92	3641.1 a	±	1791.2	988.6 b	±	702.6	343.8 b	±	417.5	2649.0 ab	±	2121.9	0.009	fruit, onion
3,4-Dimethyl-2-hexanol	789.78	98.8	±	148.0	367.4	±	127.0	112.9	±	103.8	433.0	±	497.8	0.155	
1-Octen-3-ol	985.90	59.3 a	±	31.1	27.5 ab	±	8.9	13.8 b	±	2.6	31.6 ab	±	27.7	0.027	Herb, earth, molds
Benzyl alcohol	1040.49	299.5	±	108.0	276.4	±	78.3	118.8	±	53.9	239.0	±	145.3	0.057	sweet, flower
Phenylethyl alcohol *	1120.61	324.4 a	±	203.5	136.2 ab	±	64.1	81.8 b	±	26.3	253.1 ab	±	110.4	0.024	honey, spice, rose, lilac
d-Mannose	1206.52	80.6 a	±	47.6	67.5 ab	±	12.7	20.9 b	±	7.0	41.6 ab	±	25.9	0.018	
Aldehydes															
3-Methylbutanal *	639.07	229.8	±	162.3	39.3	±	38.9	62.5	±	38.7	149.4	±	150.0	0.066	cocoa, almond, malt
Benzaldehyde *	962.63	184.4	±	151.1	84.0	±	54.5	86.2	±	43.3	170.3	±	228.5	0.555	almond, burnt sugar
Benzeneacetaldehyde	1048.85	914.7	±	698.7	530.6	±	118.9	311.8	±	152.7	748.4	±	363.0	0.142	Benzeneacetaldehyde
Nonanal *	1111.01	77.2	±	33.5	261.1	±	328.6	40.9	±	18.0	72.8	±	39.5	0.187	fat, citrus, green
Ketones															
3-Hydroxybutanone	686.92	848.9	±	568.8	662.1	±	210.3	258.3	±	145.8	659.1	±	407.9	0.124	
3-Octanone *	1028.26	1570.6	±	1132.3	1382.1	±	686.3	345.4	±	205.5	890.2	±	604.5	0.074	herb, butter, resin
1-Phenylethan-1-one	1073.10	248.1	±	121.6	243.3	±	75.8	100.8	±	41.1	167.4	±	89.9	0.050	

Table 6. *Cont.*

	LRI	Negative Control			Positive Control			Low Level			High Level			p-Value	Descriptors [a]
Lineal hydrocarbons															
2-Methylheptane	743.57	31.1	±	43.0	36.3	±	21.9	107.6	±	63.7	161.1	±	185.6	0.172	
3-Ethylhexane	752.17	33.4	±	60.9	34.4	±	31.0	136.8	±	90.1	205.3	±	242.4	0.159	
2,2,4,6,6-Pentamethylheptane	991.59	1117.6	±	954.4	1085.8	±	523.2	1896.0	±	1141.7	2561.3	±	2259.1	0.306	
Aromatic hydrocarbons															
O-Methyleugenol	1421.26	332.3 a	±	205.7	289.2 ab	±	82.2	86.5 b	±	50.3	226.3 ab	±	119.2	0.040	clove, honey
Myristicin	1544.97	250.8	±	154.9	214.6	±	62.0	77.1	±	24.2	170.0	±	93.6	0.060	spice, warm, balsamic
Elemicin	1576.72	54.8 a	±	34.9	46.9 ab	±	13.1	16.0 b	±	5.0	36.9 ab	±	19.8	0.055	spice, flower

* The identification of the compound was carried out by the mass spectrum and LRI identical with a commercial standard compound. Different letters in the same row indicate significant differences in the Tukey test ($p < 0.05$). [a] Flavor descriptors from the Cornell University Flavornet (http://www.flavornet.org/flavornet.html (accesed on 28 September 2024)) and the Good Scents Company (http://www.thegoodscentscompany.com/index.html (accesed on 28 September 2024)).

The three most abundant compounds isolated in the headspace of dry-cured sausages were diallyl disulfide (a sulfur compound), followed by two terpenes, namely α- terpinene and limonene. These three compounds did not present significant differences among formulations. According to their origin, the lack of significant differences agrees with the similar amounts of garlic, nutmeg, white pepper, and black pepper added to each formulation.

In general, the incorporation of the RGP ingredient did not affect the aroma of the dry-cured meat products, which agrees with the sensory analysis. The most abundant compounds isolated in the valorized pomace ingredient (Table 7) were 2,2,4,6,6-pentamethylheptane, ethyl decanoate, ethyl octanoate, phenylethyl alcohol, acetic acid, and 3-methyl, 1-butanol. Among these compounds, only 2,2,4,6,6-pentamethylheptane, phenylethyl alcohol, and 3-methyl, 1-butanol were also isolated in dry-cured sausages (Table 6), and they were not isolated in high quantities in the batches manufactured with the RGP ingredient.

Sulfur compounds produce a significant impact on the global meat aroma in dry-cured meat products. These compounds produce vegetable and garlic notes, but some compounds, like dimethyl disulfide, could be also derived from amino acids degradation. And it has been associated with meaty odor notes [6]. Since garlic is a potent aromatic ingredient, sulfur compounds derived from allicin, characteristic of garlic aroma [52], likely contribute significantly to the overall aroma of dry-cured sausages. Three sulfur compounds (methyl 2-propenyl disulfide (methyl allyl disulfide), allyl methyl sulfide, and dimethyldisulfide) were more abundant in PC sausages with respect to the other groups. Dimethyl disulfide is a compound that originated from the degradation of methionine and adds savory flavors [6]. Dimethyl disulfide is considered a key odorant in dry-cured ham [53], but it has been also isolated in garlic [52]. It is difficult to explain why these sulfur compounds are in the highest quantities in the PC group. In contrast to our results, other authors have reported that the use of nitrite decreased the abundance of diallyl disulfide, in heat-treated Sucuk, a type of semi-dry fermented sausage [54]. Probably, the differences in the manufacturing processes between both products would explain these dissimilarities.

The increase in sulfur compounds in PC sausages could be associated with different causes. Sulfur compounds (in the case of sulfur compounds originated from garlic) may be better preserved by the addition of ascorbic acid and nitrifying salts, since they have antioxidant properties. In addition, the incorporation of additives also affects microbial counts and amino acid degradation, which could, in turn, affect sulfur compounds originating from this pathway. In this line, other authors have also found changes in sulfur compounds (despite their origin from spices) in dry-cured meat products because of changes in the chloride salts addition. In this context, methyl 2-propenyl disulfide and diallyl disulfide were significantly affected by the replacement of chloride salts in pastırma (a traditional Turkish dry-cured meat product) [55]. So, despite these sulfur compounds originating from garlic, their stability could be probably affected by the formulation of the dry-cured products.

A wide variety of terpenes were isolated in dry-cured sausages. Terpenes have well-defined odors in the literature, so α-pinene has been described to add a pine odor. Meanwhile, limonene and carene terpene add lemon notes. α- and β-pinene are two isomers found in nature in essential oils. In 6 of 19 compounds, significant differences were presented among groups. The highest levels of terpenes were in the PC group, specifically L-β-pinene, limonene, terpinolene, and β-bisabolene, all of which originated from spices added during the manufacturing process. Similarly, NC also presented the highest values of limonene, 4-terpineol, α-terpineol, and trans-α-Bergamotene. Aquilani et al. (2018) [17] reported differences in the terpenes content abundances in Cinta Senese dry-fermented sausages with grape seed and chestnut, in combination with 3-hydroxytyrosol, as a substitute for sodium nitrite. However, these changes did not affect the overall aroma profile, since changes in individual compounds are not always globally perceived. This would be in line with our results, since the panelist did not find differences in the sensory analysis regarding odor and taste intensity in the dry-cured meat products (Figure 2). Only the PC group was noticed to be less spicy than the other sausages, which could be correlated with the highest levels of sulfur and terpenes in this group. These compounds may modulate the perception of spicy taste in the sausages.

Table 7. Volatile compounds (μg kg^{-1}) isolated in the headspace of the valorized ingredient from red grape pomace.

		LRI	Ingredient Red Grape Pomace			Descriptors [a]
Alcohols	Ethanol *	-	1014.4	±	416.6	sweet
	3-Methyl-1-butanol *	715.16	2173.3	±	1185.8	whiskey, malt, burnt
	2,3-Butanediol *	810.65	462.5	±	409.3	fruit, onion
	1-Hexanol *	873.64	524.3	±	30.5	resin, flower, green
	Benzyl Alcohol *	1034.79	328.3	±	64.6	sweet, flower
	Phenylethyl alcohol *	1113.88	3043.5	±	100.3	honey, spice, rose, lilac
Aldehydes	Hexanal	790.34	316.0	±	7.5	grass, tallow, fat
	Benzaldehyde *	958.98	70.7	±	26.2	almond, burnt sugar
	Phenyl acetaldehyde	1043.62	48.7	±	4.2	Hawthorne, honey, sweet
	Nonanal *	1103.98	76.8	±	39.2	fat, citrus, green
Esters	3-Methylbutyl acetate *	880.34	122.6	±	16.7	banana
	2-Methylbutyl acetate	882.62	13.5	±	23.4	fruit
	Ethyl hexanoate *	999.81	552.6	±	47.7	Pineapple
	Hexyl acetate *	1015.21	24.9	±	2.5	Fruit, herb
	Ethyl octanoate *	1198.16	3686.7	±	428.6	fruit, fat
	2-Phenylethyl acetate	1260.84	23.4	±	3.3	rose, honey, tobacco
	Ethyl decanoate *	1399.01	5706.5	±	629.4	grape
	Dibutyl hexanedioate	1772.74	43.9	±	7.6	
Hydrocarbons	2-Methyl heptane	743.57	875.2	±	902.3	
	3-Ethyl hexane	751.71	424.6	±	85.9	
	3-Ethyl octane	972.65	61.1	±	26.1	
	2,2,4,6,6-Pentamethyl heptane	989.05	7740.3	±	2245.9	
	2,2,4,4-Tetramethyl octane	1025.93	595.0	±	178.8	
	Dodecane, 2,6,10-trimethyl	1049.14	75.3	±	23.1	
Others	Acetic acid	692.27	2343.8	±	610.4	sour
	Methoxy-phenyl-oxime	928.85	1094.5	±	360.5	
	Eucalyptol	1029.41	15.8	±	27.5	

* The identification of the compound was carried out by the mass spectrum and LRI identical with a commercial standard compound. [a] Flavor descriptors from the Cornell University Flavornet (http://www.flavornet.org/flavornet.html (accesed on 28 September 2024)) and the Good Scents Company (http://www.thegoodscentscompany.com/index.html (accesed on 28 September 2024)).

Three alcohols were most abundantly isolated in NC samples, including ethanol, 2,3-butanediol, and D-mannose, which are compounds related to the effect of microbial fermentation. During the fermentation process, LAB produces lactic acid and small amounts of other compounds, such as ethanol and 2,3-butanediol. The highest levels of these compounds in NC, which presented the lowest LAB counts and the highest pH, could probably be attributed to less developed fermentation in this batch compared to the other groups. These compounds are the initial products of fermentation, which should be transformed into other flavor compounds. In fact, significant correlations were found between these parameters. The LAB counts were negatively correlated with pH (r = −0.611, $p < 0.01$), ethanol (r = −0.640, $p < 0.01$), and 2,3- butanediol (r = −0.571, $p < 0.01$). The proteolytic and lipolytic activities of both microorganisms are essential to the sensory quality of fermented sausages. The highest levels of sugar-like d-mannose in NC sausages would also indicate less fermentation development in this group. The flavor and aroma of fermented meats are a combination of several elements. Lactic acid bacteria produce lactic acid and other compounds. However, to ensure the sensory quality of fermented sausages, the contribution of the proteolytic and lipolytic activities of *Staphylococcus* is fundamental. In a recent paper with an ingredient from white grape pomace (obtained by the same procedure) and added to dry-cured sausages at different levels (0.5% and 3%), the compound 2,3-butanediol was the highest in the 3% pomace sausages. The high level of this compound was related to less juiciness and a defective texture in the sensory analysis of the dry-cured sausage [53]. In the current study, these changes were not noticed when the sensory analysis was performed.

On the other hand, the levels of 1-octen-3-ol and phenylethyl alcohol (both lipid-oxidation-derived compounds) were the highest in NC sausages, which agrees with the highest TBA-RS values observed in this group. In this line, significant positive correlations were found between the TBA-RS values and 1-octen-3-ol (r = +0.453, $p < 0.05$) and phenylethyl alcohol (r = +0.683, $p < 0.001$), which would support the origin of these compounds.

Aldehyde compounds probably originated from amino acids degradation (benzaldehyde and benzeneacetaldehyde) or the Maillard reaction (3-methyl, butanal) and lipid oxidation (nonanal). None of these compounds originated significant differences among formulations.

The use of the RPG ingredient did not negatively affect the volatile profile of the sausages. However, the incorporation of ascorbic acid and nitrifying salts in the PC group favored the highest contents of some terpenes or sulfur compounds. This fact was not reflected in the sensory analysis of sausages, where only significant differences in the spicy taste were reported. The effect of nitrites in flavor formation is not known in depth, but in the case of dry-cured meat products like ham, a clear effect of nitrites on the modulation of lipid-derived compound reactions has been described [48]. However, in fermented products like salchichón, which have a different volatile profile with more importance on spices and microbial fermentation, this effect is not well understood.

4. Conclusions

The valorization process for red grape pomace allowed the full utilization of the entire by-product, resulting in an ingredient rich in fiber and phenolic compounds. The ingredient from red grape pomace favored an adequate fermentation of the dry-cured salchichón, since it enhanced the development of lactic acid bacteria and the acidification of the product. Red grape pomace also provided an antioxidant effect, so lipid-oxidation development was lower in the dry-cured product. This fact has a special relevance, since the sale of the sliced product has increased as a consequence of new market trends. Due to the high surface exposure of the product, the development of oxidation reactions is crucial to reach a long shelf-life. In addition, the ingredient did not negatively affect the sensory perception of the sausages. However, the addition of red grape pomace did not provide the bright-red cured color of salchichón caused by nitrifying salts. Future studies should evaluate the possibility of a partial supply of nitrifying salts instead of a total reduction of these additives, to avoid reducing the safety and modifying the appearance of dry-cured products. Research on the pomace valorization process should prioritize preserving phenolic compounds in the ingredient, as this would enhance its effectiveness in preserving dry-cured meat products.

Author Contributions: M.D.: writing—original draft, investigation. M.J.P.: writing—review and editing, investigation. J.D.-A.: writing—review and editing, supervision. J.J.G.-P.: writing—review and editing, investigation. M.J.M.-M.: writing—review and editing, investigation. M.R.R.-B.: project administration, writing—review and editing, supervision. All authors have read and agreed to the published version of the manuscript.

Funding: The results have been financed by the project "Valorization of the by-products of the oenological industry through alternative technologies to conventional ones to improve the preservation of meat products" (IB20073) financed from the Junta de Extremadura and FEDER funds.

Institutional Review Board Statement: The study was conducted in accordance with the Declaration of Helsinki, and approved by the Institutional Review Board (or Ethics Committee) of University of Extremadura (protocol code 110//2022 and date of approval 16 June 2022).

Informed Consent Statement: Informed consent was obtained from all subjects involved in the study.

Data Availability Statement: The authors declare that the data supporting the results of this work are accessible within the article. The data presented in this study are available on request from the corresponding author.

Acknowledgments: We are grateful to Olivia Fariña and Antonia Trejo for their technical support.

Conflicts of Interest: The authors declare no conflicts of interest. The funding agencies had no function in the design of the study, in the collection, analyses, or interpretation of the data; in the writing of the manuscript; or in the decision to publish the results.

References

1. Beriain, M.J.; Peña, M.P.; Bello, J. A Study of the Chemical Components Which Characterize Spanish Saucisson. *Food Chem.* **1993**, *48*, 31–37. [CrossRef]
2. Lizaso, G.; Chasco, J.; Beriain, M.J. Microbiological and Biochemical Changes during Ripening of Salchichon, a Spanish Dry Cured Sausage. *Food Microbiol.* **1999**, *16*, 219–228. [CrossRef]
3. Chasco, J.; Lizaso, G.; Beriain, M.J. Cured Colour Development during Sausage Processing. *Meat Sci.* **1996**, *44*, 203–211. [CrossRef] [PubMed]
4. Navarro, J.L.; Nadal, M.I.; Izquierdo, L.; Flores, J. Lipolysis in Dry Cured Sausages as Affected by Processing Conditions. *Meat Sci.* **1997**, *45*, 161–168. [CrossRef]
5. Estévez, M. Protein Carbonyls in Meat Systems: A Review. *Meat Sci.* **2011**, *89*, 259–279. [CrossRef] [PubMed]
6. Flores, M. Understanding the Implications of Current Health Trends on the Aroma of Wet and Dry Cured Meat Products. *Meat Sci.* **2018**, *144*, 53–61. [CrossRef] [PubMed]
7. Lorenzo, J.M.; Franco, D. Fat Effect on Physico-Chemical, Microbial and Textural Changes through the Manufactured of Dry-Cured Foal Sausage Lipolysis, Proteolysis and Sensory Properties. *Meat Sci.* **2012**, *92*, 704–714. [CrossRef] [PubMed]
8. Berardo, A.; Claeys, E.; Vossen, E.; Leroy, F.; De Smet, S. Protein Oxidation Affects Proteolysis in a Meat Model System. *Meat Sci.* **2015**, *106*, 78–84. [CrossRef] [PubMed]
9. Lorenzo, J.M.; González-Rodríguez, R.M.; Sánchez, M.; Amado, I.R.; Franco, D. Effects of Natural (Grape Seed and Chestnut Extract) and Synthetic Antioxidants (Buthylatedhydroxytoluene, BHT) on the Physical, Chemical, Microbiological and Sensory Characteristics of Dry Cured Sausage "Chorizo". *Food Res. Int.* **2013**, *54*, 611–620. [CrossRef]
10. Majou, D.; Christieans, S. Mechanisms of the Bactericidal Effects of Nitrate and Nitrite in Cured Meats. *Meat Sci.* **2018**, *145*, 273–284. [CrossRef]
11. Higuero, N.; Moreno, I.; Lavado, G.; Vidal-Aragón, M.C.; Cava, R. Reduction of Nitrate and Nitrite in Iberian Dry Cured Loins and Its Effects during Drying Process. *Meat Sci.* **2020**, *163*, 108062. [CrossRef] [PubMed]
12. Li, L.; Shao, J.; Zhu, X.; Zhou, G.; Xu, X. Effect of Plant Polyphenols and Ascorbic Acid on Lipid Oxidation, Residual Nitrite and N-Nitrosamines Formation in Dry-Cured Sausage. *Int. J. Food Sci. Technol.* **2013**, *48*, 1157–1164. [CrossRef]
13. Hammes, W.P. Metabolism of Nitrate in Fermented Meats: The Characteristic Feature of a Specific Group of Fermented Foods. *Food Microbiol.* **2012**, *29*, 151–156. [CrossRef] [PubMed]
14. Ferysiuk, K.; Wójciak, K.M. Reduction of Nitrite in Meat Products through the Application of Various Plant-Based Ingredients. *Antioxidants* **2020**, *9*, 711. [CrossRef] [PubMed]
15. Falowo, A.B.; Fayemi, P.O.; Muchenje, V. Natural Antioxidants against Lipid-Protein Oxidative Deterioration in Meat and Meat Products: A Review. *Food Res. Int.* **2014**, *64*, 171–181. [CrossRef]

16. Hospital, X.F.; Carballo, J.; Fernández, M.; Arnau, J.; Gratacós, M.; Hierro, E. Technological Implications of Reducing Nitrate and Nitrite Levels in Dry-Fermented Sausages: Typical Microbiota, Residual Nitrate and Nitrite and Volatile Profile. *Food Control* **2015**, *57*, 275–281. [CrossRef]
17. Aquilani, C.; Sirtori, F.; Flores, M.; Bozzi, R.; Lebret, B.; Pugliese, C. Effect of Natural Antioxidants from Grape Seed and Chestnut in Combination with Hydroxytyrosol, as Sodium Nitrite Substitutes in Cinta Senese Dry-Fermented Sausages. *Meat Sci.* **2018**, *145*, 389–398. [CrossRef]
18. Shah, M.A.; Bosco, S.J.D.; Mir, S.A. Plant Extracts as Natural Antioxidants in Meat and Meat Products. *Meat Sci.* **2014**, *98*, 21–33. [CrossRef] [PubMed]
19. Beres, C.; Costa, G.N.S.; Cabezudo, I.; da Silva-James, N.K.; Teles, A.S.C.; Cruz, A.P.G.; Mellinger-Silva, C.; Tonon, R.V.; Cabral, L.M.C.; Freitas, S.P. Towards Integral Utilization of Grape Pomace from Winemaking Process: A Review. *Waste Manag.* **2017**, *68*, 581–594. [CrossRef] [PubMed]
20. Dwyer, K.; Hosseinian, F.; Rod, M.R.M. The Market Potential of Grape Waste Alternatives. *J. Field Robot.* **2014**, *3*, 91–106. [CrossRef]
21. García-Lomillo, J.; González-SanJosé, M.L. Applications of Wine Pomace in the Food Industry: Approaches and Functions. *Compr. Rev. Food Sci. Food Saf.* **2017**, *16*, 3–22. [CrossRef] [PubMed]
22. Ilyas, T.; Chowdhary, P.; Chaurasia, D.; Gnansounou, E.; Pandey, A.; Chaturvedi, P. Sustainable Green Processing of Grape Pomace for the Production of Value-Added Products: An Overview. *Environ. Technol. Innov.* **2021**, *23*, 101592. [CrossRef]
23. Antonić, B.; Jančíková, S.; Dordević, D.; Tremlová, B. Grape Pomace Valorization: A Systematic Review and Meta-Analysis. *Foods* **2020**, *9*, 1627. [CrossRef] [PubMed]
24. Rondeau, P.; Gambier, F.; Jolibert, F.; Brosse, N. Compositions and Chemical Variability of Grape Pomaces from French Vineyard. *Ind. Crops Prod.* **2013**, *43*, 251–254. [CrossRef]
25. Makris, D.P. Green Extraction Processes for the Efficient Recovery of Bioactive Polyphenols from Wine Industry Solid Wastes—Recent Progress. *Curr. Opin. Green. Sustain. Chem.* **2018**, *13*, 50–55. [CrossRef]
26. Toepfl, S.; Mathys, A.; Heinz, V.; Knorr, D. Review: Potential of High Hydrostatic Pressure and Pulsed Electric Fields for Energy Efficient and Environmentally Friendly Food Processing. *Food Rev. Int.* **2006**, *22*, 405–423. [CrossRef]
27. Cheftel, J.C.; Culioli, J. Effects of High Pressure on Meat: A Review. *Meat Sci.* **1997**, *46*, 211–236. [CrossRef]
28. Corrales, M.; García, A.F.; Butz, P.; Tauscher, B. Extraction of Anthocyanins from Grape Skins Assisted by High Hydrostatic Pressure. *J. Food Eng.* **2009**, *90*, 415–421. [CrossRef]
29. Ramírez, R.; Delgado, J.; Rocha-Pimienta, J.; Valdés, M.E.; Martín-Mateos, M.J.; Ayuso-Yuste, M.C. Preservation of White Wine Pomace by High Hydrostatic Pressure. *Heliyon* **2023**, *9*, e21199. [CrossRef]
30. D'Arrigo, M.; Delgado-Adámez, J.; Rocha-Pimienta, J.; Valdés-Sánchez, M.E.; Ramírez-Bernabé, M.R. Integral Use of Red Wine Pomace after Hydrostatic High Pressure: Application of Two Consecutive Cycles of Treatment. *Foods* **2024**, *13*, 149. [CrossRef] [PubMed]
31. Martín-Mateos, M.J.; Delgado-Adámez, J.; Moreno-Cardona, D.; Valdés-Sánchezm, M.E.; Ramírez-Bernabé, M.R. Application of an Ingredient Made of White Wine Pomace for the Preservation of Fresh Pork Burgers. *Foods* **2023**, *12*, 4468. [CrossRef]
32. Pateiro, M.; Bermúdez, R.; Lorenzo, J.M.; Franco, D. Effect of Addition of Natural Antioxidants on the Shelf-Life of 'Chorizo', a Spanish Dry-Cured Sausage. *Antioxidants* **2015**, *4*, 42–67. [CrossRef]
33. Pini, F.; Aquilani, C.; Giovannetti, L.; Viti, C.; Pugliese, C. Characterization of the Microbial Community Composition in Italian Cinta Senese Sausages Dry-Fermented with Natural Extracts as Alternatives to Sodium Nitrite. *Food Microbiol.* **2020**, *89*, 103417. [CrossRef]
34. Kurt, S. The Effects of Grape Seed Flour on the Quality of Turkish Dry Fermented Sausage (Sucuk) during Ripening and Refrigerated Storage. *Korean J. Food Sci. Anim. Resour.* **2016**, *36*, 300–308. [CrossRef] [PubMed]
35. AOAC. *Official Methods of Analysis of AOAC International*, 20th ed.; George, W.L., Jr., Ed.; Official Methods of Analysis of AOAC International: Rockville, MA, USA, 2016; ISBN 0935584870.
36. Folch, J.; Lees, M.; Sloane Stanley, G.H. A Simple Method for the Isolation and Purification of Total Lipides from Animal Tissues. *J. Biol. Chem.* **1957**, *226*, 497–509. [CrossRef] [PubMed]
37. Villanueva, M.J.; Barragán, R. Determinación Cuantitativa de La Fracción Hidrocarbonada En Alimentos. *Anal. Bromatol.* **1985**, *XXXVII*, 61–77.
38. Sørensen, G.; Jørgensen, S.S. A Critical Examination of Some Experimental Variables in the 2-Thiobarbituric Acid (TBA) Test for Lipid Oxidation in Meat Products. *Z. Fur Lebensm. Unters. Und-Forsch.* **1996**, *202*, 205–210. [CrossRef]
39. Oliver, C.N.; Ahn, B.-W.; Moerman, E.J.; Goldstein, S.; Stadtman, E.R. Age-Related Changes in Oxidized Proteins. *J. Biol. Chem.* **1987**, *262*, 5488–5491. [CrossRef]
40. Dewanto, V.; Xianzhong, W.; Adom, K.K.; Liu, R.H. Thermal Processing Enhances the Nutritional Value of Tomatoes by Increasing Total Antioxidant Activity. *J. Agric. Food Chem.* **2002**, *50*, 3010–3014. [CrossRef] [PubMed]
41. Sheng, K.; Qu, H.; Liu, C.; Yan, L.; You, J.; Shui, S.; Zheng, L. A Comparative Assess of High Hydrostatic Pressure and Superfine Grinding on Physicochemical and Antioxidant Properties of Grape Pomace. *Int. J. Food Sci. Technol.* **2017**, *52*, 2106–2114. [CrossRef]

42. Corrales, M.; Toepfl, S.; Butz, P.; Knorr, D.; Tauscher, B. Extraction of Anthocyanins from Grape By-Products Assisted by Ultrasonics, High Hydrostatic Pressure or Pulsed Electric Fields: A Comparison. *Innov. Food Sci. Emerg. Technol.* **2008**, *9*, 85–91. [CrossRef]
43. Javier, H.; Siles, J.A.; Aida, G.; del Carmen, G.M.; de los Ángeles, M.M. Revalorization of Grape Marc Waste from Liqueur Wine: Biomethanization. *J. Chem. Technol. Biotechnol.* **2019**, *94*, 1499–1508. [CrossRef]
44. Hoz, L.; D'Arrigo, M.; Cambero, I.; Ordóñez, J.A. Development of an N-3 Fatty Acid and α-Tocopherol Enriched Dry Fermented Sausage. *Meat Sci.* **2004**, *67*, 485–495. [CrossRef] [PubMed]
45. Moretti, V.M.; Madonia, G.; Diaferia, C.; Mentasti, T.; Paleari, M.A.; Panseri, S.; Pirone, G.; Gandini, G. Chemical and Microbiological Parameters and Sensory Attributes of a Typical Sicilian Salami Ripened in Different Conditions. *Meat Sci.* **2004**, *66*, 845–854. [CrossRef] [PubMed]
46. Toldrá, F.; Flores, M. Dry and Semidry. In *Encyclopedia of Meat Sciences*, 2nd ed.; Academic Press: Cambridge, MA, USA, 2014; ISBN 9780123847317.
47. Delgado Adámez, J.; Gamero Samino, E.; Valdés Sánchez, E.; González-Gómez, D. In Vitro Estimation of the Antibacterial Activity and Antioxidant Capacity of Aqueous Extracts from Grape-Seeds (*Vitis vinifera* L.). *Food Control* **2012**, *24*, 136–141. [CrossRef]
48. Toldrá, F.; Aristoy, M.-C.; Flores, M. Relevance of Nitrate and Nitrite in Dry-Cured Ham and Their Effects on Aroma Development. *Grasas Y Aceites* **2009**, *60*, 291–296. [CrossRef]
49. Morita, H.; Yoshikawa, H.; Suzuki, T.; Hisamatsu, S.; Kato, Y.; Sakata, R.; Nagata, Y.; Yoshimura, T. Anti-Microbial Action against Verotoxigenic *Escherichia Coli* O157:H7 of Nitric Oxide Derived from Sodium Nitrite. *Biosci. Biotechnol. Biochem.* **2004**, *68*, 1027–1034. [CrossRef] [PubMed]
50. Feng, X.; Li, C.; Jia, X.; Guo, Y.; Lei, N.; Hackman, R.M.; Chen, L.; Zhou, G. Influence of Sodium Nitrite on Protein Oxidation and Nitrosation of Sausages Subjected to Processing and Storage. *Meat Sci.* **2016**, *116*, 260–267. [CrossRef]
51. Yu, H.; Qin, C.; Wu, X.; Ge, Q.; Wu, M.; Wu, J.; Wang, M.; Wang, Z. Effect of Grape Seed and Rosemary Phenolics on Protein Oxidation in Chinese-Style Sausage. *J. Food Agric. Environ.* **2013**, *11*, 231–236.
52. Kilic-Buyukkurt, O.; Kelebek, H.; Bordiga, M.; Keskin, M.; Selli, S. Changes in the Aroma and Key Odorants from White Garlic to Black Garlic Using Approaches of Molecular Sensory Science: A Review. *Heliyon* **2023**, *9*, e19056. [CrossRef] [PubMed]
53. Carrapiso, A.I.; Martín-Mateos, M.J.; D'Arrigo, M.; Delgado-Adámez, J.; Saraiva, J.A.; Ramírez-Bernabé, M.R. High-Hydrostatic-Pressure-Stabilized White Grape Pomace to Improve the Oxidative Stability of Dry-Cured Sausages ("Salchichón"). *Foods* **2024**, *13*, 687. [CrossRef]
54. Sallan, S.; Kaban, G.; Kaya, M. The Effects of Nitrite, Sodium Ascorbate and Starter Culture on Volatile Compounds of a Semi-Dry Fermented Sausage. *LWT* **2022**, *153*, 112540. [CrossRef]
55. Yalınkılıç, B.; Kaban, G.; Kaya, M. Effect of Sodium Replacement on the Quality Characteristics of Pastırma (a Dry-Cured Meat Product). *Food Sci. Hum. Wellness* **2022**, *12*, 266–274. [CrossRef]

foods

MDPI

Article

Enhancing Deer Sous Vide Meat Shelf Life and Safety with *Eugenia caryophyllus* Essential Oil against *Salmonella enterica*

Miroslava Kačániová [1,2,*], Stefania Garzoli [3], Anis Ben Hsouna [4,5], Zhaojun Ban [6], Joel Horacio Elizondo-Luevano [7], Maciej Ireneusz Kluz [2], Rania Ben Saad [4], Peter Haščík [8], Natália Čmiková [1], Božena Waskiewicz-Robak [2], Ján Kollár [9] and Alessandro Bianchi [10]

1 Institute of Horticulture, Faculty of Horticulture and Landscape Engineering, Slovak University of Agriculture, Trieda Andreja Hlinku 2, 94976 Nitra, Slovakia; n.cmikova@gmail.com
2 School of Medical & Health Sciences, University of Economics and Human Sciences in Warsaw, Okopowa 59, 01043 Warszawa, Poland; m.kluz@vizja.pl (M.I.K.); b.waszkiewicz-robak@vizja.pl (B.W.-R.)
3 Department of Chemistry and Technologies of Drug, Sapienza University, P. le Aldo Moro, 5, 00185 Rome, Italy; stefania.garzoli@uniroma1.it
4 Laboratory of Biotechnology and Plant Improvement, Centre of Biotechnology of Sfax, B.P "1177", Sfax 3018, Tunisia; benhsounanis@gmail.com (A.B.H.); raniabensaad@gmail.com (R.B.S.)
5 Department of Environmental Sciences and Nutrition, Higher Institute of Applied Sciences and Technology of Mahdia, University of Monastir, Monastir 5000, Tunisia
6 School of Biological and Chemical Engineering, Zhejiang University of Science and Technology, Zhejiang Provincial Key Laboratory of Chemical and Biological Processing Technology of Farm Products, Zhejiang Provincial Collaborative Innovation Center of Agricultural Biological Resources Biochemical Manufacturing, Hangzhou 310023, China; banzhaojun@zust.edu.cn
7 Faculty of Agronomy, Universidad Autónoma de Nuevo León (UANL), Av. Francisco Villa S/N, Col. Ex Hacienda el Canadá, General Escobedo, Nuevo León 66050, Mexico; joel.elizondolv@uanl.edu.mx
8 Institute of Food Technology, Faculty of Biotechnology and Food Sciences, Slovak University of Agriculture, Trieda Andreja Hlinku 2, 94976 Nitra, Slovakia; peter.hascik@uniag.sk
9 Institute of Landscape Architecture, Faculty of Horticulture and Landscape Engineering, Slovak University of Agriculture, Tulipánová 7, 94976 Nitra, Slovakia; jan.kollar@uniag.sk
10 Department of Agriculture, Food and Environment, University of Pisa, Via del Borghetto 80, 56124 Pisa, Italy; alessandro.bianchi@phd.unipi.it
* Correspondence: miroslava.kacaniova@gmail.com or m.kacaniova@vizja.pl

Abstract: Modern lifestyles have increased the focus on food stability and human health due to evolving industrial goals and scientific advancements. Pathogenic microorganisms significantly challenge food quality, with *Salmonella enterica* and other planktonic cells capable of forming biofilms that make them more resistant to broad-spectrum antibiotics. This research examined the chemical composition and antibacterial and antibiofilm properties of the essential oil from *Eugenia caryophyllus* (ECEO) derived from dried fruits. GC-MS analyses identified eugenol as the dominant component at 82.7%. Additionally, the study aimed to extend the shelf life of sous vide deer meat by applying a plant essential oil and inoculating it with *S. enterica* for seven days at 4 °C. The essential oil demonstrated strong antibacterial activity against *S. enterica.* The ECEO showed significant antibiofilm activity, as indicated by the MBIC crystal violet test results. Data from MALDI-TOF MS analysis revealed that the ECEO altered the protein profiles of bacteria on glass and stainless-steel surfaces. Furthermore, the ECEO was found to have a beneficial antibacterial effect on *S. enterica.* In vacuum-packed sous vide red deer meat samples, the anti-*Salmonella* activity of the ECEO was slightly higher than that of the control samples. These findings underscore the potential of the ECEO's antibacterial and antibiofilm properties in food preservation and extending the shelf life of meat.

Keywords: clove essential oil; chemical composition; antimicrobial activity; antibiofilm activity; insecticidal activity; game meat; pathogenic bacteria; chili cook

1. Introduction

The preservation of food product quality during storage is a crucial step for reducing food waste and enhancing sustainability [1,2]. Ensuring food safety is equally important, as it protects consumers from potential health hazards caused by microbial contamination [3]. Enhancing food longevity with modern preservation techniques, including the application of natural essential oils, helps ensure the retention of nutritional value and safety in food products [3–5]. This approach not only minimizes waste but also supports sustainable practices in the food industry, aligning with modern health and environmental goals [6].

During the production and storage of food, certain bacterial contaminants have the ability to proliferate or endure. Salmonellosis, one of the most common food-borne infections worldwide, is caused by the presence of *Salmonella*, posing a significant risk associated with food products [7]. *Salmonella*'s adaptability to various temperatures and its robust heat-stress resistance enable it to evade host defenses and establish infections. Furthermore, thermal stress responses can activate genes associated with virulence and overall stress resilience [8].

Foods, whether raw or processed, are susceptible to contamination during manufacturing, sale, and distribution processes [9]. Consequently, the food industry requires preservatives to inhibit the growth of microorganisms responsible for food spoilage [10]. Before the early 1990s, there was a limited amount of research on the impact of essential oils (EOs) in food, even though EOs were included in a few commonly used food preservatives [11]. Generally, bacteria are more vulnerable to the antibacterial action of EOs when food pH is lowered, storage temperature is raised, and packing oxygen content is increased. The antibacterial activity of EOs may be restricted by the physical properties of food. Moreover, research has shown that a number of EOs are more effective as bactericidal agents than commonly used preservatives in meat applications [12].

Clove oil has long been utilized in the food industry both as a flavoring agent and for its antimicrobial properties. Its biological activities include antibacterial, antifungal, insecticidal, and antioxidant effects [13]. In addition to its use in food, clove oil serves as an antiseptic for treating oral infections [14]. This essential oil effectively inhibits the growth of molds, yeasts, and bacteria [15]. It has demonstrated efficacy against *Listeria monocytogenes* and *Salmonella* Enteritidis in both tryptone soy broth and cheese [16]. Essential oils are known for their strong anti-biofilm properties, especially when used in combination with antibacterial agents [17]. Plants are a rich source of secondary metabolites like tannins, terpenoids, alkaloids, and flavonoids, many of which have shown antibacterial properties in various studies [18].

Synthetic chemical products from various toxicological classes are used to control insects in stored grains. Although these treatments are highly effective, frequent use can lead to several issues, including increased production costs, food residue buildup, insect resistance, harm to human health, and environmental contamination [19]. The plants can have their parts processed into powders, extracts, or oils for application. As noted by Mazzonetto and Vendramim [20], these products are cost-effective, readily accessible, easy to apply without the need for specialized personnel, and are environmentally and medically safe, posing no adverse effects.

This study examined the heat resistance of five different *Salmonella* strains in teriyaki-marinated chicken breasts using a sous vide method. The chicken breasts, packaged and inoculated, were submerged entirely in a circulating water bath. Temperature settings of 55, 57.5, or 60 °C were achieved within an hour and maintained for specific durations. Based on linear regression analysis, *Salmonella* D-values ranged from 47.65 min at 55 °C to 7.48 min at 60 °C in chicken breast samples. Post-marination, the bacteria exhibited increased susceptibility to heat lethality, which is critical for ensuring the microbiological safety of sous vide-processed marinated meats [21]. To enhance the efficacy of sous vide, EOs were combined with *S. enterica* in another experiment. Samples were vacuum-sealed, inoculated with *S. enterica*, and cooked sous vide at temperatures ranging from 50 to 65 °C for specified periods. After a 20 min heat treatment at 65 °C, significant reductions in bacteria and coliforms were observed in the samples. The integration of EOs with sous vides effectively contributed to the meat's stabilization and safety [22].

To our knowledge, this is the first study on the use of ECEOs in meat preservation, specifically in game meat. Clove is a highly prized spice used as a food preservative and for a variety of therapeutic reasons. The ECEO and its principal active component, eugenol, indicate antibacterial and antifungal action, aromaticity, and safety as promising and valuable antiseptics in the food sector. The ECEO can help improve the flavor of deer meat. In addition, it can improve the palatability of venison by adding sweetness and tenderness to the meat. However, the effects of the ECEO on venison have never been investigated in detail. The objective of this research was to examine the chemical composition, antimicrobial properties, and effectiveness of the *Eugenia caryophyllus* essential oil against the biofilm formation of *Salmonella enterica in vitro*. Additionally, this study investigated the survival of *S. enterica* inoculated onto red deer meat processed using the sous vide cook–chill method and subsequently stored at 4 °C for 7 days, aiming to extend its shelf life.

2. Materials and Methods

2.1. Essential Oil Characteristics

The essential oil of clove (*Eugenia caryophyllus*) ECEO used in this research was procured from Hanus s.r.o. in Nitra. It was extracted by steam distillation of dried flower buds sourced from Sri Lanka.

2.2. Chemical Analysis of ECEO

The analysis was conducted using a Perkin Elmer Clarus 500 (Waltham, MA, USA) gas chromatograph equipped with a mass spectrometer and a flame ionization detector. A Varian Factor Four VF-5 capillary column (Lambda Life s.r.o., Bratislava, Slovakia)was housed in the GC oven, with helium serving as the carrier gas flowing at a rate of 1 mL/min. The temperature program for the GC oven began with an initial isothermal phase at 60 °C for 2 min, followed by a gradual increase to 220 °C at a rate of 6 °C/min, maintaining the final temperature for 20 min. Mass spectra were obtained in electron impact mode at 70 eV, scanning from 35 to 450 m/z. Identification of volatile compounds involved comparing their mass spectra with entries in the Wiley 2.2 and Nist 02 databases and determining their linear retention indices (LRIs) relative to C_8–C_{25} n-alkanes analyzed under similar conditions described in the literature. The relative amounts of compounds, expressed as percentages, were calculated by normalizing peak areas without relying on internal standards or any corrective measures. Each experimental run was conducted in triplicate to ensure robustness and consistency of the findings [23,24].

2.3. Antimicrobial Activity

2.3.1. Bacteria Strain Preparation

This experiment focused on using *Salmonella enterica* to investigate the effects of sous vide cooking on deer meat and its impact on extending shelf life by reducing microbial counts and antimicrobial activity. The *Salmonella enterica* subsp. *enterica* CCM 4420 strain used was obtained from a microbial collection, the Czech Collection of Microorganisms in Brno, Czech Republic. The bacteria were cultured on Mueller Hinton agar (MHA) from Oxoid in Basingstoke, UK, and were incubated for 24 h at 37 °C. Once the bacterial culture reached an optical density adjusted to the 0.5 McFarland standard (equivalent to 1.5×10^8 CFU/mL), 100 μL of the inoculum was added to samples of deer thigh flesh. To ensure even distribution of the pathogen, the deer meat samples were thoroughly mixed for three minutes at room temperature following inoculation with *Salmonella enterica* [3,22].

2.3.2. Disk Diffusion Method

To assess the antimicrobial efficacy of the ECEO, we employed the disk diffusion method. Bacterial cultures were grown in Mueller–Hinton Broth (MHB, Oxoid, Basingstoke,

UK) at 37 °C for 24 h. Following incubation, the bacterial density was adjusted to 0.5 McFarland standard (1.5 × 10^8 CFU/mL) using distilled water. Subsequently, 100 μL of the bacterial suspension was spread evenly on Mueller–Hinton Agar (MHA, Oxoid, Basingstoke, UK). Sterile 6 mm disks saturated with 10 μL of the ECEO were placed on the agar plates. After incubating at 37 °C for 24 h, the zones of inhibition were measured from three different directions around each disk. Ciprofloxacin (30 μg per disc) was included as the control antibiotic. Each experiment to evaluate antimicrobial activity was conducted in triplicate to ensure a thorough and consistent assessment of results [22,25].

2.3.3. Minimal Inhibitory Concentration (MIC)

Bacterial cultures were incubated for 24 h in Mueller–Hinton broth (MHB, Oxoid, Basingstoke, UK) at 37 °C. The cultures were adjusted to an optical density corresponding to 0.5 McFarland standard and then added in 150 μL volumes to each well of a 96-well microplate. The ECEO was also added in 150 μL volumes to achieve final concentrations ranging from 10 mg/mL to 0.00488 mg/mL. The microplate was then incubated for 24 h at 37 °C. Negative controls consisted of MHB with the ECEO, while positive controls included MHB with bacterial inoculum. Following incubation, absorbance at 570 nm was measured using a Glomax spectrophotometer (Promega Inc., Madison, WI, USA). The MIC_{50} was defined as the lowest EO concentration inhibiting 50% of bacterial growth, and the MIC_{90} as the concentration inhibiting 90% of growth. To ensure accuracy and reliability, the experiment was conducted in triplicate [26].

2.4. Research on Biofilm Growth

2.4.1. Crystal Violet Study

Kačániová et al. [25] conducted a comprehensive study on the Minimal Biofilm Inhibitory Concentration (MBIC). Bacterial suspensions were cultured in Mueller–Hinton broth (MHB, Oxoid, Basingstoke, UK) at 37 °C under aerobic conditions throughout the day. After incubation, an inoculum was prepared to achieve an optical density equivalent to the 0.5 McFarland standard. A 96-well microtiter plate was set up by adding 100 μL of the bacteria and 100 μL of the ECEO per well. Starting from the first column, 100 μL of the ECEO was added, followed by a two-fold dilution using a pipette to achieve concentrations ranging from 10 mg/mL to 0.00488 mg/mL. Maximal growth control was maintained using MHB with bacterial inoculum, while MHB with the ECEO served as the negative control. After a 24 h incubation period at 37 °C, the supernatant was discarded, and the wells were washed three times with 250 μL of saline solution before drying at room temperature for 30 min. The wells were then stained with 200 μL of 0.1% *w*/*v* crystal violet for 15 min, followed by several washes with distilled water and subsequent drying. The samples were solubilized with 200 μL of 33% acetic acid, and absorbance at 570 nm was measured using a Glomax spectrophotometer (Promega Inc., Madison, USA). The MBIC was determined as the concentration where the absorbance was equal to or less than the negative control. $MBIC_{50}$ and $MBIC_{90}$ were defined as the concentrations inhibiting 50% and 90% of biofilm development, respectively.

2.4.2. MALDI-TOF MS Biotyper for Biofilm Formation Detection

The Bruker Daltonics MALDI-TOF MicroFlex instrument (Bremen, Germany)was employed to assess protein degradation during biofilm formation. Initially, 100 μL of *S. enterica* bacterial inoculum and 20 mL of MHB were combined in 50 mL polypropylene tubes containing small glass and stainless-steel slides. Experimental tubes were treated with the ECEO to achieve a final concentration of 0.1%, while control tubes remained untreated. Over seven days at 37 °C, tubes were agitated at 170× *g*. Each day, biofilms from glass and steel surfaces were collected using sterile cotton swabs and transferred to target plates. Planktonic cells from untreated control samples were also analyzed. After adding 300 μL of culture material, control bacterial cultures were centrifuged for one minute at 12,000× *g*. Pellets underwent three washes in ultrapure water before being centrifuged again and transferred to target plates for analysis. Reconstituted pellets and swabs (1 μL each) were applied to plates with 10 mg/mL of α-cyano-4-hydroxycinnamic acid matrix. Plates were dried and subjected to MALDI-TOF analysis in linear positive mode, with mass-to-charge ratios calibrated between 2000 and 20,000. Eighteen standard global spectra (MSPs), as described by Kačániová et al. [25,26], were analyzed using automated methods to calculate Euclidean distances and construct dendrograms.

2.5. Extending the Shelf Life of Deer Sous Vide Meat

2.5.1. Preparation of Samples of Deer Meat

This study focused on examining deer meat samples obtained from the biceps femoris muscle of a 5-year-old deer originating from Slovak hunting grounds. The analysis of the thigh meat revealed its composition per 100 g: 71.97 g of water, 0.75 g of fat, 21.85 g of protein, and 0.035 g of cholesterol. A total of 4 kg of thigh meat was collected and initially stored in a refrigerator before being transferred to a microbiological laboratory for further analysis. The meat was then sliced into 5 g portions using a sterile knife, resulting in 723 individual samples. These samples were allocated across different time points as follows: three raw deer meat samples on day 0 and 240 samples each on days 1, 7, and 14 for both control and treated groups. Each 5 g portion of deer meat was divided into control and treatment groups. For the treatment group, the meat was mixed with a 1% (v/w) solution of the ECEO dissolved in sunflower oil. Following this, all samples underwent vacuum packing using a Concept vacuum packer from Chocen, Czech Republic. Control samples were packed in polyethylene bags, while the treatment groups were vacuum-packed after mixing with the ECEO solution.

During the preparation process, 100 μL of *Salmonella enterica* was added to each sample, along with the ECEO solution. Careful precautions were taken to prevent contamination during the brief mixing period, which lasted approximately one minute prior to vacuum sealing. During our trial, we explored various methods for preparing fresh deer meat:

1. Fresh deer meat was stored in polyethylene bags at 4 °C and then cooked at temperatures between 50 °C and 65 °C for 5 to 25 min;
2. Control vacuum: Deer meat, vacuum-sealed in polyethylene bags at 4 °C, underwent cooking in a water bath at temperatures from 50 °C to 65 °C for 5 to 25 min;
3. Essential oil treatment: Deer meat treated with a 1% ECEO solution, vacuum-packed, and kept at 4 °C was cooked in a water bath at temperatures from 50 °C to 65 °C for 5 to 25 min;
4. *Salmonella enterica* contamination: Deer meat inoculated with *Salmonella enterica*, vacuum-packed, stored at 4 °C until exposed, then cooked in a water bath at temperatures from 50 °C to 65 °C for 5 to 25 min;
5. *Salmonella enterica* and essential oil treatment: Deer meat treated with both *Salmonella enterica* and a 1% ECEO solution, vacuum-packed, stored at 4 °C, and subsequently cooked in a water bath at temperatures from 50 °C to 65 °C for 5 to 25 min.

On day zero, raw deer meat samples were processed as controls. These samples were mixed with either the ECEO or *Salmonella enterica* and allowed to rest for 24 h before undergoing sous vide cooking using the CASO SV1000 machine from Arnsberg, Germany. The meat was packed in polyethylene high-barrier bags known for their durability, resistance to moisture, and capability to withstand temperatures ranging from −30 °C to +100 °C. These bags are specifically designed without plasticizers like bisphenol A or microplastics, ensuring food safety during prolonged refrigeration.

2.5.2. Microbial Analyses

Microbiological evaluations were conducted periodically throughout the experiment. Following a 24 h storage period at 4 °C, the samples underwent heat treatment and were then assessed at scheduled intervals. Initially, 5 g of red deer meat samples were placed in sterile stomacher bags and diluted with 45 mL of peptone water to achieve a 1:10 dilution ratio. The samples were homogenized using a stomacher apparatus for 20 min. After homogenization, 0.1 mL aliquots from appropriate dilutions were spread onto standard plate count agar medium and incubated in a shaking incubator for 30 min. For culturing coliform bacteria, Violet Red Bile Lactose Agar (VRBL; Oxoid, Basingstoke, UK) was utilized and incubated at 37 °C for 24 to 48 h. Plate Count Agar (PCA; Oxoid, Basingstoke, UK) was employed for Total Viable Count (TVC) and incubated at 30 °C for 48 to 72 h. Viable counts were determined based on visible growth on these media. *Salmonella* spp. were detected by culturing on Xylose Lysine Deoxycholate Agar (XLD; Oxoid, Basingstoke, UK) and incubating for 24 to 48 h.

2.5.3. Identification of Microorganisms Using Mass Spectrometry

Microorganisms derived from deer thigh tissue samples were identified using the MALDI-TOF MS Biotyper system from Bruker Daltonics in Bremen, Germany, employing established reference libraries. To prepare the matrix solution, an initial stock was created comprising 50% acetonitrile, 47.5% water, and 2.5% trifluoroacetic acid. This stock solution was formulated by combining 500 μL of pure acetonitrile, 475 μL of filtered water, and 25 μL of 10% trifluoroacetic acid. Subsequently, the "HCCA matrix solution" was prepared in a 250 μL Eppendorf flask, thoroughly mixed with the organic solvent, and sourced from Aloqence Science in Vrable, Slovakia, based on prior guidance [22]. Eight distinct colonies from the Petri dishes were then processed accordingly. Biological material from these colonies was transferred into an Eppendorf flask with 300 μL of distilled water, mixed thoroughly, and centrifuged at 10,000× *g* for two minutes using a ROTOFIX 32A centrifuge from ITES in Vranov, Slovakia. Following centrifugation, 900 μL of ethanol was added, and after removal of the supernatant, the pellet was air-dried at room temperature (20 °C). Finally, 30 μL of 70% formic acid and 30 μL of acetonitrile were added to the pellet. Scores obtained from the MALDI-TOF analysis were interpreted based on the following criteria: scores below 1.700 were considered unreliable, scores between 2.300 and 3.000 indicated highly probable species identification, scores between 2.000 and 2.299 suggested genus identification with potential species identification, and scores between 1.700 and 1.999 indicated likely genus identification.

2.6. Statistic Analysis

Each evaluation was performed three times, and the findings are expressed as mean values ± the standard deviation (SD). Statistical analysis was conducted using a one-way ANOVA (CoStat version 6.451, CoHort Software, Pacific Grove, CA, USA) followed by Duncan's multiple range test (MRT), with significance set at $p \leq 0.05$ for sample differentiation.

Graphical representation was generated using JMP Pro 17.0 software (SAS Institute, Cary, NC, USA).

3. Results

3.1. Chemical Composition of ECEO

The composition of the ECEO was analyzed using GC-MS methodology. A total of 11 compounds were detected, as detailed in Table 1. Eugenol, identified as a predominant phenolic compound, constituted 82.7% of the composition, with β-caryophyllene (9.9%), acetyl eugenol (3.4%), and humulene (1.4%) identified as the other major constituents. Additionally, a chromatogram is depicted in Figure 1 to illustrate the compound's elution profiles.

Table 1. Chemical composition (percentages mean values ± standard deviation) of the ECEO.

N°	Component [1]	LRI [2]	LRI [3]	*Eugenia caryophyllus* EO [4]
1	eugenol	1360	1363	82.7 ± 2.15
2	α-copaene	1365	1368	0.5 ± 0.02
3	isoeugenol	1441	1439	0.1 ± 0.01
4	β-caryophyllene	1460	1457	9.9 ± 0.18
5	humulene	1470	1466	1.3 ± 0.03

Table 1. *Cont.*

N°	Component [1]	LRI [2]	LRI [3]	*Eugenia caryophyllus* EO [4]
6	acetyl eugenol	1522	1525	3.4 ± 0.04
7	δ-cadinene	1530	1533	0.6 ± 0.03
8	trans-calamenene	1540	1536	0.5 ± 0.02
9	α-calacorene	1555	1560	0.1 ± 0.01
10	caryophyllene oxide	1608	1613	0.8 ± 0.02
11	humulene epoxide II	1618	1620	0.1 ± 0.00
	SUM			100.0
	Oxygenated Sesquiterpenes			0.9
	Hydrocarbon Sesquiterpenes			12.9
	Others			86.2

[1] The components are reported according to their elution order on the apolar column; [2] Linear Retention indices measured on the apolar column; [3] Linear Retention indices from literature; [4] Percentage values of *Eugenia caryophyllus* components.

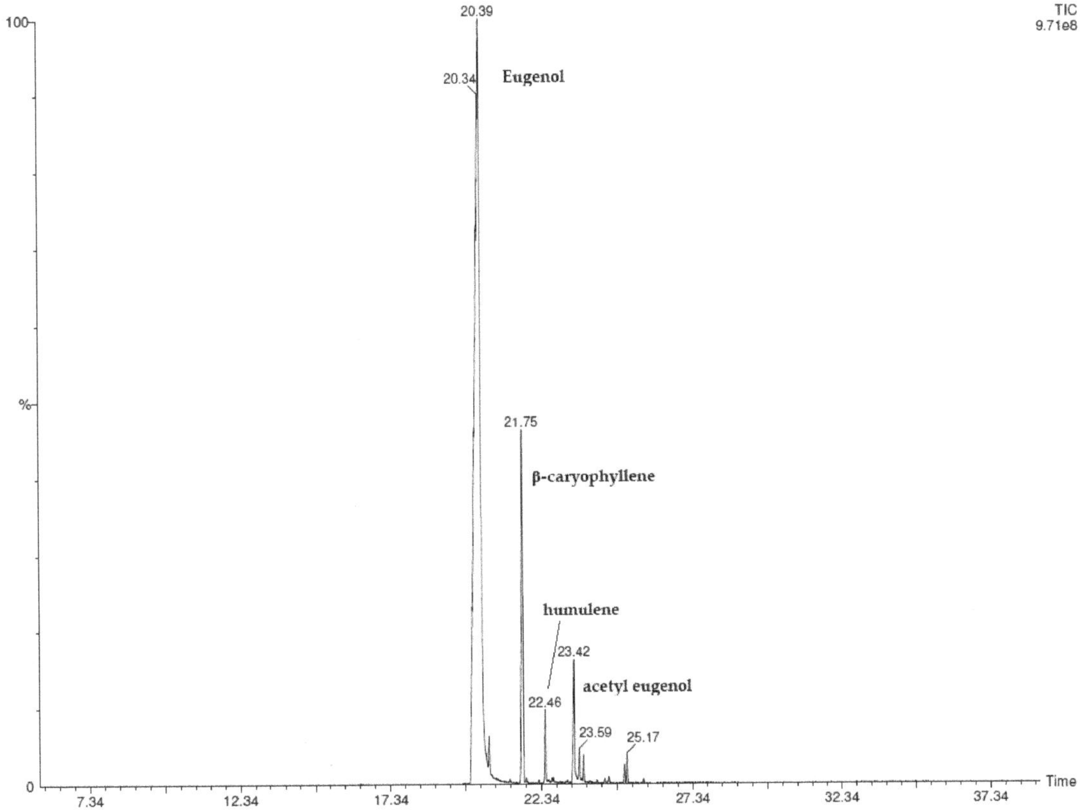

Figure 1. GC-MS chromatogram of the ECEO.

3.2. Antimicrobial Activity of ECEO

Table 2 presents the results of the antimicrobial activity evaluation of the ECEO using the disc diffusion method and minimal inhibitory concentration (MIC). The ECEO exhibited significant antimicrobial activity against *S. enterica*, as evidenced by a zone of inhibition measuring 15.67 mm. In contrast, the antibiotic ciprofloxacin showed superior efficacy with a zone of inhibition measuring 29.67 mm. The MIC values were determined as MIC_{50} at 0.328 ± 0.06 mg/mL and MIC_{90} at 0.384 ± 0.01 mg/mL. Furthermore, the minimal biofilm inhibition concentration, assessed using the crystal violet biofilm assay, was found to be $MBIC_{50}$ at 0.377 ± 0.05 mg/mL and $MBIC_{90}$ at 0.396 ± 0.03 mg/mL.

Table 2. Antimicrobial and antibiofilm activity of the *Eugenia caryophyllus* essential oil (ECEO) against *Salmonella enterica*. Data are presented as mean values ± standard deviation (SD) of three tests.

Inhibition Zone (mm)	ECEO	Ciprofloxacin
Salmonella enterica	15.67 ± 0.58	29.67 ± 0.56
Minimal inhibition concentration (mg/mL)	MIC_{50}	MIC_{90}
Salmonella enterica	0.328 ± 0.06	0.384 ± 0.01
Minimal biofilm inhibition concentration (mg/mL)	$MBIC_{50}$	$MBIC_{90}$
Salmonella enterica	0.377 ± 0.05	0.396 ± 0.03

Figure 2A–F depicts the MS spectra of different stages of *S. enterica* development treated with the ECEO on glass and stainless-steel surfaces. The spectra of planktonic cells, used as controls, are also presented. On the third day of treatment (SEPC 3, SEG 3, and SES 3), differences in protein spectra numbers were observed between the experimental and control groups, indicating some variation. However, similarities in spectrum evolution suggested that both groups were synthesizing similar proteins. By the fifth day (SEPC 5, SEG 5, and SES 5), distinct differences in mass spectrum evolution were noted between biofilms on both surfaces, suggesting the ECEO's impact on biofilm stability. Significant variations were evident by the seventh day, particularly in the spectra of biofilms on plastic and stainless steel, indicating effective disruption by the ECEO. Nonetheless, some similarities in spectrum evolution persisted until the experiment's conclusion. These findings demonstrate the ECEO's ability to disrupt *S. enterica* biofilm homeostasis, with notable effects observed from day 3 to day 7 on both surfaces. This suggests that higher concentrations of the ECEO could potentially suppress biofilm development effectively over extended periods.

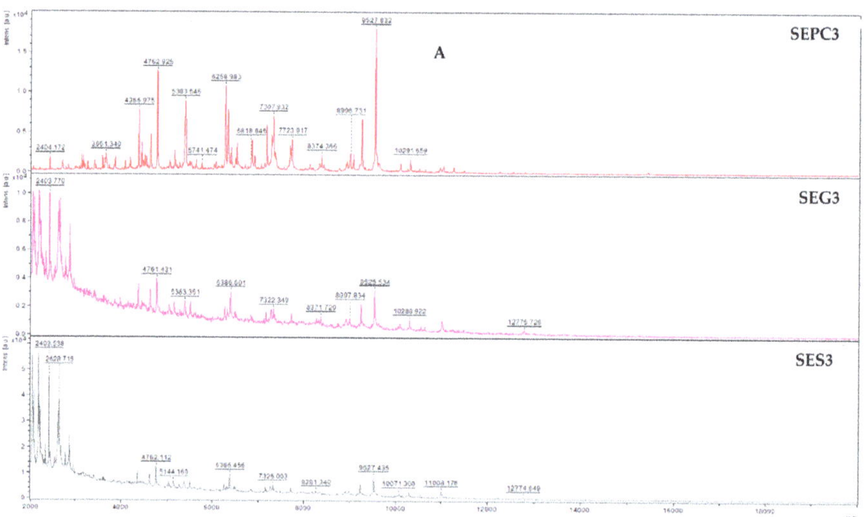

Figure 2. *Cont.*

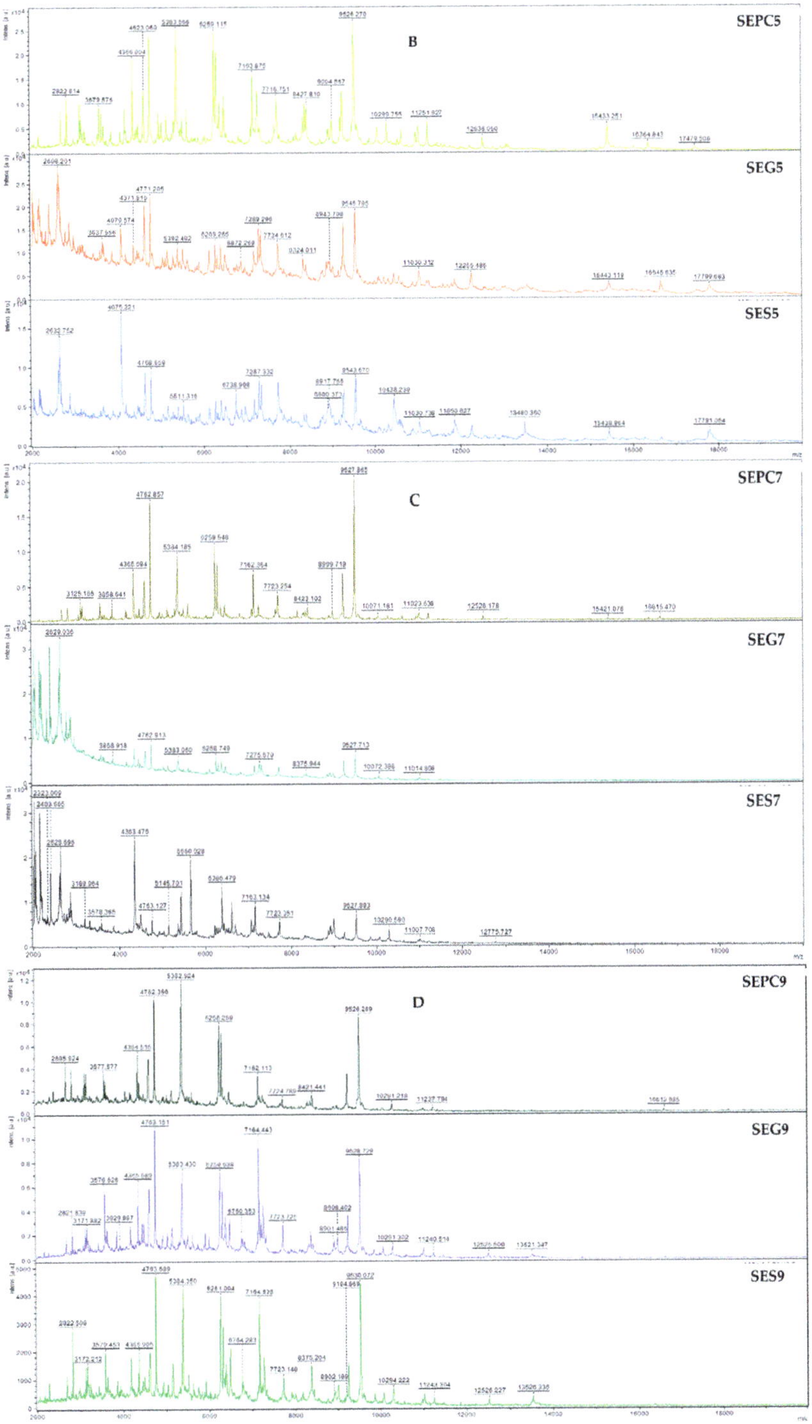

Figure 2. *Cont.*

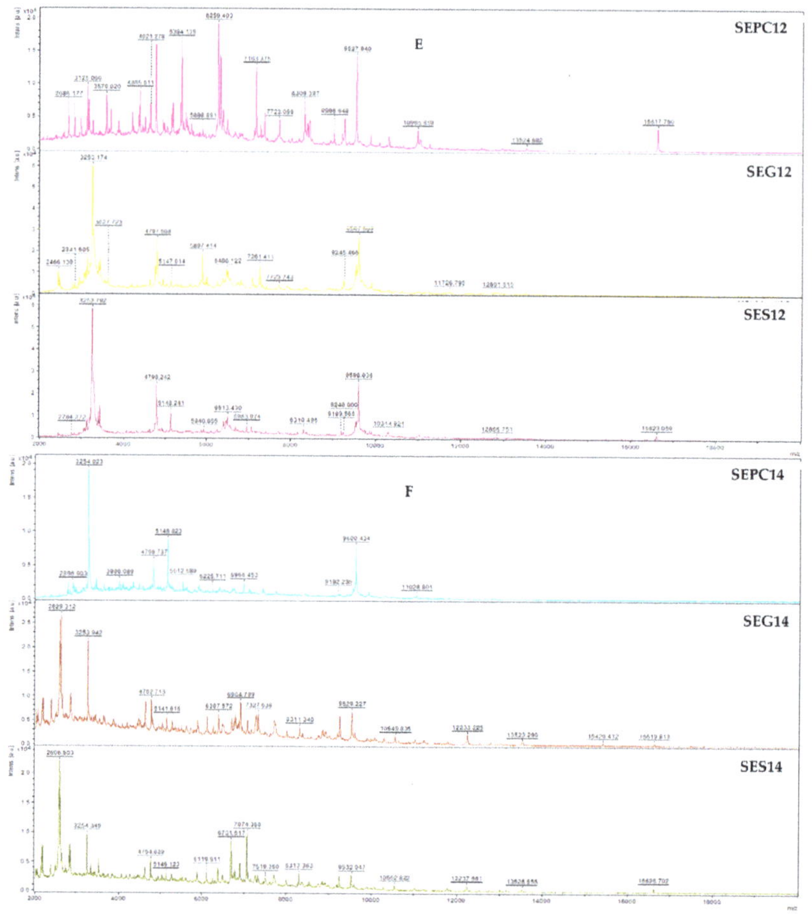

Figure 2. Representative MALDI-TOF mass spectra of *S. enterica*: (**A**) 3rd day; (**B**) 5th day; (**C**) 7th day; (**D**) 9th day; (**E**) 12th day; (**F**) 14th day. SE = *S. enterica*; G = glass; S = stainless-steel; and PC = planktonic cells.

The dendrogram depicted in Figure 3 illustrates that on days 3 and 9, as well as during the early stages of biofilm formation on the experimental glass surface on days 5 and 7, the control groups exhibited the smallest MSP distances. Comparatively, the MSP distance was higher for the stainless-steel surface than for the glass surface, indicating that the ECEO had a more pronounced inhibitory effect on *S. enterica* biofilms on stainless steel. However, both glass and stainless-steel surfaces showed the greatest increase in MSP distance on days 9 and 14 of the experiment. Another aspect investigated was the minimal spectral peak (MSP) distances between planktonic cells and controls. Throughout the study, the MSP distances in the experimental group increased. Specifically, on the third day of the trial with the stainless-steel surface, the MSP distance in the experimental group was the shortest. By days 9 and 14, particularly on the plastic surface, the MSP distance for the experimental group had reached its maximum length. Similar trends were observed on days 5 and 7. Our study's findings highlight the inhibitory and detrimental effects of the ECEO on *S. enterica* biofilm growth on both stainless steel and glass surfaces.

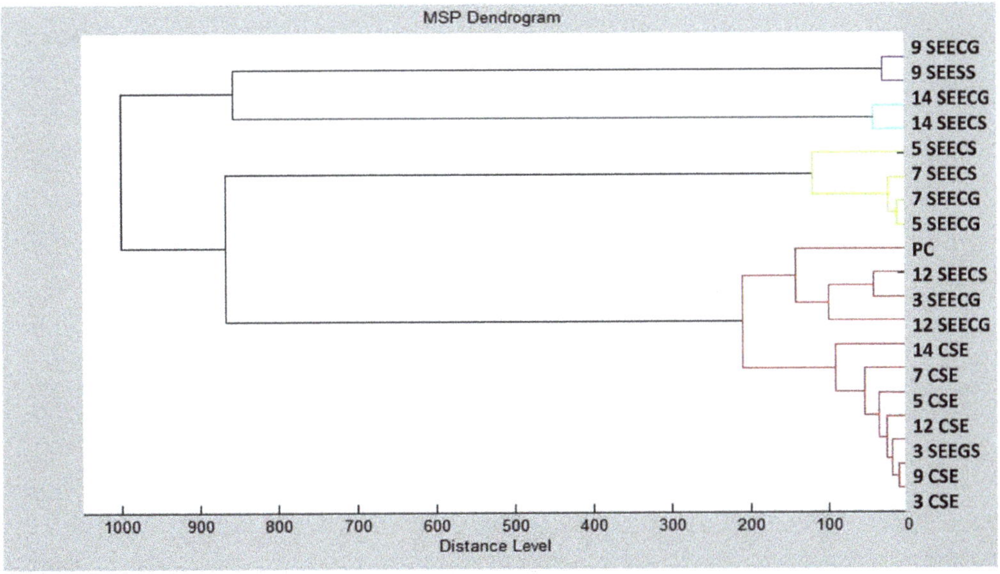

Figure 3. Dendrogram of *S. enterica* generated using MSPs of the planktonic cells and the control. SE = *S. enterica*; C = glass; S = stainless-steel; and PC = planktonic cells.

3.3. Sous Vide Red Deer Meet Microbiological Analyses

Figure 4 and Table S1 present the total viable count (TVC) of sous vide red deer meat samples subjected to different temperatures, times, the ECEO, and *S. enterica* treatments. Raw, uncooked, and unpackaged red deer meat served as control samples. Initial assessments on day 0 showed a TVC of 2.95 ± 0.06 log CFU/g with no presence of coliform bacteria.

On day 1, the TVC ranged from 2.00 log CFU/g (55 °C for 20 min) to 3.21 log CFU/g (50 °C for 5 min) in the sous vide red deer meat control group. Vacuum-packaged samples exhibited lower TVC compared to non-vacuum-packaged ones (Table S1). Specifically, vacuum-packaged samples ranged from 1.58 log CFU/g (55 °C for 20 min) to 3.08 log CFU/g (50 °C for 5 min), the ECEO-treated samples ranged from 1.20 log CFU/g (55 °C for 20 min) to 2.89 log CFU/g (50 °C for 5 min), and samples treated with both the ECEO and *S. enterica* ranged from 1.51 log CFU/g (60 °C for 5 min) to 3.44 log CFU/g (50 °C for 5 min). These results were recorded on the first day of storage.

By day 7, the TVC in the control group ranged from 1.14 log CFU/g (65 °C for 10 min) to 3.76 log CFU/g (50 °C for 5 min). Vacuum-packaged sous vide deer meat ranged from 1.75 log CFU/g (55 °C for 20 min) and the ECEO-treated samples ranged from 1.82 log CFU/g (55 °C for 10 min) to 3.16 log CFU/g (50 °C for 5 min) (Figure 4, Table S1). Samples treated solely with the ECEO on day 7 ranged from 1.82 log CFU/g (55 °C for 5 min) to 3.16 log CFU/g (50 °C for 5 min). For samples with S. enterica application, the TVC ranged from 1.26 log CFU/g (60 °C for 20 min) to 3.68 log CFU/g (50 °C for 5 min), and those treated with both the ECEO and *S. enterica* ranged from 1.67 log CFU/g.

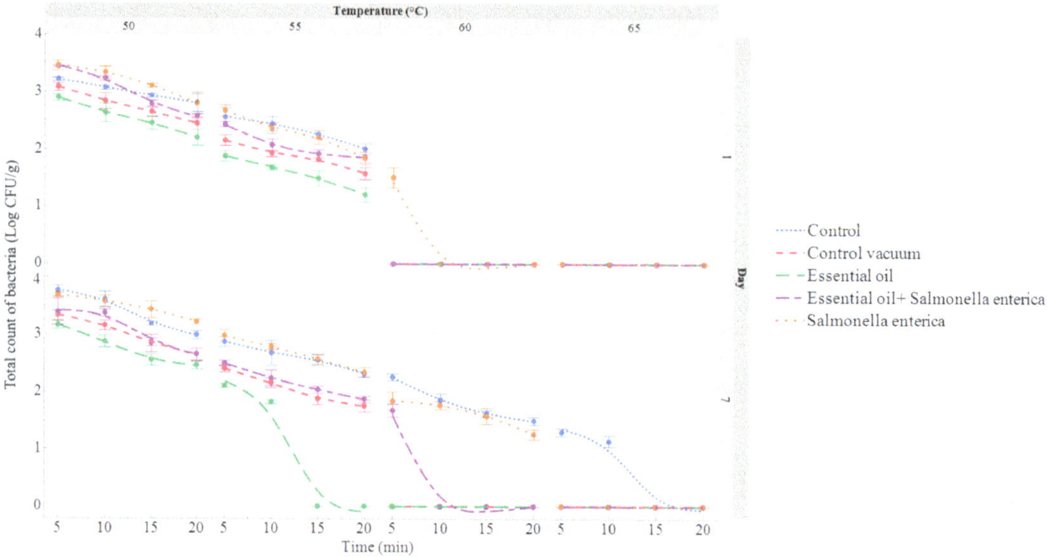

Figure 4. Total viable count (log CFU/g) of sous vide deer meat samples after 1 and 7 days of storage, treated in a water bath at temperatures between 50 and 65 °C for 5 to 20 min. Data are the mean (bars indicate ± SD) of 3 deer meat samples. Control: deer meat samples placed in polyethylene bags without vacuum. Control vacuum: deer meat samples vacuum-packed in polyethylene bags. Essential oil: deer meat samples treated with 1% ECEO and vacuum-packed. *Salmonella enterica*: deer meat samples inoculated with *S. enterica* and vacuum-packed. Essential oil + *Salmonella enterica:* deer meat samples treated with 1% ECEO and inoculated with *S. enterica* and vacuum-packed.

The number of coliform bacteria (CB) in samples of sous vide red deer meat is shown in Figure 5 and Table S2. On day 0, CB counts were zero. In the control group, packaged under aerobic conditions using polyethylene bags, CB were detected at 2.05 log CFU/g only in the first treatment of temperature and time. Coliform bacteria were first observed in sous vide deer meat samples on day seven. In the group where *S. enterica* was applied, CB ranged from 1.71 log CFU/g at a temperature treatment of 50 °C for 20 min to 2.69 log CFU/g at 50 °C for 5 min. In the group treated with the ECEO along with *S. enterica*, CB ranged from 1.52 log CFU/g at 50 °C for 20 min to 2.52 log CFU/g at 50 °C for 5 min. By day 7, CB counts in the control group ranged from 1.85 log CFU/g for the group treated for 5 min at 55 °C to 3.04 log CFU/g for the group treated for 5 min at 50 °C. In the vacuum-packed control group, CB counts ranged from 1.33 log CFU/g for the group treated at 55 °C for 5 min to 2.87 log CFU/g for the group treated at 50 °C for 5 min (Figure 5 and Table S2). In the group where the ECEO was applied, the number of CB was zero. On day 7 in the group with *S. enterica* application, CB ranged from 2.41 log CFU/g at 50 °C for 20 min to 3.12 log CFU/g at 50 °C for 5 min. CB counts ranged from 2.52 log CFU/g at 50 °C for 15 min to 3.25 log CFU/g in the group treated with both the ECEO and *S. enterica* application.

As depicted in Figure 6 and Table S3, *S. enterica* counts were only detected in the last two groups throughout the storage period. On day 1, counts in the group inoculated with *S. enterica* ranged from 1.97 log CFU/g (50 °C for 20 min) to 2.94 log CFU/g (50 °C for 5 min). In contrast, the group treated with the ECEO and inoculated with *S. enterica* showed counts ranging from 1.66 log CFU/g (50 °C for 20 min) to 2.66 log CFU/g (50 °C for 5 min) (Table S3). By day 7, counts in the *S. enterica* group ranged from 2.24 log CFU/g (50 °C for 20 min) to 2.94 log CFU/g (50 °C for 5 min), while in the group treated with the ECEO and inoculated with *S. enterica*, counts ranged from 1.78 log CFU/g (50 °C for 15 min) to 2.85 log CFU/g (50 °C for 5 min).

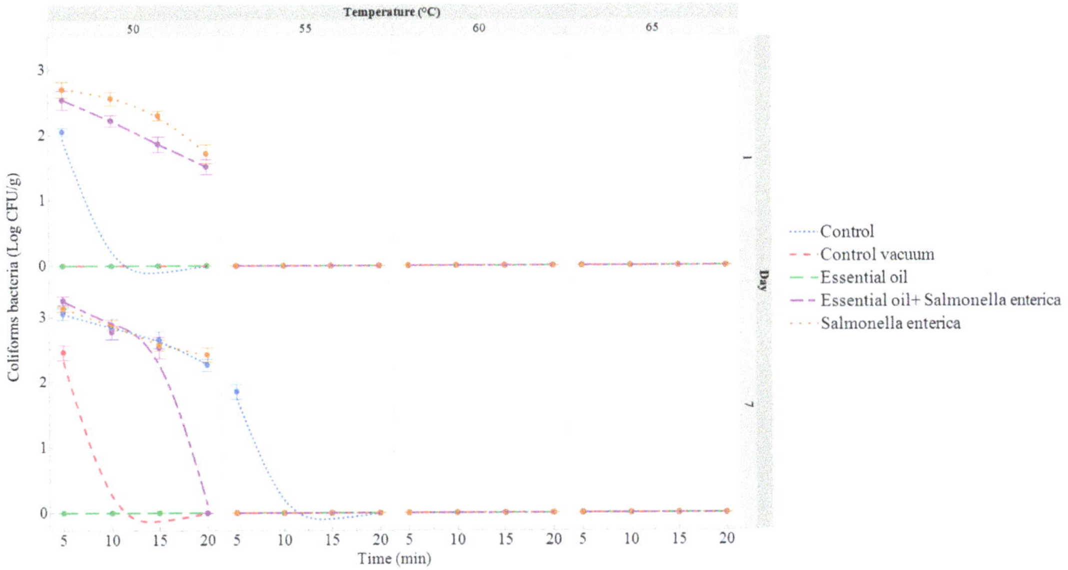

Figure 5. Total coliform bacteria (log CFU/g) of sous vide deer meat samples after 1 and 7 days of storage, treated in a water bath at temperatures between 50 and 65 °C for 5 to 20 min. Data are the mean (bars indicate ± SD) of 3 deer meat samples. Control: deer meat samples placed in polyethylene bags without vacuum. Control vacuum: deer meat samples vacuum-packed in polyethylene bags. Essential oil: deer meat samples treated with 1% ECEO and vacuum-packed. *Salmonella enterica*: deer meat samples inoculated with *S. enterica* and vacuum-packed. Essential oil + *Salmonella enterica:* deer meat samples treated with 1% ECEO and inoculated with *S. enterica* and vacuum-packed.

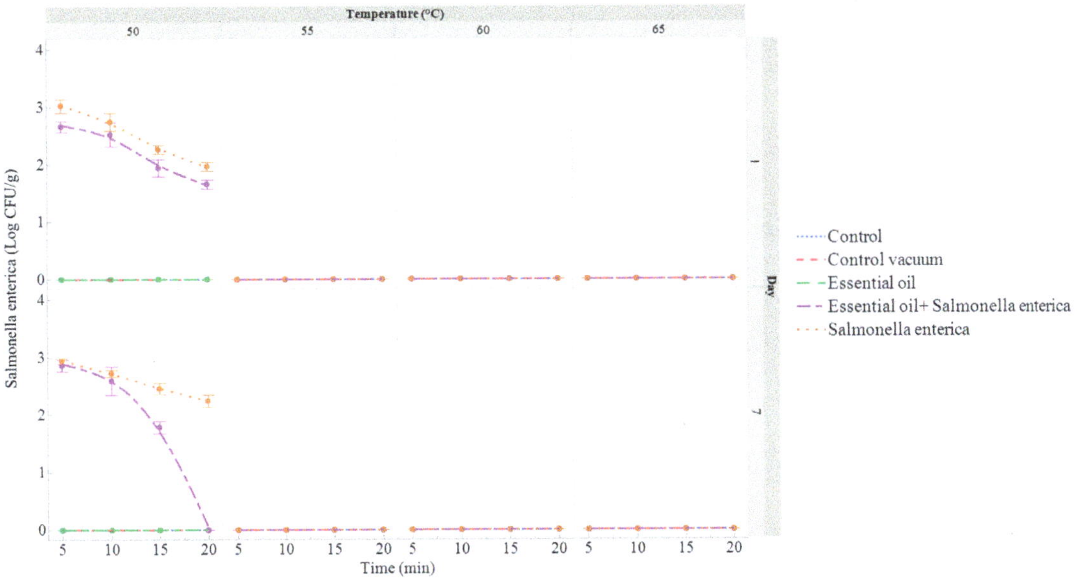

Figure 6. *Salmonella enterica* count (log CFU/g) of sous vide deer meat samples after 1 and 7 days of storage, treated in a water bath at temperatures between 50 and 65 °C for 5 to 20 min. Data are the mean

(bars indicate ± SD) of 3 deer meat samples. Control: deer meat samples placed in polyethylene bags without vacuum. Control vacuum: deer meat samples vacuum-packed in polyethylene bags. Essential oil: deer meat samples treated with 1% ECEO and vacuum-packed. *Salmonella enterica*: deer meat samples inoculated with *S. enterica* and vacuum-packed. Essential oil + *Salmonella enterica:* deer meat samples treated with 1% ECEO and inoculated with *S. enterica* and vacuum-packed.

Figure 7 displays species, genera, and families isolated from red deer sous vide meat samples on the first day of storage. A total of 377 isolates were identified using mass spectrometry with scores up to 2. These isolates belonged to 13 families, 17 genera, and 32 species. The most frequently identified species were from the families Pseudomonadaceae and Bacillaceae. The predominant species isolated from the sous vide deer meat samples on the first day included *S. enterica* (13%), which was intentionally inoculated into the meat, followed by *Pseudomonas fragi*, *Pseudomonas gessardii*, *Pseudomonas graminis*, *Pseudomonas libanensis*, and *Pseudomonas lundensis*, each with a frequency of 6%. *Hafnia alvei* and *Bacillus cereus* were also isolated with a frequency of 5% each.

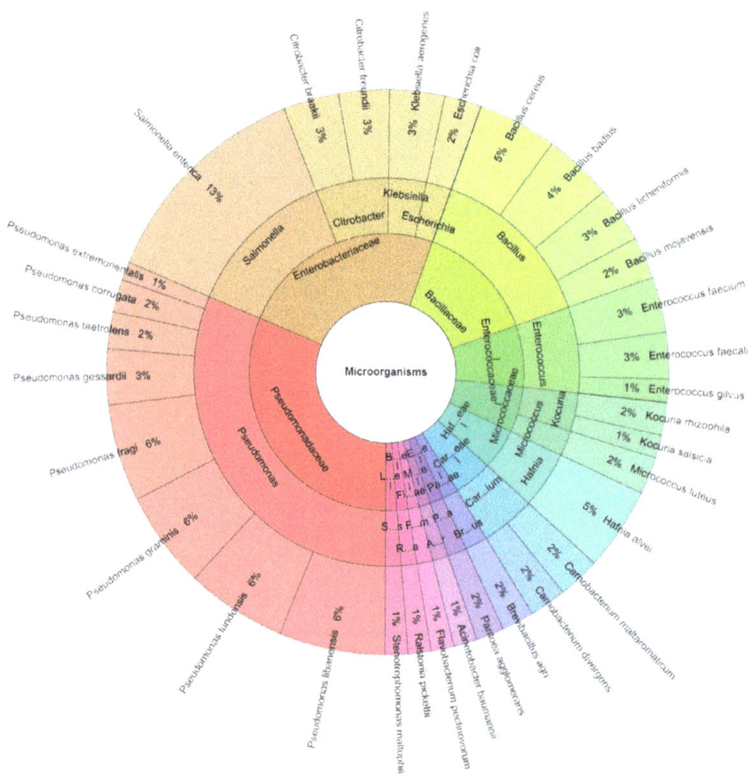

Figure 7. Krona chart: Isolated species, genera, and families from deer sous vide meat at 1 day.

Figure 8 shows species, genera, and families isolated from red deer sous vide meat samples after seven days of storage. A total of 388 isolates were identified using MALDI-TOF MS Biotyper with high scores. These isolates belonged to eleven families, thirteen genera, and twenty-seven species. The most frequently isolated family was Pseudomonadaceae, comprising nine species. The most commonly isolated species was *S. enterica* (17%) in the groups inoculated with this bacterium, followed by *Pseudomonas fragi*, *Pseudomonas lundensis*, and *Pseudomonas taetrolens*, each accounting for 7%.

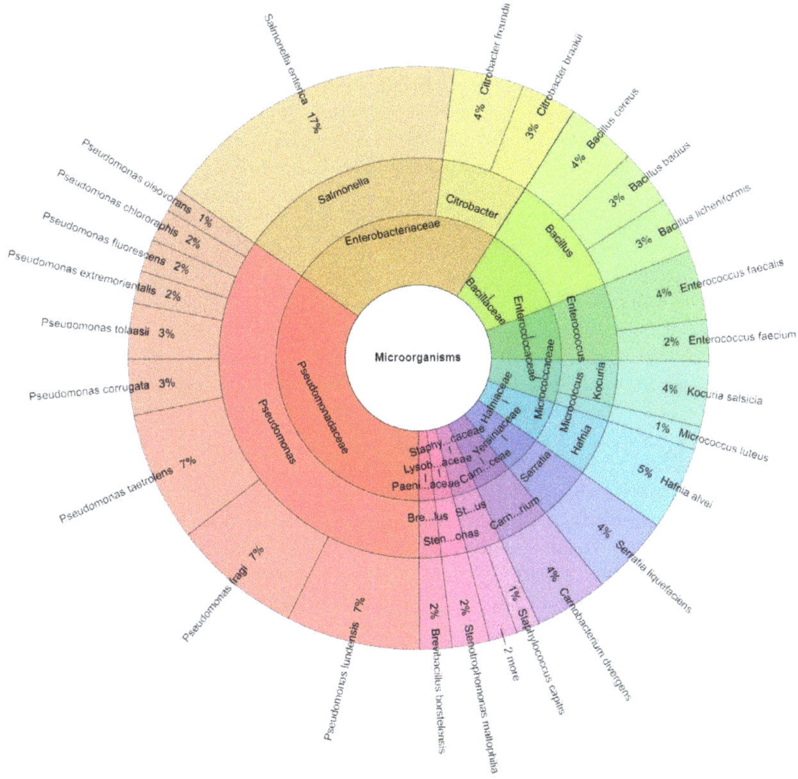

Figure 8. Krona chart: Isolated species, genera, and families from deer sous vide meat after 7 days.

4. Discussion

In our work, the ECEO consisted mainly of eugenol (82.7%). Comparing published data becomes challenging due to variations in EO composition, extraction methods, types of organisms tested, growth phases, inoculum volumes, culture media, pH levels, and incubation conditions, all of which can influence experimental outcomes [27–29]. Despite methodological differences, reported inhibition zones and minimum inhibitory concentrations (MICs) often show similarity across studies. Notably, much of the data focuses on *S. enterica*, a commonly used model organism [30]. Our study investigated the antimicrobial potential of clove essential oil against *S. enterica* using both the disc diffusion method and MIC determination. A different study indicates that the ECEO and its components are eugenol, β-caryophyllene, and acetyl eugenol, which exhibit significant antibacterial activity against a broad spectrum of microorganisms, encompassing Gram-positive and Gram-negative bacteria including *Streptococcus* species, *Staphylococcus aureus*, *Salmonella typhimurium*, *Escherichia coli*, *Klebsiella pneumoniae*, *Pseudomonas aeruginosa*, and *Agrobacterium* spp. Additionally, it effectively inhibits yeasts such as *Aspergillus* species, *Penicillium*, and *Candida albicans* [31]. Clove essential oil, extracted from the dried flower buds of the clove tree, is renowned for its antibacterial and antioxidant characteristics, primarily due to phenolic compounds such as eugenol. It has demonstrated effectiveness against foodborne pathogens, including *Salmonella*, *Escherichia coli*, *Staphylococcus*, and *Listeria* [32]. In our study, a 15.67 mm inhibition zone was observed for clove's antimicrobial activity against *S. enterica*. In a different investigation, the zones of inhibition against *S. aureus*, *E. coli*, *L. monocytogenes*, and *S.* Typhimurium were measured as 2.83 cm, 2.81 cm, 2.47 cm, and 2.22 cm, respectively [33]. The existence and dimensions of inhibition zones serve as indicators of bacteria sensitivity to the essential oil. Typically, if the inhibition zone measures less than 0.7 cm, the sample is deemed inactive against the bacteria. Conversely, an inhibition zone diameter exceeding 1.2 cm indicates effective inhibitory efficacy [34]. Therefore, clove essential oil demonstrated effective inhibition against all tested bacteria in our study, consistent with its well-documented broad spectrum of inhibitory effects. The lipophilic properties of clove essential oil are likely responsible for disrupting bacterial cell membranes, thereby affecting their permeability [35]. Furthermore, our study revealed that clove essential oil exhibited inhibitory effects at concentrations as low as 0.328 mg/mL against *S. enterica.* According to Duarte et al. [36], essential oils with MICs up to 0.5 mg/mL are classified as having strong antimicrobial activity, while those with MICs between 0.6 and 1.5 mg/mL are considered moderate, and MICs above 1.6 mg/mL indicate weak activity. The MIC value of 0.328 mg/mL reported in our study aligns with the potent antibacterial activity noted by Silvestri et al. [37], who reported an MIC of 0.3 mg/mL for clove essential oil. For *Escherichia coli*, the MIC of clove essential oil in our study was 0.3047 mg/mL, lower than the range of 0.400–0.600 mg/mL reported by Silvestri et al. [37]. Similarly, the MIC for *S.* Typhimurium was lower in our study (0.0400 mg/mL) compared to values reported by Beraldo et al. [38]. MICs *for L. monocytogenes* were also lower than those reported by Beraldo et al. [38], where the MIC was 0.800 mg/mL. Discrepancies in MIC values across different studies may be attributed to variations in sample culture conditions, concentrations of components, and techniques used for essential oil extraction [39].

Microorganisms can develop heightened resistance to antimicrobial treatments when they form biofilms and intricate and spatially organized communities [40]. According to Zhao et al. [41], many pathogenic bacteria, including the *S. enterica* studied here, can cause illness through biofilm formation. Consequently, preventing the production and growth of biofilms poses a significant challenge in managing pathogenic bacteria and has become a pressing therapeutic concern [42,43]. Our study demonstrated strong antibiofilm activity using violet crystal assays. Another study using crystal violet staining highlighted the robust biofilm-forming capabilities of *Salmonella* Derby. Further investigation is required to explore the antibiofilm potential of EOs at sub-inhibitory concentrations from an economic perspective, and the specific mechanisms through which different EOs hinder *Salmonella* biofilm development remain unclear [44]. Clove essential oil, rich in eugenol, has been extensively researched for its ability to inhibit bacterial biofilms, including those formed by *Salmonella* Typhimurium, *Escherichia coli* O157, *Listeria monocytogenes*, and *Staphylococcus aureus* [42,45–47]. Although research has mainly focused on *S.* Typhimurium and *Salmonella* Enteritidis, there is limited exploration into the anti-biofilm effects of EOs on *Salmonella* Derby [48,49].

In our study, the ECEO, which has eugenol, β-caryophyllene, and acetyl eugenol as the main components, demonstrated effectiveness in disrupting the equilibrium of biofilms. However, from the beginning to the end of the experiment, there was a consistent resemblance in the progression of mass spectra between the experimental and control groups. Nonetheless, the data clearly indicate that the ECEO had a significant impact on disturbing the homeostasis of *S. enterica* biofilms. The sustained effectiveness of the ECEO on both surfaces suggests long-lasting benefits. In another study, the effect of the ECEO on the biofilms of *Bacillus subtilis* and *Stenotrophomonas maltophilia* was investigated using MALDI-TOF MS Biotyper to detect changes in molecular structures associated with growth suppression. To enhance clarity, the planktonic cell spectrum was substituted for the control spectrum, as the planktonic and biofilm spectra in the control group developed similarly. Each day during the experiment, spectra depicting planktonic growth in the control group and experimental spectra from various surfaces (wood and glass) were illustrated [25].

In our study, we assessed the microbiological quality of sous vide deer meat samples throughout a 7-day shelf-life period. Specifically, we assessed total counts and the presence of coliform bacteria and *Salmonella.* Our findings indicated that the microbial load decreased with higher temperatures and longer heat treatment times, with the lowest counts observed in groups treated with the ECEO. Clove essential oil demonstrated significant antibacterial activity against *S. enterica*, particularly under more intense heat treatments. Badei et al. [50] previously highlighted its effectiveness in inhibiting microbial growth in cookies, and it has been shown to combat both Gram-positive and Gram-negative bacteria, including reducing *E. coli* levels in ground beef and fermented sausages [51]. Recently, the ECEO has been employed as a natural food preservative and coloring agent due to its antibacterial properties and health benefits [52,53]. It has also been suggested as an alternative to acetic acid, sodium bicarbonate, and chlorine-based disinfectants for washing fresh-cut vegetables to mitigate microbial risks and extend shelf life [52,54]. Eugenol, constituting over 80% of the ECEO, primarily contributes to its antiseptic properties [55,56]. Studies by Latifah-Munirah et al. [55] and Rajkowska et al. [57] have demonstrated the effectiveness of eugenol and the ECEO against various pathogenic bacteria, including *S. enterica*, *S. aureus*, *E. coli*, *Proteus mirabilis*, *Pseudomonas aeruginosa*, and *Streptococcus mutans*. However, detailed investigations into the molecular antibacterial mechanisms of eugenol and the ECEO are limited, with current studies mainly focusing on their initial antibacterial activities [43,58]. Comprehensive studies, specifically on the effects of eugenol on foodborne microorganisms, are currently lacking. Nonetheless, due to its bactericidal, analgesic, antioxidant, and potential anti-cancer properties, the ECEO holds promise for diverse applications in the food and health sectors [59–61].

5. Conclusions

Our findings demonstrate the great effectiveness of 1.0% ECEO applied to deer meat combined with vacuum packaging against *Salmonella enterica*, coliform bacteria, and total viable count. Food shelf life is positively impacted, and food safety is improved when microorganisms in food are rendered inactive. As a natural antibacterial with a mild flavor, the ECEO can be used to keep vacuum-packaged deer meat fresher longer. Additional research is required to improve the inhibition of total viable numbers. This study concludes by highlighting the potential antibacterial and antibiofilm qualities of the ECEO, both *in vitro* and in relation to the preservation of meat. These characteristics point to its possible use in food preservation to prevent food spoiling and guarantee food safety, especially when paired with other modern packaging technologies and processing methods. *Salmonella enterica* contamination can be prevented by using the ECEO, which could prolong the shelf life of sous vide red deer meat while maintaining quality and safety standards.

Supplementary Materials: The following supporting information can be downloaded at: https://www.mdpi.com/article/10.3390/foods13162512/s1, Table S1. Total viable count (log CFU/g) of sous vide red deer meat samples after storage 1, and 7 days treated in a water bath at temperatures between 50 and 65 °C for 5 to 20 min. Data are the mean (bars indicate ± SD) of 3 red deer meat samples; Table S2. Total coliform bacteria (log CFU/g) of sous vide red deer meat samples after storage 1, and 7 days treated in a water bath at temperatures between 50 and 65 °C for 5 to 20 min. Data are the mean (bars indicate ± SD) of 3 red deer meat samples; Table S3. *S. enterica* count (log CFU/g) of sous vide red deer meat samples after storage for 1 and 7 days treated in a water bath at temperatures between 50 and 65 °C for 5 to 20 min. Data are the mean (bars indicate ± SD) of 3 red deer meat samples.

Author Contributions: Conceptualization, M.K., N.Č., and A.B., methodology M.K., S.G., A.B.H., Z.B., J.H.E.-L., M.I.K., R.B.S., P.H., N.Č., B.W.-R., J.K., and A.B.; investigation, M.K.; writing-original draft M.K., S.G., A.B.H., Z.B., J.H.E.-L., M.I.K., R.B.S., P.H., N.Č., B.W.-R., J.K., and A.B.; formal analyses, M.K., S.G., A.B.H., Z.B., J.H.E.-L., M.I.K., R.B.S., P.H., N.Č., J.K., and A.B.; writing-review & editing, M.K., S.G., A.B.H., Z.B., J.H.E.-L., M.I.K., R.B.S., P.H., N.Č., B.W.-R., J.K., and A.B.; resources, M.K; supervision, M.K. and S.G.; project administration, M.K.; funding acquisition, M.K.; visualization, N.Č., software, A.B.; data curation, A.B. All authors have read and agreed to the published version of the manuscript.

Funding: This research was funded by the grant APVV-20-0058. The potential of the essential oils from aromatic plants for medical use and food preservation.

Data Availability Statement: The original contributions presented in the study are included in the article/Supplementary Materials, further inquiries can be directed to the corresponding author.

Acknowledgments: This research was supported by the grant VEGA 1/0059/24, Chemical properties and biological activity (in vitro, in vivo, and in situ) of plant volatile mixtures, their main components, and inclusion systems.

Conflicts of Interest: The authors declare no conflicts of interest.

References

1. Sagar, N.A.; Pareek, S. Safe Storage and Preservation Techniques in Commercialized Agriculture. In *Natural Remedies for Pest, Disease and Weed Control*; Elsevier: Amsterdam, The Netherlands, 2020; pp. 221–234. ISBN 978-0-12-819304-4.
2. Bianchi, A.; Taglieri, I.; Zinnai, A.; Macaluso, M.; Sanmartin, C.; Venturi, F. Effect of Argon as Filling Gas of the Storage Atmosphere on the Shelf-Life of Sourdough Bread—Case Study on PDO Tuscan Bread. *Foods* **2022**, *11*, 3470. [CrossRef] [PubMed]
3. Kačániová, M.; Čmiková, N.; Vukovic, N.L.; Verešová, A.; Bianchi, A.; Garzoli, S.; Ben Saad, R.; Ben Hsouna, A.; Ban, Z.; Vukic, M.D. *Citrus Limon* Essential Oil: Chemical Composition and Selected Biological Properties Focusing on the Antimicrobial (In Vitro, In Situ), Antibiofilm, Insecticidal Activity and Preservative Effect against *Salmonella Enterica* Inoculated in Carrot. *Plants* **2024**, *13*, 524. [CrossRef] [PubMed]
4. Santiesteban-López, N.A.; Gómez-Salazar, J.A.; Santos, E.M.; Campagnol, P.C.B.; Teixeira, A.; Lorenzo, J.M.; Sosa-Morales, M.E.; Domínguez, R. Natural Antimicrobials: A Clean Label Strategy to Improve the Shelf Life and Safety of Reformulated Meat Products. *Foods* **2022**, *11*, 2613. [CrossRef] [PubMed]
5. Tongnuanchan, P.; Benjakul, S. Essential Oils: Extraction, Bioactivities, and Their Uses for Food Preservation. *J. Food Sci.* **2014**, *79*, R1231–R1249. [CrossRef] [PubMed]
6. Alsaffar, A.A. Sustainable Diets: The Interaction between Food Industry, Nutrition, Health and the Environment. *Food Sci. Technol. Int.* **2016**, *22*, 102–111. [CrossRef]
7. Heredia, N.; García, S. Animals as Sources of Food-Borne Pathogens: A Review. *Anim. Nutr.* **2018**, *4*, 250–255. [CrossRef] [PubMed]
8. Dawoud, T.M.; Davis, M.L.; Park, S.H.; Kim, S.A.; Kwon, Y.M.; Jarvis, N.; O'Bryan, C.A.; Shi, Z.; Crandall, P.G.; Ricke, S.C. The Potential Link between Thermal Resistance and Virulence in Salmonella: A Review. *Front. Vet. Sci.* **2017**, *4*, 93. [CrossRef] [PubMed]
9. Pakdel, M.; Olsen, A.; Bar, E.M.S. A Review of Food Contaminants and Their Pathways Within Food Processing Facilities Using Open Food Processing Equipment. *J. Food Prot.* **2023**, *86*, 100184. [CrossRef]
10. Teshome, E.; Forsido, S.F.; Rupasinghe, H.P.V.; Olika Keyata, E. Potentials of Natural Preservatives to Enhance Food Safety and Shelf Life: A Review. *Sci. World J.* **2022**, *2022*, 9901018. [CrossRef]
11. Hanková, K.; Lupoměská, P.; Nový, P.; Všetečka, D.; Klouček, P.; Kouřimská, L.; Hlebová, M.; Božik, M. Effect of Conventional Preservatives and Essential Oils on the Survival and Growth of *Escherichia Coli* in Vegetable Sauces: A Comparative Study. *Foods* **2023**, *12*, 2832. [CrossRef]
12. Falleh, H.; Ben Jemaa, M.; Saada, M.; Ksouri, R. Essential Oils: A Promising Eco-Friendly Food Preservative. *Food Chem.* **2020**, *330*, 127268. [CrossRef] [PubMed]
13. Huang, Y.; Ho, S.-H.; Lee, H.-C.; Yap, Y.-L. Insecticidal Properties of Eugenol, Isoeugenol and Methyleugenol and Their Effects on Nutrition of *Sitophilus Zeamais* Motsch. (Coleoptera: Curculionidae) and *Tribolium Castaneum* (Herbst) (Coleoptera: Tenebrionidae). *J. Stored Prod. Res.* **2002**, *38*, 403–412. [CrossRef]
14. Nuñez, L.; D' Aquino, M. Microbicide Activity of Clove Essential Oil (*Eugenia Caryophyllata*). *Braz. J. Microbiol.* **2012**, *43*, 1255–1260. [CrossRef] [PubMed]
15. Matan, N.; Rimkeeree, H.; Mawson, A.J.; Chompreeda, P.; Haruthaithanasan, V.; Parker, M. Antimicrobial Activity of Cinnamon and Clove Oils under Modified Atmosphere Conditions. *Int. J. Food Microbiol.* **2006**, *107*, 180–185. [CrossRef] [PubMed]
16. Smith-Palmer, A.; Stewart, J.; Fyfe, L. The Potential Application of Plant Essential Oils as Natural Food Preservatives in Soft Cheese. *Food Microbiol.* **2001**, *18*, 463–470. [CrossRef]
17. Slobodníková, L.; Fialová, S.; Rendeková, K.; Kováč, J.; Mučaji, P. Antibiofilm Activity of Plant Polyphenols. *Molecules* **2016**, *21*, 1717. [CrossRef]
18. Seyedtaghiya, M.; Nayeri Fasaei, B.; Peighambari, S.M. Antimicrobial and Antibiofilm Effects of *Satureja Hortensis* Essential Oil against *Escherichia Coli* and *Salmonella* Isolated from Poultry. *Iran. J. Microbiol.* **2021**, *13*, 74–80. [CrossRef] [PubMed]
19. Campos, F.M.; Couto, J.A.; Hogg, T.A. Influence of Phenolic Acids on Growth and Inactivation of *Oenococcus Oeni* and Lactobacillus Hilgardii. *J. Appl. Microbiol.* **2003**, *94*, 167–174. [CrossRef]
20. Mazzonetto, F.; Vendramim, J.D. Efeito de Pós de Origem Vegetal Sobre *Acanthoscelides Obtectus* (Say) (Coleoptera: Bruchidae) Em Feijão Armazenado. *Neotrop. Entomol.* **2003**, *32*, 145–149. [CrossRef]
21. Karyotis, D.; Skandamis, P.N.; Juneja, V.K. Thermal Inactivation of *Listeria Monocytogenes* and *Salmonella* Spp. in Sous-Vide Processed Marinated Chicken Breast. *Food Res. Int.* **2017**, *100*, 894–898. [CrossRef]

22. Kačániová, M.; Čmiková, N.; Kluz, M.I.; Akacha, B.B.; Ben Saad, R.; Mnif, W.; Waszkiewicz-Robak, B.; Garzoli, S.; Ben Hsouna, A. Anti-*Salmonella* Activity of *Thymus* Serpyllum Essential Oil in Sous Vide Cook–Chill Rabbit Meat. *Foods* **2024**, *13*, 200. [CrossRef] [PubMed]
23. Garzoli, S.; Cicaloni, V.; Salvini, L.; Trespidi, G.; Iriti, M.; Vitalini, S. SPME-GC-MS Analysis of the Volatile Profile of Three Fresh Yarrow (*Achillea Millefolium* L.) Morphotypes from Different Regions of Northern Italy. *Separations* **2023**, *10*, 51. [CrossRef]
24. Garzoli, S.; Laghezza Masci, V.; Franceschi, S.; Tiezzi, A.; Giacomello, P.; Ovidi, E. Headspace/GC–MS Analysis and Investigation of Antibacterial, Antioxidant and Cytotoxic Activity of Essential Oils and Hydrolates from *Rosmarinus Officinalis* L. and *Lavandula Angustifolia* Miller. *Foods* **2021**, *10*, 1768. [CrossRef] [PubMed]
25. Kačániová, M.; Galovičová, L.; Valková, V.; Ďuranová, H.; Borotová, P.; Štefániková, J.; Vukovic, N.L.; Vukic, M.; Kunová, S.; Felsöciová, S.; et al. Chemical Composition and Biological Activity of *Salvia Officinalis* Essential Oil. *Acta Hortic. Regiotect.* **2021**, *24*, 81–88. [CrossRef]
26. Kačániová, M.; Vukovic, N.L.; Čmiková, N.; Galovičová, L.; Schwarzová, M.; Šimora, V.; Kowalczewski, P.Ł.; Kluz, M.I.; Puchalski, C.; Bakay, L.; et al. *Salvia Sclarea* Essential Oil Chemical Composition and Biological Activities. *Int. J. Mol. Sci.* **2023**, *24*, 5179. [CrossRef] [PubMed]
27. Burt, S. Essential Oils: Their Antibacterial Properties and Potential Applications in Foods—A Review. *Int. J. Food Microbiol.* **2004**, *94*, 223–253. [CrossRef] [PubMed]
28. Kalemba, D.; Kunicka, A. Antibacterial and Antifungal Properties of Essential Oils. *Curr. Med. Chem.* **2003**, *10*, 813–829. [CrossRef] [PubMed]
29. Kewlani, A.; Nayak, V.; Gohel, T.; Gupta, P. Phytochemical Analysis and Antimicrobial Activity of Selected Medicinal Plants. *Sci. Prepr.* **2022**. [CrossRef]
30. Deng, X.; Liao, Q.; Xu, X.; Yao, M.; Zhou, Y.; Lin, M.; Zhang, P.; Xie, Z. Analysis of Essential Oils from Cassia Bark and Cassia Twig Samples by GC-MS Combined with Multivariate Data Analysis. *Food Anal. Methods* **2014**, *7*, 1840–1847. [CrossRef]
31. Haro-González, J.N.; Castillo-Herrera, G.A.; Martínez-Velázquez, M.; Espinosa-Andrews, H. Clove Essential Oil (*Syzygium Aromaticum* L. Myrtaceae): Extraction, Chemical Composition, Food Applications, and Essential Bioactivity for Human Health. *Molecules* **2021**, *26*, 6387. [CrossRef]
32. Chaieb, K.; Hajlaoui, H.; Zmantar, T.; Kahla-Nakbi, A.B.; Rouabhia, M.; Mahdouani, K.; Bakhrouf, A. The Chemical Composition and Biological Activity of Clove Essential Oil, *Eugenia Caryophyllata* (*Syzigium Aromaticum* L. Myrtaceae): A Short Review. *Phytother. Res.* **2007**, *21*, 501–506. [CrossRef] [PubMed]
33. Radünz, M.; Dos Santos Hackbart, H.C.; Camargo, T.M.; Nunes, C.F.P.; De Barros, F.A.P.; Dal Magro, J.; Filho, P.J.S.; Gandra, E.A.; Radünz, A.L.; Da Rosa Zavareze, E. Antimicrobial Potential of Spray Drying Encapsulated Thyme (*Thymus Vulgaris*) Essential Oil on the Conservation of Hamburger-like Meat Products. *Int. J. Food Microbiol.* **2020**, *330*, 108696. [CrossRef]
34. Arora, D.S.; Kaur, J. Antimicrobial Activity of Spices. *Int. J. Antimicrob. Agents* **1999**, *12*, 257–262. [CrossRef]
35. Radünz, M.; Da Trindade, M.L.M.; Camargo, T.M.; Radünz, A.L.; Borges, C.D.; Gandra, E.A.; Helbig, E. Antimicrobial and Antioxidant Activity of Unencapsulated and Encapsulated Clove (*Syzygium Aromaticum*, L.) Essential Oil. *Food Chem.* **2019**, *276*, 180–186. [CrossRef] [PubMed]
36. Duarte, M.C.T.; Figueira, G.M.; Sartoratto, A.; Rehder, V.L.G.; Delarmelina, C. Anti-Candida Activity of Brazilian Medicinal Plants. *J. Ethnopharmacol.* **2005**, *97*, 305–311. [CrossRef] [PubMed]
37. Silvestri, J.D.F.; Paroul, N.; Czyewski, E.; Lerin, L.; Rotava, I.; Cansian, R.L.; Mossi, A.; Toniazzo, G.; Oliveira, D.D.; Treichel, H. Perfil Da Composição Química e Atividades Antibacteriana e Antioxidante Do Óleo Essencial Do Cravo-Da-Índia (*Eugenia Caryophyllata* Thunb.). *Rev. Ceres* **2010**, *57*, 589–594. [CrossRef]
38. Beraldo, C.; Daneluzzi, N.S.; Scanavacca, J.; Doyama, J.T.; Fernandes Júnior, A.; Moritz, C.M.F. Eficiência de Óleos Essenciais de Canela e Cravo-Da-Índia Como Sanitizantes Na Indústria de Alimentos. *Pesqui. Agropecuária Trop.* **2013**, *43*, 436–440. [CrossRef]
39. Ríos, J.L.; Recio, M.C. Medicinal Plants and Antimicrobial Activity. *J. Ethnopharmacol.* **2005**, *100*, 80–84. [CrossRef] [PubMed]
40. Hou, J.; Miao, L.; Wang, C.; Wang, P.; Ao, Y.; Qian, J.; Dai, S. Inhibitory Effects of ZnO Nanoparticles on Aerobic Wastewater Biofilms from Oxygen Concentration Profiles Determined by Microelectrodes. *J. Hazard. Mater.* **2014**, *276*, 164–170. [CrossRef]
41. Zhao, Z.; Xue, W.; Wang, J.; Zhang, C.; Zhou, D. The Role of Trace P-Hydroxybenzoic Acid to *Chlorella* for Advanced Wastewater Treatment: Mitigating Bacterial Contamination and Boosting Biomass Recovery. *Resour. Conserv. Recycl.* **2023**, *199*, 107229. [CrossRef]
42. Kim, Y.-G.; Lee, J.-H.; Gwon, G.; Kim, S.-I.; Park, J.G.; Lee, J. Essential Oils and Eugenols Inhibit Biofilm Formation and the Virulence of *Escherichia Coli* O157:H7. *Sci. Rep.* **2016**, *6*, 36377. [CrossRef] [PubMed]
43. Wang, F.; Wei, F.; Song, C.; Jiang, B.; Tian, S.; Yi, J.; Yu, C.; Song, Z.; Sun, L.; Bao, Y.; et al. *Dodartia Orientalis* L. Essential Oil Exerts Antibacterial Activity by Mechanisms of Disrupting Cell Structure and Resisting Biofilm. *Ind. Crops Prod.* **2017**, *109*, 358–366. [CrossRef]
44. Liu, Y.; Yan, Y.; Dong, P.; Ni, L.; Luo, X.; Zhang, Y.; Zhu, L. Inhibitory Effects of Clove and Oregano Essential Oils on Biofilm Formation of Salmonella Derby Isolated from Beef Processing Plant. *LWT* **2022**, *162*, 113486. [CrossRef]
45. Zhang, C.; Li, C.; Abdel-Samie, M.A.; Cui, H.; Lin, L. Unraveling the Inhibitory Mechanism of Clove Essential Oil against *Listeria Monocytogenes* Biofilm and Applying It to Vegetable Surfaces. *LWT* **2020**, *134*, 110210. [CrossRef]
46. Cui, H.; Zhang, C.; Li, C.; Lin, L. Inhibition of *Escherichia Coli* O157:H7 Biofilm on Vegetable Surface by Solid Liposomes of Clove Oil. *LWT* **2020**, *117*, 108656. [CrossRef]

47. Mohan, A.; Purohit, A.S. Anti-Salmonella Activity of Pyruvic and Succinic Acid in Combination with Oregano Essential Oil. *Food Control* **2020**, *110*, 106960. [CrossRef]
48. Kadam, S.R.; Den Besten, H.M.W.; Van Der Veen, S.; Zwietering, M.H.; Moezelaar, R.; Abee, T. Diversity Assessment of Listeria Monocytogenes Biofilm Formation: Impact of Growth Condition, Serotype and Strain Origin. *Int. J. Food Microbiol.* **2013**, *165*, 259–264. [CrossRef] [PubMed]
49. Tokam Kuaté, C.R.; Bisso Ndezo, B.; Dzoyem, J.P. Synergistic Antibiofilm Effect of Thymol and Piperine in Combination with Aminoglycosides Antibiotics against Four *Salmonella Enterica* Serovars. *Evid. Based Complement. Altern. Med.* **2021**, *2021*, 1567017. [CrossRef] [PubMed]
50. Badei, A.Z.M.A.; Faheid, S.M.M.; El-Akel, A.T.M.; Mahmoud, B.S.M. Application of Some Spices in Flavoring and Preservation of Cookies: 2-Antimicrobial and Sensory Properties of Cardamom, Cinnamon and Clove. *Dtsch. Lebensm.-Rundsch.* **2002**, *98*, 261–265.
51. Ji, J.; Shankar, S.; Royon, F.; Salmieri, S.; Lacroix, M. Essential Oils as Natural Antimicrobials Applied in Meat and Meat Products—A Review. *Crit. Rev. Food Sci. Nutr.* **2023**, *63*, 993–1009. [CrossRef]
52. Ju, J.; Xie, Y.; Yu, H.; Guo, Y.; Cheng, Y.; Qian, H.; Yao, W. A Novel Method to Prolong Bread Shelf Life: Sachets Containing Essential Oils Components. *LWT* **2020**, *131*, 109744. [CrossRef]
53. Kumar Pandey, V.; Shams, R.; Singh, R.; Dar, A.H.; Pandiselvam, R.; Rusu, A.V.; Trif, M. A Comprehensive Review on Clove (*Caryophyllus Aromaticus* L.) Essential Oil and Its Significance in the Formulation of Edible Coatings for Potential Food Applications. *Front. Nutr.* **2022**, *9*, 987674. [CrossRef] [PubMed]
54. Park, J.-B.; Kang, J.-H.; Song, K.B. Clove Bud Essential Oil Emulsion Containing Benzethonium Chloride Inactivates *Salmonella* Typhimurium and *Listeria Monocytogenes* on Fresh-Cut Pak Choi during Modified Atmosphere Storage. *Food Control* **2019**, *100*, 17–23. [CrossRef]
55. Latifah-Munirah, B.; Himratul-Aznita, W.H.; Mohd Zain, N. Eugenol, an Essential Oil of Clove, Causes Disruption to the Cell Wall of *Candida Albicans* (ATCC 14053). *Front. Life Sci.* **2015**, *8*, 231–240. [CrossRef]
56. Wongsawan, K.; Chaisri, W.; Tangtrongsup, S.; Mektrirat, R. Bactericidal Effect of Clove Oil against Multidrug-Resistant Streptococcus Suis Isolated from Human Patients and Slaughtered Pigs. *Pathogens* **2019**, *9*, 14. [CrossRef] [PubMed]
57. Rajkowska, K.; Otlewska, A.; Kunicka-Styczyńska, A.; Krajewska, A. Candida Albicans Impairments Induced by Peppermint and Clove Oils at Sub-Inhibitory Concentrations. *Int. J. Mol. Sci.* **2017**, *18*, 1307. [CrossRef]
58. Hu, Q.; Zhou, M.; Wei, S. Progress on the Antimicrobial Activity Research of Clove Oil and Eugenol in the Food Antisepsis Field. *J. Food Sci.* **2018**, *83*, 1476–1483. [CrossRef]
59. Das, M.; Roy, S.; Guha, C.; Saha, A.K.; Singh, M. In Vitro Evaluation of Antioxidant and Antibacterial Properties of Supercritical CO2 Extracted Essential Oil from Clove Bud (*Syzygium Aromaticum*). *J. Plant Biochem. Biotechnol.* **2021**, *30*, 387–391. [CrossRef]
60. El-Darier, S.M.; El-Ahwany, A.M.D.; Elkenany, E.T.; Abdeldaim, A.A. An in Vitro Study on Antimicrobial and Anticancer Potentiality of Thyme and Clove Oils. *Rendiconti Lincei Sci. Fis. E Nat.* **2018**, *29*, 131–139. [CrossRef]
61. Kheawfu, K.; Pikulkaew, S.; Rades, T.; Müllertz, A.; Okonogi, S. Development and Characterization of Clove Oil Nanoemulsions and Self-Microemulsifying Drug Delivery Systems. *J. Drug Deliv. Sci. Technol.* **2018**, *46*, 330–338. [CrossRef]

foods

MDPI

Review

Effect of Heat Pasteurization and Sterilization on Milk Safety, Composition, Sensory Properties, and Nutritional Quality

Ahmad Rabbani [1], Mutamed Ayyash [1], Crystal D. C. D'Costa [1], Gang Chen [2], Yajun Xu [3] and Afaf Kamal-Eldin [1,4,*]

1 Department of Food Science, College of Agriculture and Veterinary Medicine, United Arab Emirates University, Al Ain P.O. Box 15551, United Arab Emirates; 202090507@uaeu.ac.ae (A.R.); mutamed.ayyash@uaeu.ac.ae (M.A.)

2 Key Laboratory of Geriatric Nutrition and Health (Beijing Technology and Business University), Ministry of Education, 100048, China; gang.chen@btbu.edu.cn

3 Beijing Key Laboratory of Toxicological Research and Risk Assessment for Food Safety, Department of Nutrition and Food Hygiene, School of Public Health, Peking University, No. 38 Xueyuan Road, Beijing 100083, China; xuyajun@bjmu.edu.cn

4 National Water and Energy Center, United Arab Emirates University, Al Ain P.O. Box 15551, United Arab Emirates

* Correspondence: afaf.kamal@uaeu.ac.ae; Tel.: +971-37136563

Abstract: Milk pasteurization and sterilization by heat treatment have an exciting history, which followed steady steps. The main aim of these treatments is to extend the shelf life of milk by destroying pathogenic and milk spoilage bacteria. With developments in pasteurization techniques, the assurance of milk safety, and extended shelf life, pasteurized bovine milk has become a staple food, especially in Western diets. However, some concerns have recently been raised about the effect of pasteurization on the sensory properties and nutritional quality of milk, and alternative methods, such as high-pressure processing, are being investigated. The primary purpose of milk pasteurization and sterilization is summarized in this review article. The associated changes that affect the compositional, sensory, and nutritional quality of milk are discussed, with particular emphasis on protein structure and function. The review is concluded by considering alternative methods, their advantages and limitations, along with future prospects.

Keywords: pasteurization; sterilization; milk safety; vitamins; minerals; nutritional quality

1. Introduction

After the French revolution, Nicolas Appert made a breakthrough in food preservation in 1809 by describing the appertization process that combines canning and heating [1]. Louis Pasteur later set a theoretical framework for how heat treatment preserves foods by destroying spoilage microorganisms. In 1863, Pasteur solved the spoilage problems of wine and beer by carrying out heating at temperatures and times sufficiently long to inactivate spoilage microorganisms in a process later named pasteurization [2]. Milk pasteurization on a commercial scale started in Denmark and Sweden after Strauss and Monrad led a campaign promoting pasteurization throughout the USA in 1889 [2]. Koplik [3] found that the consumption of raw milk contributed to pathogen transmission from animals to humans. In the same year, milk pasteurization became common in Denmark to reduce the risk of spreading tuberculosis [4]. Sheffield Dairy Farms installed the first pasteurization equipment in Bloomfield, New Jersey, USA, in 1891. Soxhlet suggested that milk used to feed infants must be heated for public health reasons [2]. The first law mandating milk pasteurization was issued in Chicago in 1909. Thereafter, pasteurization became standard practice in the commercial milk industry [4]. A narrative history of the conceptualization and implementation of milk pasteurization is provided in Figure 1.

Several pasteurization techniques have been developed with variable effects on milk quality and shelf life. Due to its increased shelf life, pasteurized milk has become a staple food and an integral part of the diet, especially in the West. However, in recent years, there has been some resurgence in the consumption of raw milk, largely driven by the belief that unprocessed milk offers better health benefits than heat-treated milk. This perception has led to an increase in the consumption of raw milk by sensitive groups, including infants, the elderly, and immunocompromised individuals, as well as those adhering to specific dietary preferences [5]. This trend has ignited ongoing debates about the safety and nutritional merits of raw and pasteurized milk for direct consumption. While "natural" food products are often viewed favorably by the public, scientific evidence does not inherently equate naturalness with safety, healthiness, or taste. Between 2007 and 2012, various milk-borne outbreaks were reported in the European Union, with raw milk being implicated in several cases. In response, institutions such as the European Food Safety Authority (EFSA), the U.S. Food and Drug Administration (FDA) [6], and the Centers for Disease Control and Prevention (CDC) have issued evaluations and advisories highlighting the potential risks associated with raw milk consumption [7].

Simultaneously, interest in the sensory properties and health implications of milk consumption has increased, with some believing that the nutritional and sensory values of milk are adversely affected by pasteurization. Thus, alternative methods, such as high-pressure processing (HPP), power ultrasonics (PU), pulsed electric fields (PEFs) and microfiltration (MF), are being investigated [8–11]. This review article aims to summarize the primary purpose of milk pasteurization and discuss the associated changes that may affect not only the safety but also the compositional, sensory, and nutritional quality of milk. Specifically, particular emphasis is placed on the effects of milk pasteurization on protein structure and function. Finally, alternative methods to pasteurization, their advantages, and their limitations, along with future prospects, are discussed in the conclusion of this article.

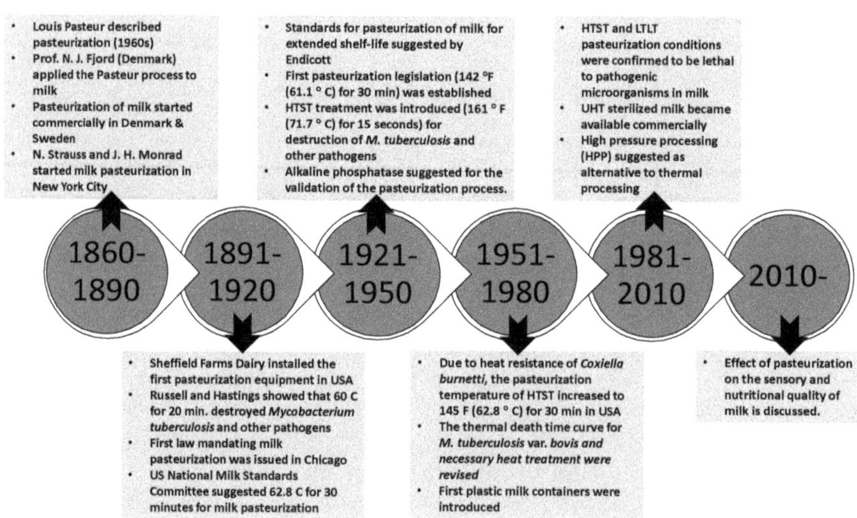

Figure 1. A summary of historical developments in the conceptualization and application of milk pasteurization [2,6].

2. Pasteurization and Sterilization Processes and Bacterial Destruction

Dr. Richard Seligman described pasteurization in 1923 as an energy-efficient process by which microorganisms were destroyed, with minor damage to the physicochemical properties of milk [12]. According to the FAO/WHO (2004) [13], pasteurization can be defined as a heat treatment designed to reduce the number of pathogenic microorganisms found in milk and liquid milk products. The quality of raw milk is determined by its bacterial count, which depends on both the health of the animal and potential contamination after milking. Thus, the final quality of the milk is significantly influenced by the somatic cell count (SCC), which reflects the immune response of the animal, particularly in cases of mastitis (udder inflammation). Enzymes associated with a high SCC in milk cause protein and fat degradation and may produce off-flavors during refrigerated storage. Healthy cows generally have SCC $< 50{,}000$ cells/mL of milk. In contrast, SCC in bulk tank milk may exceed 200,000 cells/mL due to the contribution of samples of high SCC from a few animals in the herd [14,15].

Milk pasteurization aims to serve two primary purposes, i.e., killing pathogenic bacteria and reducing the number of live spoilage bacteria, leading to the prolongation of the shelf life of the milk. After establishing the importance of pasteurization on milk safety, scientists focused on the details of pasteurization and sterilization processes. The initial pasteurization conditions, known as flash pasteurization, in which the milk was heated to up to ~80 °C, continued to be used until the 1960s. Enright [16] showed that older pasteurization processes were inadequate to inactivate *Coxiella burnetti*, which causes Q-fever in humans, and suggested new pasteurization conditions, i.e., heating at 62.8 °C for 30 min for batch processes and heating at 71.7 °C for 1 s for continuous processes. Different heat treatments, such as thermal treatment (65 °C for 15 s), low-temperature long-time pasteurization (LTLT, 65 °C for 30 min), high-temperature short-time pasteurization (HTST, 72 °C for 15 s), extended shelf life pasteurization (120–130 °C for 1–4 s), ultra-high-temperature sterilization (UHT, 136–145 °C for 2–8 s), in-container or vat sterilization (112 °C for 15 min), and innovative steam injection (ISI) treatments, are currently available (Table 1). These processes target variable microbial species and result in milk with different shelf lives [17]. Thermal treatments destroy heat-sensitive spoilage bacteria; pathogenic bacteria are mainly eliminated by pasteurization, while sterilization kills all bacteria and spores [2,18,19].

The primary goal of pasteurization conditions is to destroy *Mycobacterium tuberculosis* and *C. burnett* (i.e., the most temperature-resistant milk pathogens). The effects of certain temperature–time combinations on different pathogenic species are described in Figure 2. About 99.9% of pathogens are killed by pasteurization through HTST, and viable *Mycobacterium avium* populations are effectively reduced. All vegetative pathogens, such as human pathogenic *Escherichia coli*, *Listeria* spp., *Salmonella* spp., *Campylobacter jejuni*, and *Clostridium botulinum*, are destroyed by HTST pasteurization. However, heat-resistant spores of *Bacillus cereus* or *C. botulinum* are not destroyed by pasteurization. Instead, the germination of these spores can be induced while refrigerating pasteurized milk. Vegetative and most sporulating pathogens, especially *C. botulinum* and *B. cereus*, but not the spores of some extreme heat-resistant nonpathogens, such as *Bacillus thermodurans*, are destroyed by higher-temperature treatments (e.g., UHT, sterilization, and ISI). The efficiency of pasteurization processes can be determined by measuring the activity of alkaline phosphatase, which is a natural enzyme in mammalian kinds of milk with higher thermal tolerance than the most heat-resistant, non-spore-forming pathogens commonly found in milk [20,21]. Positive phosphatase activity indicates that the milk has not been sufficiently pasteurized or has been contaminated with raw milk or bacteria after processing [22].

Table 1. Different heat treatments used in the milk and dairy industries [23].

Heat Treatments	Temperature–Time Combination Required	Time	Pathogens Destroyed
Thermization	57–68 °C	5 s–30 min	Non-spore-forming pathogens and psychrotropic spoilage bacteria
Flash pasteurization	72–80 °C	15–30 s	Non-spore-forming pathogens and psychrotropic spoilage bacteria
Extended shelf life pasteurization (ESLP)	125–140 °C	1–10 s	Psychrotropic, mesophilic, and non-spore-forming bacteria
HTST	72–74 °C	15–20 s	*Coxiella burnetii*, the most heat-resistant pathogen in raw milk
Ultra-high-temperature (UHT) indirect heating	130–145 °C	5–20 s	*Clostridium Botulinum* and target *Coxiella burnetii;* bacterial endospore
Ultra-high-temperature (UHT) direct heating	142–150 °C	2–6 s	Heat-resistant spore formers without excessive chemical damage
Sterilization	110–120 °C or 125 °C	10–20 min 5 min	All non-spore-forming bacteria except heat-resistant spore-forming bacteria
Innovative steam injection (ISI)	160–180 °C	0.1 s	Heat-resistant spores

The spoilage of pasteurized milk is caused by post-pasteurization contamination with Gram-negative psychrotrophic bacteria, such as *Enterobacter, Serratia, Hafnia, Citrobacter, Pseudomonas, Alcaligenes,* and *Flavobacterium* [24]. The destruction of lactic acid bacteria by pasteurization increases milk's shelf life. However, it may also lead to unintended consequences, such as the increased growth of bacterial spores (e.g., *Bacillus* spores) and vegetative bacteria that survive pasteurization in the absence of lactic acid bacteria. Thus, when implementing the pasteurization of milk, there is a need for septic packaging to prevent recontamination post-pasteurization. Recyclable glass bottles were initially used until Ruben Rausing introduced paperboard containers in Sweden [25]. The Tetra Classic triangular (1951) was created, followed by Tetra Brick (1963), and finally, the Tetrapack packages currently used [26]. It has been proven in follow-up studies that milk pasteurization has contributed enormously to the public health and safety of populations worldwide [27]. However, the recent identification of heat-resistant microorganisms of public health significance, such as *Listeria monocytogenes* and *Mycobacterium avium* subsp. *paratuberculosis*, may question the adequacy of pasteurization heat treatment [28].

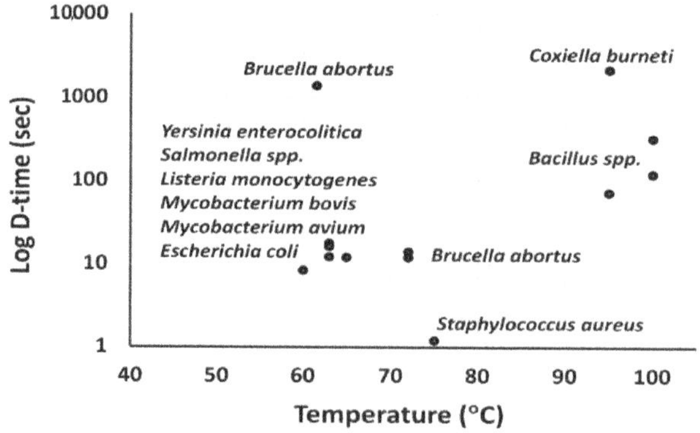

Figure 2. The relationship between pasteurization temperature and log D-time (the time required to reduce the number of microorganisms in one log cycle) [29].

3. Effects of Pasteurization and Sterilization on Milk Safety

Interest and popularity in the consumption of raw unpasteurized milk are increasing day by day as it is suggested in some reports that healthy microflora and bioactive components present in milk are also destroyed by pasteurization. The main reasons used to advocate the consumption of raw milk are its enhanced nutritional value, health benefits, overall quality, and better taste [27]. However, these claims have not been substantiated due to a lack of scientific evidence and the high cost of the required investigations. Instead, various epidemiological studies have confirmed that the contamination of raw milk by various pathogens causes the transmission of infectious diseases in humans [9,30,31]. The prevalence level of different pathogens available in raw milk has been reported in various research studies; for example, bacteria like *L. monocytogenes* and *C. jejuni* are found in unpasteurized milk at a prevalence level of about 12–13% [32]. The prevalence of pathogens in milk is influenced by many factors, including milking facilities, farm hygiene, different seasons, farm management practices, the type of utensils used to store milk, and the number of animals available on the farm [33]. One of the crucial sources of foodborne pathogens includes dairy farms, and the level of contamination in raw milk depends to a large extent upon the milking and farming practices used. Pathogens can contaminate raw milk even if it appears to be of acceptable quality and from healthy animals [34]. The possible mechanisms by which pathogens can contaminate raw milk are systematic infections, udder infections, feed contamination, and human transmission during milking.

Various foodborne pathogens causing outbreaks are present in raw milk, like *Salmonella* spp., *Campylobacter*, and some verocytotoxin-producing human pathogens like *E. coli*, *Y. enterocolitis*, *E. coli* O157:H7, and *L. monocytogenes* [35]. Nearly 2–6% of foodborne bacterial diseases in many industrialized countries are caused by pathogens in raw unpasteurized milk. Human pathogens found in raw milk are responsible for causing foodborne illnesses and outbreaks. When a high number of *Staphylococcus aureus* is present in raw milk due to contamination, it produces an enterotoxin, which is harmful to humans. Listeriosis is a condition that occurs when there is a high infectious load of *L. monocytogenes* in milk [36]. Common symptoms of milk-borne infections include abdominal cramps, diarrhea, nausea, vomiting, fever, etc. More severe symptoms, such as hemolytic uremic syndrome caused by *E. coli* O157:H7 and Guillain-Barré syndrome caused by *Campylobacter* spp., can be seen in people with severe illnesses. These symptoms may also result in chronic problems such as arthritis or, in some cases, even death.

Raw milk has been identified as a frequent source of various foodborne outbreaks and illnesses (Table 2). Statistical analysis data for milk-related human outbreaks in the United States have been reported and reviewed [37]. According to these data, consuming raw milk products was responsible for 121 outbreaks, which resulted in 1571 confirmed cases, including two deaths and 202 hospitalizations. A high number of outbreaks (55) were reported in 21 states where the sale of raw milk was allowed compared to a smaller number of cases and outbreaks in states where raw milk sale was banned and prohibited [37]. The outbreaks, illnesses, and health issues reported due to the consumption of raw milk reflect only a small proportion of the actual numbers [38,39]. For example, between 2001 and 2010, it was shown in data from Minnesota that 3.7% of patients acquired sporadic infections due to the consumption of raw unpasteurized milk [40]. In Minnesota, it was estimated that 20,500 patients (17%) had sporadic infection and enteric pathogen infection due to the consumption of raw milk. This resulted in a raw-milk-associated outbreak and illnesses. Children under the age of five were adversely affected, and it was observed that 76% of them were given raw unpasteurized milk from their farms [40].

S. aureus is the primary pathogen associated with contagious mastitis, which occurs at 24.4–37% in bulk tank milk samples. Contamination by coagulase-negative *Staphylococcus* and *Streptococcus* spp. was also shown in colostrum and bulk tank milk. The prevalence rates for *S. aureus* and *S. agalactiae* isolation were 31% and 10%, respectively [41]. Sixteen enterotoxin genes (seg–seq, sea–see), along with the toxic shock syndrome toxin gene (tsst-1), isolated from *S. aureus* in milk from mastitis-affected cows, were compared for their prevalence. A total of 73 out of 78 *S. aureus* isolates were positive for one or more enterotoxin genes. Along with the enterotoxin genes sed and tsst-1, some new *Staphylococcus* enterotoxin genes (e.g., sem, sen, and sei) were also reported in most *S. aureus* isolates. The high epidemiological prevalence of enterotoxin and *S. aureus* in raw milk is a concern since this species is a common pathogen isolated from raw milk and is responsible for foodborne infections and outbreaks [42].

Table 2. Outbreaks related to raw milk and raw milk products from available epidemiological data.

Year	Pathogen	State	Outbreak Information	Reference
2015	*Campylobacter jejuni*	Italy	*Campylobacter jejuni* contamination of raw milk across several Italian regions was estimated to cause between 230,776 and 301,785 cases per year (D–R I model) and up to 5.25 million cases per year under worst-case assumptions (D–R II model) depending on storage conditions.	[43]
2014	*Staphylococcus aureus*	Italy	There were no reported outbreaks, but an estimated 485 servings per year contained ≥20 ng enterotoxin A.	[44]
2013	*Campylobacter* spp. *L. monocytogenes* *Salmonella* spp.	New Zealand	A total of 93 cases per 100,000 servings contained *Campylobacter* spp., 201 cases contained *Shiga toxin*-producing *E. coli* (*STEC*), and 15 cases contained *Salmonella* spp. for *Listeria monocytogenes*.	[45]
2011	*Listeria monocytogenes* *E. coli O157:H7,* *Campylobacter,* *Salmonella*	New York	A quantitative risk assessment in New York estimated *Listeria monocytogenes* infections from raw milk consumption to range from 2.7×10^{-7} to 1.0×10^{-4} cases per person per year.	[46]
2007–2011	Campylobacter jejuni *E. coli* O157:H7	Italy	Between 2007 and 2011, an estimated 6.3–7.2 cases of HUS (Hemolytic Uremic Syndrome) were linked to raw milk consumption in Italy, caused by *E. coli* O157:H7. Additionally, outbreaks of *Campylobacter jejuni* were reported in the Veneto and Marche regions during the 2008–2009 period, and two *E. coli* O157:H7 outbreaks occurred in Emilia Romagna over the same period.	[47]
2009	*S.aureus* *Staphylococcus enterotoxin A*	California	A total of 25.3% of 51,963 raw milk samples tested positive for *Staphylococcus aureus*, indicating a substantial contamination rate. Additionally, Staphylococcal Enterotoxin A (SEA) exposure levels at these high percentiles could reach 94 ng/serving.	[48]
2008	*Campylobacter* spp.	California	Of 16 cases, 4 cases were CC for *Campylobacter*; 3/4 drank raw milk; and the rest were employees. Two individuals were hospitalized, including one with a form of Guillain–Barré syndrome.	[49]

Table 2. *Cont.*

Year	Pathogen	State	Outbreak Information	Reference
2007	*C. jejuni*	Kansas	Of 25 cases, 7 cases were CC, 18 probably occurred over several months; 16/28 persons who consumed raw milk at a gathering became ill.	[50]
2007	*Salmonella typhimurium*	Pennsylvania	There were 29 cases, with an age range of 5 months–76 years; 16/29 were <7 years, 29 cases were CC, there were identical PFGE patterns, and two individuals were hospitalized.	[51]
2007	*C. jejuni*	Kansas	There were 68 cases, and 4 cases were CC for *C. jejuni*; two individuals were hospitalized.	[52]
2006–2007	*Salmonella*		There were 85 cases, primarily including Hispanic people. A total of 85 cases were CC, with identical PFGE patterns; 36 individuals were hospitalized.	[53]
2006	*E. coli O157:H7*	California	There were 6 cases, with 5 CC with identical PFGE patterns. There was one non-CC case, HUS; three individuals were hospitalized.	[54]
2005	*E. coli* O157:H7	Washington	Of 18 cases, 8 cases were CC; 7/8 had identical PFGE patterns. Five people were hospitalized, and four had HUS.	[53]
2002–2003	*S. typhimurium*	Multi-State in the USA	There were 62 cases, and 62 were CC, with identical PFGE patterns and an epidemiologic link to an implicated dairy outbreak strain isolated from milk, cream, and butter samples.	[55]
2002	*C. jejuni*	Utah	Of 13 cases, 5/6 cases were CC; six individuals sought medical attention and none were hospitalized.	[56]
2001	*Salmonella*	-	A total of 26 cases were CC for MDR-SN; 23 individuals were treated with antibiotics, and 8 were hospitalized.	[57]
2001	*C. jejuni*	Wisconsin	Of 75 cases, 28 cases were CC; the PFGE patterns of 21 tested individuals were identical.	[58]

CC: culture confirmed; PFGE: pulsed-field gel electrophoresis, HUS: hemolytic uremic syndrome, MDR: multiple drug resistant.

4. Effect of Pasteurization and Sterilization on Milk Constituents

4.1. Effects on Milk Protein Structure and Functionality

The heat treatment of milk leads to various protein alterations by denaturation, aggregative interactions, Maillard reactions, and a loss of nutritional value [59]. Different chemical reactions may occur during pasteurization, including denaturation, hydrolysis, glycation, β-elimination reaction, iso-peptide bond formation, and racemization. The milk protein denaturation range by heat falls between 62 °C and 72 °C [60]. The series of reactions during the heating of milk involves different amino acids, especially lysine, tryptophan, asparagine, threonine, phosphoserine, and glutamine, and can affect various milk characteristics. The amount of lysine is reduced by heat treatment, mainly due to Maillard reactions with lactose [61]. Amino acids are building blocks and key components of milk proteins, contributing to nutrition, bioactivity, and functional properties like emulsification and gelling. Heat-sensitive amino acids such as lysine and tryptophan undergo chemical changes during thermal processing, including Maillard reactions and oxidation. These reactions reduce amino acid availability, impair protein digestibility, and affect milk's flavor, color, and stability [62].

Montilla [63] reported that changes in pH at different temperatures caused the denaturation of milk proteins. Partial denaturation of the globular structure of native whey proteins at temperatures above 60 °C causes the unfolding and exposure of their hydrophobic residues and disulfide bonds. Although these reactions may be reversible at low temperatures, new irreversible hydrophobic interactions may be formed by them at high temperatures [64]. Unfolded proteins can also aggregate with other proteins through disulfide linkages and sulphydryl-disulfide interchanges, as shown in Figure 3 [65].

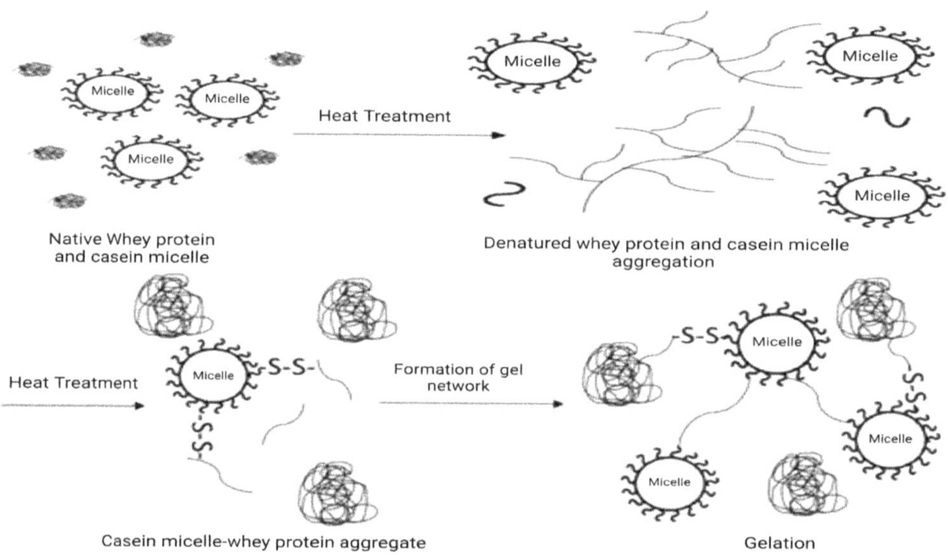

Figure 3. Schematic diagram showing possible interactions in heat-treated milk system. Modified from [66].

Heating milk above 60 °C may also lead to interactions between the casein micelles and the denatured whey proteins and cause their conversion from a soluble form to a colloidal state [67]. Heated casein micelles increase in size and associate with unfolded whey proteins, leading to the formation of adhesive hard spheres and an increase in viscosity [68]. The association between caseins and whey proteins is caused by hydrophobic interactions at temperatures <70 °C and disulfide bonds at higher temperatures [65]. For example, complexes between β-lactoglobulin and κ-casein aggregates are formed in bovine milk upon heating [65]. Upon heating at low temperatures for an extended time, β-lactoglobulin has enough time to unfold and associate with the micelle. Still, it does not unfold appropriately at high temperatures and may refold into a new structure and form aggregates with molecular species other than κ-casein. The formation of β-lactoglobulin/κ-casein complexes increases with the increased proportion of β-lactoglobulin in milk. Upon prolonged heating at low temperatures, α-lactalbumin forms complexes with β-lactoglobulin and, consequently, with κ-casein (Table 3) [59,69,70].

Table 3. Free sulfhydryl (SH) group and disulfide (S-S) bonds in milk whey proteins.

Protein	-SH Groups	S-S Bonds	Reference
α-LA	None	4 (Cys6-Cys120, Cys28-Cys111, Cys61-Cys77, and Cys73-Cys91)	[71]
β-LG	1 (Cys121)	2 (Cys66-Cys160 and Cys106-Cys119)	[72]
BSA	1 (Cys34)	17	[65]

Some heat-stable indigenous enzymes, such as plasmin and cathepsin, are activated by milk heating, leading to proteolysis during storage [73]. Bovine milk β-casein is hydrolyzed by plasmin (EC 3.4.21.7) to produce three C-terminal fragments [γ1- (fractions 29–209), γ2- (fractions 106–209), and γ3- (fractions 108–209)], αs1-casein is hydrolyzed to produce 14 peptides, and to a lesser extent, αs2-casein is hydrolyzed to release several fragments [74–76]. For example, five peptides were identified in heat-treated bovine milk resulting from the enzyme hydrolysis of α_{s1}- and β-caseins (Table 4). However, plasmin activity in bovine milk is strongly inhibited by native and denatured β-lactoglobulin [77]. This might explain the lower prevalence of hydrolytic peptides in bovine milk compared to camel milk, which lacks β-lactoglobulin [78]. Plasmin activity is believed to be enhanced by heat treatment and to contribute significantly to age gelation in UHT-treated milk [75].

Table 4. Peptides generated from bovine α_{s1}- and β-caseins after heat treatment [73].

Precursor	Peptide Sequence (Position)	Peptide (*m*/*z*)	Released by
αS1-Casein	IPNPIGSENSEKTTMPLW (182–199)	2014.0	Heat
	SDIPNPIGSENSEKTTMPLW (180–199)	2216.1	Cathepsin G
	RPKHPIKHQGLPQEVLNENLLRFF (1–24)	2910.6	Cathepsin B, Cathepsin D
β-Casein	EMPFPKYPVEPFTESQSL (108–125)	2126.0	Plasmin, Cathepsin D
	HKEMPFPKYPVEPFTESQSL (106–125)	2391.2	Plasmin, Cathepsin D

The Maillard reaction (nonenzymatic glycation) is a chemical reaction between the amino and carbonyl groups. Upon heating, lactose reacts with the ε-amino groups of lysine residues in milk proteins to form the Amadori product (lactulosyl-lysine-R), followed by the elimination of galactose moiety from lactose through the 4-deoxyosone pathway and the formation of an amino-reductone structure [79]. A small reduction (1–4%) in lysine concentration has been observed in pasteurized milk [80]. In the final stages of this reaction, hydroxymethylfurfural and other pigments are formed, causing browning of the milk, especially when heated at high temperatures (Figure 4). Tryptophan is destroyed during pasteurization, and mutagenic derivatives can be formed [81]. Arginine may also be converted into ornithine and citrulline by severe heat treatment, and deamination occurs when excess heat is applied to glutamine and asparagine [82]. These reactions cause a loss of nutritional value in protein because the altered amino acids might become unavailable for the metabolic process or because they cause the formation of toxic end products.

The heat stabilities of milk proteins expressed in terms of Arrhenius kinetics are shown in Table 5. Immunoglobulins and bovine serum albumin are the least stable, while β-lactoglobulin has intermediate stability [59,83]. Because α-lactalbumin is more sensitive to heat than β-lactoglobulin, it denatures at ~62 °C. However, its unfolding is reversible and does not form aggregates at temperatures below 80 °C [69]. Bovine whey proteins are subject to significant alterations at temperatures >130 °C, with β-lactoglobulin experiencing more changes than α-lactalbumin [63].

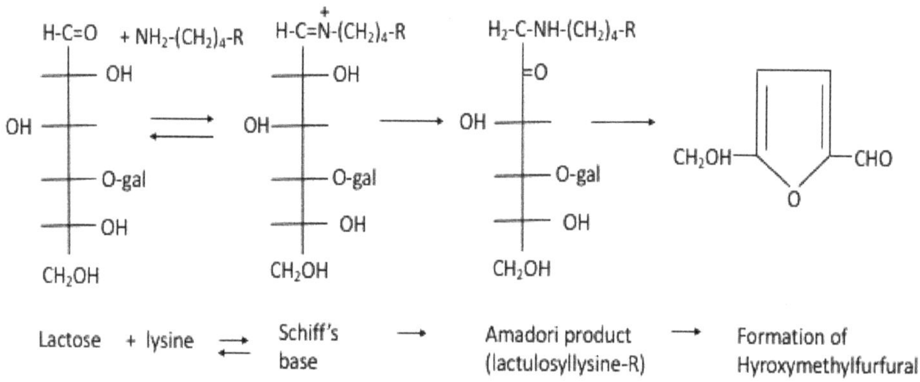

Figure 4. Maillard reaction between lactose and lysine residues in milk.

Table 5. Arrhenius kinetics describing the heat-induced degradation of milk whey proteins and lysine.

Component	Temperature Range (°C)	ln k_0	E_a ($kJmol^{-1}$)	*n*	Reference
Bovine serum albumin	82–150	13.18	49	1	[84]
Immunoglobulin	60–76	90.38	275	1	[85]
	76–82	54.21	170	1	
α-Lactalbumin	70–85	84.92	269	1	[86]
	85–150	16.95	69	1	
β-Lactoglobulin	70–90	89.43	280	1.5	[86,87]
	90–150	12.66	48	1.5	
	75–85	120.64	374	1.8	
Lysine (AA)	130–160	8.77	109	2	[88]

In conclusion, pasteurization induces a range of structural and chemical changes in milk proteins, including denaturation, aggregation, Maillard reactions, and enzymatic modifications, which can lead to both positive and negative effects. On the one hand, the heat-induced denaturation of whey proteins enhances digestibility and can improve certain functional properties such as viscosity and emulsification. On the other hand, excessive or prolonged heating may result in the loss of essential amino acids (e.g., lysine and tryptophan), a reduced nutritional value, and the formation of potentially harmful by-products. Additionally, heat-activated enzymes like plasmin can contribute to post-pasteurization proteolysis and shelf life challenges such as age gelation. Thus, while pasteurization is vital for ensuring microbial safety, its impact on protein quality is a balance between desirable functional outcomes and potential nutritional drawbacks, which are highly dependent on the processing conditions applied.

4.2. *Effect on Antimicrobial Systems*

While pasteurization is intended to destroy the most heat-tolerant pathogens, it might destroy essential nutrients, enzymes, and microorganisms in milk. Raw milk has been suggested to contain numerous antimicrobial systems that prevent the growth of pathogens and contribute to immunity, including lysozyme, xanthine oxidase, and lactoperoxidase. During storage, the activity of bacteria and other organisms is restricted due to cold or very low temperatures, along with the pasteurization effect [33]. While evaluating the antimicrobial activity of raw milk against pasteurized milk, Pitt [89] found that *Salmonella enteritidis* and *S. aureus* increased in both kinds of milk at 37 °C. However, the level of the pathogen in raw milk decreased after 32 h of growth, suggesting that the inactivation of antimicrobial activity in raw milk was caused by pasteurization. The prevalence of *S. aureus* and *S. enteritidis* after 70–72 h of inoculation in pasteurized milk was almost 100 and 1000 times higher than in raw milk, respectively [33]. *L. monocytogenes* inoculated at 37 °C in raw milk resulted in an initial bacterial population of about 10^4 CFU/mL after 12 h, after which it lost its ability to grow [90]. No viable cells of *L. monocytogenes* were found 56 h after inoculation in raw milk, suggesting that the microorganism had been "killed" by raw milk. In another study, the impact of unpasteurized milk on the level of *L. monocytogenes* at 15 °C was investigated [91]. During the study of the inhibitory role of the lactoperoxidase mechanism, six different strains of *L. monocytogenes*, which were isolated from unpasteurized milk, were used. The level of *L. monocytogenes* in heat-treated milk was found to increase by 2–3.8 log cycles after 65 h in static conditions [90]. In another study, the population of *L. monocytogenes* in raw milk increased by 0.8–2.3 log cycles under the same conditions [33]. Interest in components of the milk fat globule membrane (MFGM), such as lipid antimicrobial and antiviral properties (e.g., sphingomyelin, phosphatidylcholine, and phosphatidylethanolamine) and the peptide fragments of casein, has been increasing over time [92,93]. Although nearly all antimicrobial agents are destroyed by UHT treatment, their presence is no longer necessary since UHT-treated milk is practically sterile [94].

Raw milk contains various enzymes, many of which are unknown for their biological function or beneficial effects, which are affected by heating (Table 6). There are also some indigenous enzymes in raw milk, and excessive heat treatment can usually destroy and denature beneficial enzymes such as alkaline phosphatase and xanthine oxidase. Alkaline phosphatase (EC 3.1.3.1) indicates an effective pasteurization process due to its inactivation under pasteurization conditions [95]. Xanthine is a milk enzyme that activates lactoperoxidase with the help of hydrogen peroxide. It produces various products that destroy oxidative stress, thus having antimicrobial properties [96]. Enzyme activity is affected by temperature, thermal conductivity, pH, and the presence of substrates, inhibitors, and activators [97].

Table 6. The effect of pasteurization on antimicrobial milk proteins and enzymes.

Component	Role in Milk	Effect of Pasteurization	References
Alkaline Phosphatase (EC 3.1.3.1)	Potent anti-inflammatory enzyme	Since this enzyme is destroyed by heat, it is used as sensitive indicator for adequate pasteurization of milk	[98]
Bovine immunoglobulin	Immunogenic proteins	59–76% of activity is retained after HTST treatment	[99]
Bacteriocins	Antimicrobial peptides produced by certain milk bacteria with narrow spectrum of antimicrobial activity mainly against Gram-positive bacteria	No effect	[100,101]
Lactoferrin	Broad-spectrum antibacterial agent that binds to iron and reduces free iron supply for proliferation of bacteria, fungi, and protozoa	No effect	[102]
Lactoperoxidase(LPO, E.C. 1.11.1.7)	Acts together with hydrogen peroxide and thiocyanate ions as antibacterial agents	70–90% of enzyme activity is retained after HTST treatment; activity is gradually lost during refrigeration of pasteurized milk	[103,104]

Table 6. *Cont.*

Component	Role in Milk	Effect of Pasteurization	References
Lysozyme	Breaks cell walls primarily affecting Gram-positive bacteria	>75% of enzyme activity is retained after heating (80 °C, 15 s)	[105]
Plasmin (EC 3.4.21.7)	Milk protease causes alterations in protein structure and function	Survives pasteurization but may be destroyed at high temperature	[106]
Xanthine oxidase	Claimed to have antimicrobial properties by supplying hydrogen peroxidetolactoperoxidase	No effect	[96,107]

In summary, while pasteurization enhances milk safety by eliminating pathogens, it also compromises the natural antimicrobial systems inherent in raw milk. Key enzymes and bioactive components such as lactoperoxidase, lysozyme, and xanthine oxidase, known to inhibit pathogen growth, are significantly reduced or inactivated by heat treatment. This inactivation may diminish the milk's natural defense mechanisms, making it more susceptible to microbial proliferation during storage. Therefore, understanding and preserving these antimicrobial properties, where possible, is crucial for balancing the safety and functional quality of milk.

5. The Effects of Pasteurization and Sterilization on the Physical Properties and Sensory Quality of Milk

The sensory quality of milk is defined by its appearance, texture in the mouth, odor, flavor, and taste [108]. During pasteurization, milk undergoes a variety of reactions that might affect its color, flavor, and organoleptic properties, such as the denaturation of proteins, lipid degradation, and Maillard reactions [109,110]. Raw bovine milk is characterized by a yellow color mainly due to its all-*trans*-β-carotene content and small amounts of lutein, zeaxanthin, and β-cryptoxanthin that are associated with the fat globules [111]. Thus, the intensity of the yellow color of milk is determined by the amount of fat and the size of the fat globules, which are also affected by milk standardization and homogenization processes. Heat treatments may destroy carotenoids by oxidation.

The milk viscosity was not significantly affected by mild heat treatments. Still, it increases in severe heat treatments due to changes in protein structures and the formation of larger particles and aggregates. Pasteurization at 60–65 °C causes a slight decrease in viscosity, but pasteurization at 70 °C and higher temperatures may cause significant increases in viscosity [112]. The denaturation of β-lactoglobulin exposes its free sulfhydryl groups, causing its dimerization/oligomerization and aggregation with other whey and casein proteins through sulfhydryl–disulfide interchange reactions [113]. These interactions may lead to the formation of weak three-dimensional structures with liquid-like behavior, contributing to increased viscosity [113]. Age gelation occurs during the storage of UHT-treated milk (~>12 weeks at 20–25 °C) through the formation of a robust extended solid-like gel [114]. The viscosity of mildly processed milk increases with an increased fat content and the presence of more and larger fat globules, which increase the resistance of milk to flow [115,116].

Heat treatment significantly influences the flavor profile of milk by promoting chemical changes such as protein denaturation, Maillard reactions, and lipid oxidation. These reactions are especially prominent in ultra-high-temperature (UHT) processing and ultra-pasteurization, which exceed 90 °C. One of the effects of heat is the unfolding of whey proteins, such as β-lactoglobulin, which leads to the release of volatile sulfur compounds like hydrogen sulfide (H_2S), methional, and dimethyl sulfide, which in turn causes a cooked flavor in milk [117]. UHT treatment causes ketone and a cooked flavor in milk due to the presence of methyl ketone and the oxidation of sulfur and lactone compounds, which originate in the lipid part of milk. This flavor remains in milk for some time and may disappear within a week depending mainly on the type of UHT process applied. Flavor compounds, such as sulfur and nitrogen-containing compounds, diacetyl, Strecker aldehydes, and maltol, are produced during thermal processing when amino acids and lactose in milk undergo Maillard reactions [118]. Methyl ketone is generated due to the β-oxidation of free fatty acids, which is induced by the degradation of lipids during pasteurization [119]. Ultra-pasteurized milk is characterized by its various distinct flavors, such as cooked, caramelized, and sulfurous flavors, which make it different from HTST milk. These distinctive flavors are a drawback of ultra-pasteurized milk [120]. Due to the presence of sulfur-containing compounds (e.g., dimethyl sulfide, hydrogen sulfide, methional, carbon disulfide, dimethyl trisulfide, and Maillard compounds, i.e., furfural, benzaldehyde, 2- and 3-methylbutanal, and 2-acetyl-1-pyrroline), ultra-pasteurized milk is different from HTST milk, which also has distinctive sulfur and cooked flavors. The amount of various sulfur and Maillard-reacting compounds is also influenced by the fat concentration in milk, which, in turn, affects the flavor and taste of milk [118].

Zhao [121] demonstrated that HTST milk (63 °C/30 min or 72 °C/15 s) contains higher levels of low-molecular-weight compounds like 2-butanone and dimethyl ketone, whereas UHT processing using Direct Steam Injection (DSI) at 150 °C for 0.1 s leads to a notable rise in high-molecular-weight aldehydes such as hexanal, pentanal, and nonanal. These aldehydes contribute to grassy, fruity, and floral notes in UHT-treated milk in contrast to the lighter, less complex flavor of HTST milk. Benzaldehyde, an aromatic compound from the Maillard reaction, adds sweet almond-like notes, while dimethyl sulfone is primarily linked to sulfurous off-flavors [122]. Consumer preferences vary widely; Liem [123] reported that in China, where 60% of the population consumes long-life milk, the cooked and sulfurous flavors are better accepted, whereas Australian consumers, who primarily consume fresh HTST milk, prefer milder flavors, highlighting regional differences in flavor tolerance and product expectations.

Meng [124] further evaluated DSI and DSIJ (Direct Steam Injection Jet) treatments using the solvent-assisted flavor evaporation (SAFE) and solid-phase microextraction (SPME) techniques. They identified 59 volatile compounds across both methods, including nonanal, 2-undecanone, 2-decanone, δ-decanolactone, 3-hydroxy-2-butanone, and dimethyl sulfone. SAFE favored the detection of alcohols and aldehydes, while SPME excelled in extracting esters and sulfur-containing volatiles. DSIJ milk exhibited slightly more diverse compounds (52) compared to DSI milk (50), suggesting subtle differences in flavor complexity based on thermal intensity and duration. Ultimately, while UHT-treated milk is thermally stable and has an extended shelf life, its cooked, caramelized, and sulfur-rich profile remains a sensory drawback in markets favoring a fresh dairy flavor. The fat content and storage time further modulate these effects by enhancing the concentration of thermally derived volatiles over time.

In summary, pasteurization influences the physical properties and sensory quality of milk through various biochemical and structural changes. While mild heat treatments have a minimal impact on viscosity, higher temperatures can lead to protein denaturation, increased viscosity, and gel formation over time. Additionally, thermal processing induces flavor alterations, including the development of cooked, caramelized, and sulfurous notes, which are more pronounced in ultra-pasteurized milk. Despite these changes, pasteurization remains essential for ensuring milk safety and quality.

6. The Effects of Pasteurization and Sterilization on the Nutritional Quality of Milk

6.1. *Effects on Vitamins and Minerals*

Milk has a balanced nutritional value and digestible elements necessary for the development of babies. About 87% of milk is water, and the remaining 13% constitutes nutritionally valuable components, including carbohydrates, minerals, proteins, lipids, and vitamins. Milk is also one of the best sources of essential amino acids, calcium, phosphorous, and high-quality proteins that contribute to human nutrition, reproduction, growth, and the promotion of the development of bones and muscles. The increase in the global consumption of milk and milk products often underlines the critical question of how different pasteurization processes may affect the nutritional constituents of raw milk. Different pasteurization conditions may influence the physicochemical characteristics of milk and milk products differently. The pasteurization process decreases the amount of total fat in raw milk. For example, the total fat in raw milk was found to decrease from 3.58% to 3.07% after pasteurization and to decrease further after UHT treatment [125]. However, the fat contents of milk sold on the market are standardized to 3.5% in full-fat milk, 1.5–1.8% in semi-skimmed milk, and 0.5% in skimmed milk by adding or removing cream (Council Regulation (EC) 2597/97) [126].

Fat- and water-soluble vitamins are destroyed by excessive heat treatment or pasteurization. A drastic reduction in the levels of some essential vitamins due to pasteurization has been observed in an evaluation of the effect of pasteurization on different vitamins (e.g., A, B1, B2, B6, B12, C, and E) in a meta-analysis based on 40 different studies [127]. It was observed that pasteurization leads to some decline in the levels of vitamins, especially vitamin B2 and folate. Minor differences between raw and pasteurized milk were observed in thiamine (vitamin B1) and pyridoxine (vitamin B6) concentrations. However, the concentration of vitamin A was found to increase after pasteurization [127]. Minerals are generally heat-stable during pasteurization [128]. Pasteurization does not seem to affect the bioavailability of milk calcium and phosphorous [129]. In conclusion, pasteurization can lead to moderate losses of certain heat-sensitive vitamins, particularly B-group vitamins and vitamin C. However, essential minerals like calcium and phosphorus remain largely unaffected, preserving milk's core nutritional value. Despite minor nutrient reductions, pasteurized milk continues to serve as a vital source of high-quality proteins, essential vitamins, and minerals, supporting human growth and development across all age groups.

6.2. *Effect on Milk Digestibility and Gut Health*

Bovine milk is naturally rich in psychotropic lactic acid bacteria (*Lactobacillus*, *Lactococcus*, *Leuconostoc*, *Streptococcus*, *Enterococcus*, and small proportions of *Acinetobacter* and *Pseudomonas*), which cause spoilage in milk. The importance of the gut microbiome can influence psychological functioning and affect different aspects of mental as well as physical health. The composition of the gut microbiome mainly depends on the diet of the person. Enhancing growth and maintaining healthy gut microbiota in the diet is challenging. It was demonstrated in epidemiological studies that ingesting raw milk enhances the growth and composition of the gut microbiome [130]. Researchers evaluated the outcome of different dietary intakes on the gut microbiomes of several people who went through a 12-week research course on a farm and consumed unpasteurized milk and dairy products produced from a herd of Jersey cows. Most people who participated in this study did not consume unpasteurized dairy products before this course. The study concluded that there was a significant change in the microbiome, with *Lactobacilli* levels dramatically increasing in the fecal samples from pre-course to post-course [131]. The consumption of unpasteurized milk and dairy products was identified as the main reason for the increase in the levels of the gut microbiome. This increase is beneficial because *Lactobacilli* are known to support gut health by improving digestion, enhancing immune function, and inhibiting the growth of harmful bacteria.

6.3. Effects on Lactose Intolerance, Allergy, and Immunity

Lactose, the predominant carbohydrate available in mammalian milk, has several health benefits, including energy production, low glycemic levels, facilitation of the absorption of magnesium and calcium, and prebiotic properties, especially for children [132]. The enzyme lactase (β-galactosidase) is responsible for the digestion of lactose in humans. The absence of this enzyme results in lactose intolerance manifested as symptoms, including bloating, diarrhea, flatulence, and severe abdominal pain. More than 65% of the human population is projected to suffer from lactose intolerance, for which the prevalence increases with age [133]. There are widespread anecdotal claims that raw milk has a curing effect on lactose intolerance due to its content of natural lactase, and bacteria that produce this enzyme, mainly *Lactobacillus acidophilus*, are destroyed by pasteurization [127]. Still, this hypothesis was later rejected [134]. However, there is no evidence supporting these claims. In a recent pilot cross-over intervention study, it was shown that raw milk was not different from pasteurized milk in affecting the symptoms of lactose malabsorption or intolerance in adults suffering from lactose malabsorption [135].

The increase in allergies and asthma in Western countries was associated with the consumption of pasteurized milk [136]. In this case, there is consistent evidence that children raised on farms show decreased incidences of allergic hypersensitivities, hay fever, atopy, and asthma compared to other children. However, the exact reason(s) cannot be given and may be multifactorial [137–146]. Rosenlund [147] collected data that included about 15,000 children from five European countries with different lifestyles. It was found that children living in agricultural areas and with restricted use of vaccines, antibiotics, and antipyretics showed a negative association between unpasteurized milk consumption and asthma and allergies (Table 7). In another study including about 1,000 rural children from five European countries, Loss [148] found a negative relationship between raw milk consumption and rhinitis, otitis, and respiratory tract infection. In the hygiene hypothesis, Strachan [149] suggested that farm living and related lifestyles expose children to microbes and may protect them against some allergies. However, in a study of germ-free female BALB/cByJ mice, it was shown that the pasteurization of bovine milk improves the allergenicity of β-lactoglobulin by denaturation and suppression of the effect on conformational, i.e., sequential epitopes [150].

Table 7. Studies evaluating the effect of raw milk consumption on asthma and allergic diseases.

County Where Study Was Conducted	Exposure	Results	Reference
Crete (Greece)	Unpasteurized milk products	Adj. OR (and 95% CI) of atopy and unpasteurized farm milk consumption with and without simultaneous farm animal contact: 0.32 (0.13–0.78) and 0.58 (0.34–0.98), respectively	[151]
Austria, Germany, Switzerland	Milk directly produced or purchased on a farm	Consumption of farm milk during first year of life significantly inversely associated with asthma, hayfever, and atopy independent of other farm exposures	[141]

Table 7. *Cont.*

County Where Study Was Conducted	Exposure	Results	Reference
New Zealand	Unpasteurized milk, yogurt at least weekly before age of two years	Adj. OR and (95% CI) for early yogurt consumption and hay fever 0.30 (0.1–0.7); any unpasteurized milk and atopic eczema: 0.2 (0.1–0.8); no significant association between unpasteurized milk consumption and asthma or atopy	[146]
Finland	Farm milk in infancy	Farm milk consumption not associated with atopy; no other allergic health outcomes reported	[152]
Northern Germany	Raw, unboiled farm milk	Raw milk consumption and atopy adj. OR (and 95% CI): 0.65 (0.36–1.18); in those visiting animal houses before age of 7 years, raw milk consumption and atopy: 0.35 (0.17–0.74)	[153]
England	Unpasteurized milk	Current unpasteurized milk consumption associated with less eczema adj. Or and (95% CI) of 0.59 (0.40–0.87) and atopy of 0.42 (0.10–0.53), and higher production of whole blood stimulated IFN-γ; effect independent of farming status; no effect on asthma	[140]
Sweden, Austria, the Netherlands, Germany, Switzerland	Milk directly produced or purchased on a farm	Association between farm milk and asthma varied between genotypes of CD14/-1721; similar patterns for symptoms of hay fever and pollen sensitization; CD14/-1721 also modified association between farm milk and CD14 gene expression	[154]
Sweden, the Netherlands, Austria, Germany, Switzerland	Milk directly produced or purchased on a farm	Adj. OR and (95%CI) of farm milk consumption ever in life and asthma: 0.47 (0.61–0.88), rhinoconjunctivitis: 0.56 (0.43–0.73), sensitization to pollen: 0.67 (0.47–0.96), and food mix: 0.42 (0.19–0.92); association observed in all subgroups independent of farm-related exposures	[145]
Finland, France Austria, Germany, Switzerland	Skimmed and unskimmed farm milk, farm-produced butter, and yogurt during pregnancy	Maternal consumption of farm-produced butter during pregnancy associated with increased IFN-γ and TNF-α production in cord blood, and farm-produced yogurt inversely related to these cytokines	[155]

Adj. OR: prevalence and adjusted odds ratio; CI: confidence interval; IFN-γ: interferon gamma; CD14: cluster of differentiation 14; TNF-α: tumor necrosis factor alpha.

Pasteurization and heat treatment are believed to destroy the beneficial immunoglobulins naturally present in raw milk, although this concentration is too low to have any physiological significance in human immunity [156]. Mainer [157] concluded that there was no effect on the IgG level upon LTLT pasteurization and only a 1% reduction upon HTST pasteurization. In another study, Kulczycki, [158] reported that heat-induced aggregation could be enhanced by pasteurization, increasing the receptor-binding activity of immunoglobulin IgG. This means that pasteurized milk has better immunological function than raw milk. A potent immune-regulatory molecule known as transforming growth factor-β (TGF-β) is found in raw milk, and it has been reported that pasteurization has no effect on it [159]. In raw milk, the concentrations of immune-modulatory factors, including TNF-α, IL-1β, IL-10, and IL-6, are too small to induce any physiological significance [160]. However, there is a hypothesis that higher immunity against symptomatic infections caused by pathogens can be achieved by frequent consumption of raw milk. This happens because of continuous exposure to the nonvirulent strain of the pathogen, which can develop cross-immunity. However, the only case reported in the literature is specifically with *Campylobacter* [161].

Products from the Maillard reaction have both beneficial and harmful effects on health. Depending on how the food is prepared or processed, toxic and beneficial Maillard reaction products can be formed. Diverse Maillard reaction products can act as anti-browning, antioxidant, prooxidant, bactericidal, carcinogenic, and anti-allergic agents. The majority of these characteristics are influenced by different food processing techniques. Due to pasteurization or high-temperature processing, some foods can lose their nutritional content, while others become more nutritious. Several measures may be applied in the food industry to limit or reduce the generation of Maillard reaction products. For example, acrylamide produced at high temperatures was classified as a possible human carcinogen (International Agency for Research on Cancer, 1987) [162]. Asparaginase has been effectively used in the laboratory to decrease acrylamide. Injecting carbon dioxide during the extrusion process can also help lower acrylamide levels [163]. Maillard reactions may form other compounds, such as furosine, lactulose, HMF, etc., that can negatively affect long-term health. Finally, there is concern that milk exosomes are affected negatively by pasteurization and that human exposure to these exosomes may induce risks of developing chronic diseases, including obesity, type 2 diabetes mellitus, osteoporosis, cancers, and Parkinson's disease [164,165].

While pasteurization has a minimal effect on lactose intolerance and may slightly alter allergenicity and immune factors, current evidence does not support claims that raw milk improves these conditions. Although some bioactive compounds and enzymes are reduced, pasteurized milk maintains nutritional safety and may even enhance certain immunological functions. Moreover, the loss of natural components is often outweighed by the reduction in foodborne illness risk.

6.4. Effect on Fatty Acids and Milk Fat Globule Membrane (MFGM)

Milk fat content in commercial products is typically regulated through standardization, achieved by adjusting the cream content to produce full-fat, semi-skimmed, or skimmed milk. However, heat treatments, particularly at high temperatures, can induce both physical and chemical transformations in the milk lipid fraction. Notably, these changes include an increase in free fatty acid levels. Moreover, polyunsaturated fatty acids in milk are susceptible to thermal alterations, including the formation of conjugated isomers under severe heat treatment. One such compound, conjugated linoleic acid, is recognized for its potential anti-carcinogenic effects, indicating that some thermal transformations may have beneficial implications [166]. Nevertheless, the impact of heat on the overall fatty acid profile appears limited under typical pasteurization conditions. A study by Pestana [125] found no significant changes in the total fat content or overall fatty acid composition following pasteurization, suggesting that milk lipids exhibit relative resilience to moderate heat processing.

At the same time, pasteurization disrupts the structural integrity of the milk fat globule membrane (MFGM), a bioactive interface rich in phospholipids and glycoproteins. Damage to the MFGM may reduce lipid protection against oxidation and alter fat digestibility [167]. Additionally, the loss of MFGM-associated components such as sphingomyelin could have implications for infant brain development and immune function [168]. These findings emphasize the need to consider not just fat quantity but also structural and functional aspects of milk lipids in processing evaluations.

7. Alternative Processing Methods

High-pressure processing (HPP), power ultrasonics (PU), and pulsed electric fields (PEFs) are non-thermal processing technologies with promising impacts on food processing [169]. HPP is a substitute for conventional thermal treatment methods for foods, and it has been shown to better preserve the nutrients and bioactive components in milk [170]. The key benefits of high HPP are the retention of sensory properties and the nutritional value of food by the inactivation of microorganisms. Some commercially available pressurized foods include yogurt, fruit jams, juices, avocado pulp, jellies, and sauces [171,172]. Batch or semi-continuous equipment is used to conduct treatments, usually in the range of 100–1000 MPa, for up to 30 min to inactivate the spore of the microorganism [173]. HPP parameters include the process pressure, pressure hold time, initial product temperature, time to achieve pressure, treatment temperature, water activity, pH of the product, temperature distribution in the vessel, and decompression time, among others [174]. Under the influence of HPP, nucleic acids, enzymes, polysaccharides, and proteins may be affected, but due to their simple structure and small molecular size, vitamins and amino acids remain unaffected [173].

Hite [175] studied the effect of HPP on milk bacteria and concluded that processing at 680 MPa for 10 min would lead to complete sterilization and delay the microbial spoilage of milk for seven days. The inactivation of microorganisms by HPP involves several mechanisms, such as membrane destruction, changes in bacterial enzymes and nucleic acids, and simultaneous leakage of the contents of organelles and cells [176]. Factors like the composition and pH of food, state and type of growth, and time and pressure of application can affect the efficiency of HPP, while the effect of HPP on enzymes in milk is comparatively much less than that of heat. Several researchers examined the impact of HPP on endogenous enzymes in milk, such as glutamyltransferase, alkaline phosphatase, phosphohexose isomerase, and lactoperoxidase [172,177]. The inactivation of plasmin by HPP requires a pressure of up to 600 MPa and a temperature of more than 20 °C [178,179]. It has been shown that whey protein (mainly β-lactoglobulin) and caseins are altered by HPP [178,179]. The denaturation of β-lactoglobulin occurs at a pressure of more than 100 MPa at 25 °C, while bovine serum albumin and α-lactalbumin tolerate pressures around 400 MPa without being denatured [179]. Studies show that HPP at 600 MPa for 3 min can achieve a 5-log reduction in *Listeria monocytogenes* and *E. coli*, similar to thermal pasteurization, but without the associated thermal degradation of bioactive compounds [180]. Quantitatively, HPP-treated milk retains up to 90% of vitamins B2 and C, whereas conventional pasteurization retains only around 50–60% [181]. Over the past 20 years, HPP technology has advanced dramatically, but its high installation cost and the complexity of designing a continuous system to prevent corrosion and cross-contamination are drawbacks in the industrial application of HPP for milk treatment.

Ultrasound waves, which can pass through milk at a frequency greater than 20 kHz, in milk processing may offer advantages to the dairy industry, such as improved product quality and cost-saving properties. However, ultrasound alone is not very efficient in deactivating microorganisms and enzymes; therefore, it should be combined with sonication techniques [174,182]. Ultrasound treatments are affected by several factors, including enzyme concentration, medium composition, treatment level, frequency, and energy density [183]. The inactivation of enzymes usually increases with increasing ultrasonic power, temperature, frequency, pressure, exposure time, and amplitude and decreases as the volume of the sample increases [184]. It has been reported that enzyme deactivation decreases with an increase in enzyme concentration and increases with a rise in solid waste [185]. There was no effect on enzymes in milk when ultrasound was applied without heat treatment. Studies demonstrated that sonication at 20 kHz achieved a 5-log reduction in *Escherichia coli* O157:H7 counts in liquid foods like milk and juices, aligning with FDA pasteurization standards [183]. Unlike conventional pasteurization methods, which often reduce heat-sensitive nutrients such as vitamins B1 and C by 20–50%, ultrasound preserves up to 90% of these compounds [186]. This not only improves nutritional quality but also enhances sensory properties and energy efficiency, making ultrasound a viable, sustainable alternative to traditional pasteurization.

Power ultrasonic (PU) milk treatment is an alternative to conventional thermal techniques and has more benefits than pasteurization, e.g., energy usage reduction, the potential to target particular species, and the absence of preservative requirements [185]. The deactivation of bacteria and enzymes, milk homogenization, lactose hydrolysis, and the extraction of enzymes are the main applications of PU in the dairy industry. Expansion cycles and alternating compression are formed when ultrasonic waves pass through the liquid, causing the growth of existing bubbles by high-intensity ultrasound in the expansion cycle, which violently implode when they reach a volume at which more energy is not absorbed, a process called cavitation [187]. Physical forces created by acoustic cavitation are the primary mechanism responsible for ultrasonic microbial deactivation.

The main criteria for evaluating any new technology to replace thermal treatment methods are its ability to provide the consumer with a safe, shelf-stable, cost-effective, and better-quality product. Using a pulsed electric field (PEF) could be a suitable alternative to traditional heat or thermal treatments for different liquid and semi-liquid foods as it can destroy harmful microorganisms and some enzymes while retaining the quality and freshness of food products. The application of PEF technology to pasteurize food, such as yogurt, juices, soups, liquid eggs, and milk, has been successfully demonstrated with negligible effects on the nutritional and sensory quality of food [188]. The inactivation of microbes and enzymes by PEF depends on different parameters, such as composition, ionic strength, pH, and conductivity [189]. PEF induces minimal deactivation of enzymes and bacterial spores, so it needs to be combined with other technologies, such as thermal treatment and the addition of bacteriocins or antimicrobial agents. It has been observed that PEF treatment is successful in destroying vegetative microorganisms, but the high deactivation of spoilage enzymes and microbial spores requires other combined treatments. The minimal effect on enzymes benefits the dairy industry as milk can be processed to inactivate microorganisms while retaining beneficial enzyme activities [190]. PEF processing of unpasteurized milk has been reported to have a minimal effect on the tertiary structure of whey proteins [191]; for example, the folded structure of heat-sensitive lactoferrin was retained after PEF processing regardless of the processing mode. PEF does not affect the sensory, physical, or chemical properties of milk [192,193]. Compared to traditional thermal pasteurization, PEF has demonstrated concrete microbiological and nutritional advantages. Quantitatively, PEF-treated milk and juice samples have shown microbial reductions ranging from 0.13 to 6.2 log CFU/mL for total bacteria and up to a 0.48 log reduction in coliforms depending on the treatment parameters and food matrix [194]. While traditional pasteurization may cause significant degradation of heat-sensitive nutrients and flavor compounds, studies confirmed that PEF causes negligible nutrient loss and leads to better preservation of antioxidant and sensory qualities [195]. PEF treatments provide less destruction of flavor and nutrients and retain the quality and nutritional value of processed foods. Further research is needed to attain a higher inactivation rate and to study more factors affecting PEF treatment to make it more precise.

Microfiltration (MF) is a non-thermal processing technology increasingly utilized in the dairy industry as an alternative to traditional heat pasteurization. This membrane-based technique relies on porous filters, typically with pore sizes ranging from 0.1 to 1.4 microns, to physically separate microorganisms, including bacterial spores, from milk without significantly altering its biochemical composition [196,197]. The key benefit of MF lies in its ability to retain milk's native organoleptic properties, such as its taste, aroma, and texture, while extending the shelf life, making it highly attractive for premium milk products [198]. Unlike thermal pasteurization, MF avoids protein denaturation and the loss of heat-sensitive nutrients, thereby preserving nutritional quality. This approach also enables selective component separation, for example, removing somatic cells or bacteria while allowing fat and casein micelles to pass, offering functional advantages in cheese manufacturing and protein standardization [199]. However, membrane fouling is a persistent limitation. The accumulation of proteins and fats on the membrane surface can reduce filtration efficiency and require frequent cleaning cycles, thus impacting the operational cost-effectiveness of the system. Moreover, emerging evidence suggests that certain spoilage microbes may still survive or proliferate post-filtration, highlighting the need for complementary treatments like UV or high-pressure processing in some applications [200]. Despite these challenges, MF remains a promising technique, offering flexibility, product quality enhancement, and potential for integration into hybrid preservation systems for next-generation dairy products.

8. Economic Impact of Milk Processing Methods

The choice of milk processing technology significantly influences both the economic viability and sustainability of dairy enterprises. Traditional thermal pasteurization methods, although well established, are energy-intensive and may result in higher operational costs due to prolonged heating and cooling cycles, labor, and maintenance requirements. In contrast, alternative processing methods such as high-pressure processing (HPP), pulsed electric field (PEF), and ultrasound-assisted pasteurization and microfiltration offer promising avenues for cost-efficiency and product quality retention.

However, their implementation carries varying economic implications. A study evaluating alternative dairy processing models revealed that small-scale facilities (50–500 cows) for fluid milk, yogurt, and cheese require capital investments ranging from USD 1.5 million to USD 7 million depending on product line and automation levels. Operational costs, including energy, labor, and packaging, represent a significant portion of the total expenditure, with energy savings being a potential advantage of non-thermal technologies [201].

For example, the adoption of PEF in milk processing, while requiring an upfront capital investment in specialized equipment, can lead to lower processing temperatures and shorter treatment times, translating into long-term energy savings and reduced product loss. This aligns with findings that modern processing methods can enhance efficiency and reduce waste, improving overall profitability [202]. Furthermore, an environmental and economic analysis estimated that traditional fluid milk processing contributes about 2.4 kg CO_{2e} per kg of milk produced and processed, a figure that could be reduced with energy-efficient alternatives like PEF or HPP [203]. Market-dependent variables such as raw milk price, product demand, and consumer preferences also influence the economic feasibility of processing methods. Enterprises must also consider maintenance costs, operator training, and regulatory compliance when adopting novel technologies [204]. In conclusion, while alternative milk processing technologies can offer operational and product quality advantages, careful cost–benefit analyses are essential for industrial application. Decision-makers in the dairy sector should weigh capital investment, energy efficiency, regulatory requirements, and market positioning when considering a transition from conventional to innovative milk processing methods.

9. Future Prospective

The future of milk processing technologies lies in achieving a balance between microbial safety and the preservation of nutritional and sensory quality. While conventional heat pasteurization and sterilization have played a major role in public health, consumer demand for minimally processed and nutrient-retentive foods is accelerating the transition toward non-thermal technologies. Emerging innovations such as high-pressure processing (HPP), pulsed electric fields (PEFs), ultrasound, and microfiltration (MF)-assisted non-thermal processing have shown promising results in maintaining milk's natural bioactives, proteins, and enzymes while ensuring microbial safety. These methods offer energy-efficient and sustainable alternatives that minimize the thermal degradation of heat-labile nutrients like vitamins B1 and B2 and bioactive peptides.

Moreover, recent insights into the gut microbiome have emphasized the need to preserve milk's native microflora and immunomodulatory compounds. Functional milk enriched with probiotics, prebiotics, and immune-supporting molecules like lactoferrin and transforming growth factor-β (TGF-β) may play a future role in preventing allergies and promoting gut health. Future research may also explore nanoencapsulation and milk exosome protection strategies to safeguard these components during processing. Another promising direction is precision pasteurization, where real-time sensors and AI-driven modeling optimize thermal input to ensure pathogen death while minimizing protein denaturation and flavor loss. Coupled with smart packaging and cold-chain blockchain systems, this could significantly improve the traceability and shelf life prediction of pasteurized milk products. In the coming years, interdisciplinary efforts will be critical for designing next-generation milk processing systems. These should fulfill dual objectives, namely enhanced safety and superior quality, aligning with the principles of personalized nutrition, sustainability, and consumer health consciousness.

10. Concluding Remarks

It is suggested by current research that milk pasteurization by heat treatment is adequate to ensure microbial safety but may impact the nutritional or sensory quality of the milk depending on the treatment temperature and time. The consumption of raw milk cannot be encouraged due to the safety risks associated with the possible content of pathogenic microorganisms. More studies that evaluate the differences between raw and differently pasteurized/sterilized milk samples and their relevance to milk quality are required. In addition, the development of alternative pasteurization methods that ensure hygienic collection and assessing the microbial quality of raw milk from individual animals before its inclusion in collection tanks are required. Studies on the digestibility of different kinds of heat-treated milk and their relation to gut health, intolerance, and allergy are also highly warranted.

Author Contributions: Conceptualization, A.K.-E.; writing—original draft preparation, A.K.-E. and A.R.; writing—review and editing, A.R., M.A., C.D.C.D., G.C., Y.X. and A.K.-E.; visualization, A.K.-E., C.D.C.D. and A.R.; supervision, A.K.-E. All authors have read and agreed to the published version of the manuscript.

Funding: This project is funded by United Arab Emirates University, project 31R331.

Institutional Review Board Statement: Not applicable.

Informed Consent Statement: Not applicable.

Data Availability Statement: No new data were created or analyzed in this study. Data sharing is not applicable to this article.

Conflicts of Interest: The authors declare no conflicts of interest.

References

1. Garcia, R.; Adrian, J. Nicolas Appert: Inventor and Manufacturer. *Food Rev. Int.* **2009**, *25*, 115–125. [CrossRef]
2. Westhoff, D.C. Heating Milk for Microbial Destruction: A Historical Outline and Update. *J. Food Prot.* **1978**, *41*, 122–130. [CrossRef]
3. Koplik, H. The History of the First Milk Depot or Gouttes De Lait with Consultations in America. *J. Am. Med. Assoc.* **1914**, *LXIII*, 1574–1575. [CrossRef]
4. Smith-Howard, K. *Pure and Modern Milk: An Environmental History Since 1900*; Oxford University Press: Oxford, UK, 2017.
5. Andreoletti, O.; Lau Baggesen, D.; Bolton, D.; Butaye, P.; Cook, P.; Davies, R.; Fernández Escámez, P.S.; Griffin, J.; Hald, T.; Havelaar, A.; et al. Scientific Opinion on the Public Health Risks Related to the Consumption of Raw Drinking Milk. *EFSA J.* **2015**, *13*, 3940. [CrossRef]
6. The Dangers of Raw Milk: Unpasteurized Milk Can Pose a Serious Health Risk—FDA. Available online: https://www.fda.gov/food/buy-store-serve-safe-food/dangers-raw-milk-unpasteurized-milk-can-pose-serious-health-risk (accessed on 2 April 2025).
7. Raw Milk—Food Safety—CDC. Available online: https://www.cdc.gov/food-safety/foods/raw-milk.html (accessed on 2 April 2025).
8. Haug, A.; Høstmark, A.T.; Harstad, O.M. Bovine Milk in Human Nutrition—A Review. *Lipids Health Dis.* **2007**, *6*, 25. [CrossRef]
9. Lisboa, H.M.; Pasquali, M.B.; dos Anjos, A.I.; Sarinho, A.M.; de Melo, E.D.; Andrade, R.; Batista, L.; Lima, J.; Diniz, Y.; Barros, A. Innovative and Sustainable Food Preservation Techniques: Enhancing Food Quality, Safety, and Environmental Sustainability. *Sustainability* **2024**, *16*, 8223. [CrossRef]
10. Zhao, Y.M.; de Alba, M.; Sun, D.W.; Tiwari, B. Principles and Recent Applications of Novel Non-Thermal Processing Technologies for the Fish Industry—A Review. *Crit. Rev. Food Sci. Nutr.* **2019**, *59*, 728–742. [CrossRef] [PubMed]
11. Khaliq, A.; Mishra, A.K.; Rabbani, A. Comparative Antimicrobial Evaluation of Synthetic Antibiotics and Essential Oils Against Human Pathogenic Bacteria and Fungi. *Innov. Agric.* **2024**, *7*, 1–7. [CrossRef]
12. Knipschildt, M.E.; Andersen, G.G. Drying of Milk and Milk Products. *Robinson Mod. Dairy Technol.* **1994**, *1*, 159–254. [CrossRef]
13. Food and Agriculture Organization of the United Nations/World Health Organization (FAO/WHO). Code of Hygienic Practice for Milk and Milk Products. Joint FAO/WHO Food Standards Programme-Codex Committee on Food Hygiene, 26th Session, 29 March–2 April 2004, Washington, DC, USA. 2004. Available online: http://www.fao.org/fileadmin/user_upload/livestockgov/documents/CXP_057e.pdf (accessed on 2 April 2025).
14. Barbano, D.M.; Ma, Y.; Santos, M.V. Influence of Raw Milk Quality on Fluid Milk Shelf Life. *J. Dairy Sci.* **2006**, *89*, E15–E19. [CrossRef]
15. Ma, Y.; Ryan, C.; Barbano, D.M.; Galton, D.M.; Rudan, M.A.; Boor, K.J. Effects of Somatic Cell Count on Quality and Shelf-Life of Pasteurized Fluid Milk. *J. Dairy. Sci.* **2000**, *83*, 264–274. [CrossRef] [PubMed]
16. Enright, J.; Sadler, W.; Thomas, R. *Thermal Inactivation of Coxiella Burnetii and Its Relation to Pasteurization of Milk*; US Government Printing: Washington, WA, USA, 1957.
17. Tamime, A.Y. *Milk Processing and Quality Management*; Wiley-Blackwell: Hoboken, NJ, USA, 2009; pp. 1–324. [CrossRef]
18. Hall, C.W.; Trout, G.M. *Milk Pasteurization*; AVI Publishing: Delhi, India, 1968.
19. Varnam, A.; Sutherland, J. *Milk and Milk Products: Technology, Chemistry and Microbiology*; Aspen Publishers: Boston, MA, USA, 2001.
20. Sharma, S.K.; Sehgal, N.; Kumar, A. Dry-Reagent Strips for Testing Milk Pasteurization. *LWT Food Sci. Technol.* **2003**, *36*, 567–571. [CrossRef]
21. Ludikhuyze, L.; Claeys, W.; Hendrickx, M. Combined Pressure—Temperature Inactivation of Alkaline Phosphatase in Bovine Milk: A Kinetic Study. *J. Food Sci.* **2000**, *65*, 155–160. [CrossRef]
22. Peng, Z.; Li, Y.; Yan, L.; Yang, S.; Yang, D. Correlation Analysis of Microbial Contamination and Alkaline Phosphatase Activity in Raw Milk and Dairy Products. *Int. J. Env. Res. Public. Health* **2023**, *20*, 1825. [CrossRef] [PubMed]
23. Deeth, H.; Smithers, G. Heat Treatment of Milk–Overview. Inernational Dairy Federation, IDF Factsheet. pp. 1–4. Available online: https://www.fil-idf.org/wp-content/uploads/2018/02/Factsheet-001_Heat-treatment-1-1.pdf (accessed on 17 February 2025).
24. Martin, N.H.; Boor, K.J.; Wiedmann, M. Symposium Review: Effect of Post-Pasteurization Contamination on Fluid Milk Quality. *J. Dairy Sci.* **2018**, *101*, 861–870. [CrossRef]

25. Kirwan, M. *Handbook of Paper and Paperboard Packaging Technology*; Wiley-Blackwell: Hoboken, NJ, USA, 2012.
26. Twede, D.; Selke, S.E.M.; Kamdem, D.-P.; Pira, S. *Cartons, Crates and Corrugated Board: Handbook of Paper and Wood Packaging Technology*; Destech Pubns Inc.: Lancaster, PA, USA, 2014.
27. LeJeune, J.T.; Rajala-Schultz, P.J. Unpasteurized Milk: A Continued Public Health Threat. *Clin. Infect. Dis.* **2009**, *48*, 93–100. [CrossRef]
28. Van Brandt, L.; Van der Plancken, I.; De Block, J.; Vlaemynck, G.; Van Coillie, E.; Herman, L.; Hendrickx, M. Adequacy of Current Pasteurization Standards to Inactivate Mycobacterium Paratuberculosis in Milk and Phosphate Buffer. *Int. Dairy J.* **2011**, *21*, 295–304. [CrossRef]
29. Sarkar, S. Microbiological Considerations: Pasteurized Milk. *Int. J. Dairy Sci.* **2015**, *10*, 206–218. [CrossRef]
30. Adly, E.; Hegazy, A.A.; Kamal, M.; Abu-Hussien, S.H. Midguts of *Culex Pipiens* L. (Diptera: Culicidae) as a Potential Source of Raw Milk Contamination with Pathogens. *Sci. Rep.* **2022**, *12*, 13183. [CrossRef]
31. Ullah, S.; Khan, S.U.H.; Khan, M.J.; Khattak, B.; Fozia, F.; Ahmad, I.; Wadaan, M.A.; Khan, M.F.; Baabbad, A.; Goyal, S.M. Multiple-Drug Resistant Shiga Toxin-Producing E. Coli in Raw Milk of Dairy Bovine. *Trop. Med. Infect. Dis.* **2024**, *9*, 64. [CrossRef]
32. Gume, B.; Berhanu, L.; Kassa, T.; Bediru, H.; Fikre, A.G.; Dadi, L.S.; Mereta, S.T. Bacterial Hazard Identification and Exposure Assessment of Raw Milk Consumption in Jimma Zone, South West Ethiopia. *BMC Microbiol.* **2023**, *23*, 166. [CrossRef] [PubMed]
33. Griffiths, M.W. The Microbiological Safety of Raw Milk. *Improv. Saf. Qual. Milk.* **2010**, *1*, 27–63. [CrossRef]
34. Fusco, V.; Chieffi, D.; Fanelli, F.; Logrieco, A.F.; Cho, G.S.; Kabisch, J.; Böhnlein, C.; Franz, C.M.A.P. Microbial Quality and Safety of Milk and Milk Products in the 21st Century. *Compr. Rev. Food Sci. Food Saf.* **2020**, *19*, 2013–2049. [CrossRef]
35. Murphy, M.; Buckley, J.F.; Whyte, P.; O'Mahony, M.; Anderson, W.; Wall, P.G.; Fanning, S. Surveillance of Dairy Production Holdings Supplying Raw Milk to the Farmhouse Cheese Sector for Escherichia Coli O157, O26 and O111. *Zoonoses Public. Health* **2007**, *54*, 358–365. [CrossRef] [PubMed]
36. Jørgensen, H.J.; Mathisen, T.; Løvseth, A.; Omoe, K.; Qvale, K.S.; Loncarevic, S. An Outbreak of Staphylococcal Food Poisoning Caused by Enterotoxin H in Mashed Potato Made with Raw Milk. *FEMS Microbiol. Lett.* **2005**, *252*, 267–272. [CrossRef]
37. Langer, A.J.; Ayers, T.; Grass, J.; Lynch, M.; Angulo, F.J.; Mahon, B.E. Nonpasteurized Dairy Products, Disease Outbreaks, and State Laws—United States, 1993–2006. *Emerg. Infect. Dis.* **2012**, *18*, 385. [CrossRef]
38. Lucey, J.A. Raw Milk Consumption: Risks and Benefits. *Nutr. Today* **2015**, *50*, 189. [CrossRef] [PubMed]
39. Mungai, E.A.; Behravesh, C.B.; Gould, L.H. Increased Outbreaks Associated with Nonpasteurized Milk, United States, 2007–2012. *Emerg. Infect. Dis.* **2015**, *21*, 119. [CrossRef]
40. Robinson, T.J.; Scheftel, J.M.; Smith, K.E. Raw Milk Consumption among Patients with Non–Outbreak-Related Enteric Infections, Minnesota, USA, 2001–2010. *Emerg. Infect. Dis.* **2014**, *20*, 38. [CrossRef]
41. D'Amico, D.J.; Groves, E.; Donnelly, C.W. Low Incidence of Foodborne Pathogens of Concern in Raw Milk Utilized for Farmstead Cheese Production. *J. Food Prot.* **2008**, *71*, 1580–1589. [CrossRef]
42. Jamali, H.; Paydar, M.; Radmehr, B.; Ismail, S.; Dadrasnia, A. Prevalence and Antimicrobial Resistance of Staphylococcus Aureus Isolated from Raw Milk and Dairy Products. *Food Control* **2015**, *54*, 383–388. [CrossRef]
43. Giacometti, F.; Bonilauri, P.; Amatiste, S.; Arrigoni, N.; Bianchi, M.; Losio, M.N.; Bilei, S.; Cascone, G.; Comin, D.; Daminelli, P.; et al. Human Campylobacteriosis Related to the Consumption of Raw Milk Sold by Vending Machines in Italy: Quantitative Risk Assessment Based on Official Controls over Four Years. *Prev. Vet. Med.* **2015**, *121*, 151–158. [CrossRef]
44. Crotta, M.; Rizzi, R.; Varisco, G.; Daminelli, P.; Cunico, E.C.; Luini, M.; Graber, H.U.; Paterlini, F.; Guitian, J. Multiple-Strain Approach and Probabilistic Modeling of Consumer Habits in Quantitative Microbial Risk Assessment: A Quantitative Assessment of Exposure to Staphylococcal Enterotoxin A in Raw Milk. *J. Food Prot.* **2016**, *79*, 432–441. [CrossRef] [PubMed]
45. Soboleva, T.; French, N. *Assessment of the Microbiological Risks Associated with the Consumption of Raw Milk*; Ministry for Primary Industries: Wellington, New Zealand, 2014.
46. Latorre, A.A.; Pradhan, A.K.; Van Kessel, J.A.S.; Karns, J.S.; Boor, K.J.; Rice, D.H.; Mangione, K.J.; Grohn, Y.T.; Schukken, Y.H. Quantitative Risk Assessment of Listeriosis Due to Consumption of Raw Milk. *J. Food Prot.* **2011**, *74*, 1268–1281. [CrossRef] [PubMed]
47. Giacometti, F.; Serraino, A.; Bonilauri, P.; Ostanello, F.; Daminelli, P.; Finazzi, G.; Losio, M.N.; Marchetti, G.; Liuzzo, G.; Zanoni, R.G.; et al. Quantitative Risk Assessment of Verocytotoxin-Producing Escherichia Coli O157 and Campylobacter Jejuni Related to Consumption of Raw Milk in a Province in Northern Italy. *J. Food Prot.* **2012**, *75*, 2031–2038. [CrossRef] [PubMed]
48. Heidinger, J.C.; Winter, C.K.; Cullor, J.S. Quantitative Microbial Risk Assessment for Staphylococcus Aureus and Staphylococcus Enterotoxin A in Raw Milk. *J. Food Prot.* **2009**, *72*, 1641–1653. [CrossRef]
49. California Department of Public Health, Food and Drug Branch, Emergency Response Unit. Environmental Investigation of a Campylobacter Jejuni Outbreak in 2012 Associated with Claravale Farms Raw Whole Milk Final Report. Available online: https://realrawmilkfacts.com/PDFs/2012-claravale-farms-raw-milk-outbreak-final-report.pdf (accessed on 5 March 2025).

50. Kansas Department of Health and Environment. *Outbreak of Campylobacter Jejuni Infections Associated with Consumption of Cheese Made from Raw Milk-Western Kansas;* Kansas Department of Health and Environment: Topeka, KS, USA, 2007.
51. Salmonella Typhimurium Infection Associated with Raw Milk and Cheese Consumption—Pennsylvania. 2007. Available online: https://www.cdc.gov/mmwr/preview/mmwrhtml/mm5644a3.htm (accessed on 24 February 2025).
52. Department of Kansas Government Information. *Campylobacteriosis Outbreak Associated with Unpasteurized Milk, Reno County and Butler County, August–December 2007*; [Final Report]—Health and Environment; Department of Kansas Government Information: Topeka, KS, USA. Available online: https://kgi.contentdm.oclc.org/digital/collection/p16884coll4/id/1349/ (accessed on 24 February 2025).
53. Oliver, S.P.; Boor, K.J.; Murphy, S.C.; Murinda, S.E. Food Safety Hazards Associated with Consumption of Raw Milk. *Foodborne Pathog. Dis.* **2009**, *6*, 793–806. [CrossRef]
54. Escherichia Coli 0157:H7 Infections in Children Associated with Raw Milk and Raw Colostrum from Cows—California. 2006. Available online: https://www.cdc.gov/mmwr/preview/mmwrhtml/mm5723a2.htm (accessed on 24 February 2025).
55. Multistate Outbreak of Salmonella Serotype Typhimurium Infections Associated with Drinking Unpasteurized Milk—Illinois, Indiana, Ohio, and Tennessee, 2002–2003. Available online: https://www.cdc.gov/mmwr/preview/mmwrhtml/mm5226a3.htm (accessed on 24 February 2025).
56. Davis, K.R.; Dunn, A.C.; Burnett, C.; McCullough, L.; Dimond, M.; Wagner, J.; Smith, L.; Carter, A.; Willardson, S.; Nakashima, A.K. *Campylobacter Jjejuni* Infections Associated with Raw Milk Consumption—Utah, 2014. *MMWR Morb. Mortal. Wkly. Rep.* **2016**, *65*, 301–305. [CrossRef]
57. McCarthy, K.S.; Lopetcharat, K.; Drake, M.A. Milk Fat Threshold Determination and the Effect of Milk Fat Content on Consumer Preference for Fluid Milk. *J. Dairy Sci.* **2017**, *100*, 1702–1711. [CrossRef]
58. Outbreak of Campylobacter Jejuni Infections Associated with Drinking Unpasteurized Milk Procured through a Cow-Leasing Program—Wisconsin. 2001. Available online: https://www.cdc.gov/mmwr/preview/mmwrhtml/mm5125a2.htm (accessed on 24 February 2025).
59. Sakkas, L.; Moutafi, A.; Moschopoulou, E.; Moatsou, G. Assessment of Heat Treatment of Various Types of Milk. *Food Chem.* **2014**, *159*, 293–301. [CrossRef]
60. Guinee, T.P. Effect of High-Temperature Treatment of Milk and Whey Protein Denaturation on the Properties of Rennet—Curd Cheese: A Review. *Int. Dairy J.* **2021**, *121*, 105095. [CrossRef]
61. Zhang, Y.; Yi, S.; Lu, J.; Pang, X.; Xu, X.; Lv, J.; Zhang, S. Effect of Different Heat Treatments on the Maillard Reaction Products, Volatile Compounds and Glycation Level of Milk. *Int. Dairy J.* **2021**, *123*, 105182. [CrossRef]
62. Esteghlal, S.; Gahruie, H.H.; Niakousari, M.; Barba, F.J.; Bekhit, A.E.D.; Mallikarjunan, K.; Roohinejad, S. Bridging the Knowledge Gap for the Impact of Non-Thermal Processing on Proteins and Amino Acids. *Foods* **2019**, *8*, 262. [CrossRef] [PubMed]
63. Montilla, A.; Calvo, M.M. Goat's Milk Stability during Heat Treatment: Effect of PH and Phosphates. *J. Agric. Food Chem.* **1997**, *45*, 931–934. [CrossRef]
64. Al haj, O.A.; Al Kanhal, H.A. Compositional, Technological and Nutritional Aspects of Dromedary Camel Milk. *Int. Dairy J.* **2010**, *20*, 811–821. [CrossRef]
65. Wijayanti, H.B.; Bansal, N.; Deeth, H.C. Stability of Whey Proteins during Thermal Processing: A Review. *Compr. Rev. Food Sci. Food Saf.* **2014**, *13*, 1235–1251. [CrossRef]
66. Patel, P.B.; Thakkar, V.R.; Patel, J.S. Cellular Effect of Curcumin and Citral Combination on Breast Cancer Cells: Induction of Apoptosis and Cell Cycle Arrest. *J. Breast Cancer* **2015**, *18*, 225–234. [CrossRef] [PubMed]
67. Jovanovic, S.; Barac, M.; Macej, O.; Vucic, T.; Lacnjevac, C. SDS-PAGE Analysis of Soluble Proteins in Reconstituted Milk Exposed to Different Heat Treatments. *Sensors* **2007**, *7*, 371–383. [CrossRef]
68. Chen, G.Q.; Qu, Y.; Gras, S.L.; Kentish, S.E. Separation Technologies for Whey Protein Fractionation. *Food Eng. Rev.* **2023**, *15*, 438–465. [CrossRef]
69. Donato, L.; Guyomarc'h, F. Formation and Properties of the Whey Protein/κ-Casein Complexes in Heated Skim Milk—A Review. *Dairy Sci. Technol.* **2009**, *89*, 3–29. [CrossRef]
70. Tolkach, A.; Kulozik, U.; Tolkach, A.; Kulozik, U. Reaction Kinetic Pathway of Reversible and Irreversible Thermal Denaturation of β-Lactoglobulin. *Lait* **2007**, *87*, 301–315. [CrossRef]
71. Brew, K. Milk Proteins | α-Lactalbumin. *Encycl. Dairy Sci. Second. Ed.* **2011**, 780–786. [CrossRef]
72. McSweeney, P.L.H.; Fox, P.F. *Advanced Dairy Chemistry: Volume 1A: Proteins: Basic Aspects*, 4th ed.; Springer Science & Business Media: Berlin, Germany, 2013; pp. 1–548. [CrossRef]
73. Dalabasmaz, S.; Pischetsrieder, M. Design of a Prediction Model for the Differentiation of Pasteurized Milk from Heated ESL Milk by Peptide Profiling. *Proteomics* **2019**, *19*, 1800292. [CrossRef] [PubMed]

74. Ismail, B.; Nielsen, S.S. Invited Review: Plasmin Protease in Milk: Current Knowledge and Relevance to Dairy Industry. *J. Dairy Sci.* **2010**, *93*, 4999–5009. [CrossRef]
75. Rauh, V.M.; Johansen, L.B.; Ipsen, R.; Paulsson, M.; Larsen, L.B.; Hammershøj, M. Plasmin Activity in UHT Milk: Relationship between Proteolysis, Age Gelation, and Bitterness. *J. Agric. Food Chem.* **2014**, *62*, 6852–6860. [CrossRef] [PubMed]
76. Ryskaliyeva, A.; Henry, C.; Miranda, G.; Faye, B.; Konuspayeva, G.; Martin, P. Combining Different Proteomic Approaches to Resolve Complexity of the Milk Protein Fraction of Dromedary, Bactrian Camels and Hybrids, from Different Regions of Kazakhstan. *PLoS ONE* **2018**, *13*, e0197026. [CrossRef]
77. Scollard, P.G.; Beresford, T.P.; Needs, E.C.; Murphy, P.M.; Kelly, A.L. Plasmin Activity, β-Lactoglobulin Denaturation and Proteolysis in High Pressure Treated Milk. *Int. Dairy J.* **2000**, *10*, 835–841. [CrossRef]
78. Khaliq, A.; Mishra, A.K.; Niroula, A.; Baba, W.N.; Shaukat, M.N.; Rabbani, A. An Updated Comprehensive Review of Camel Milk: Composition, Therapeutic Properties, and Industrial Applications. *Food Biosci.* **2024**, *62*, 105531. [CrossRef]
79. Shimamura, T.; Kurogi, Y.; Katsuno, S.; Kashiwagi, T.; Ukeda, H. Demonstration of the Presence of Aminoreductone Formed during the Maillard Reaction in Milk. *Food Chem.* **2011**, *129*, 1088–1092. [CrossRef]
80. Erbersdobler, H.F.; Somoza, V. Forty Years of Furosine—Forty Years of Using Maillard Reaction Products as Indicators of the Nutritional Quality of Foods. *Mol. Nutr. Food Res.* **2007**, *51*, 423–430. [CrossRef]
81. Jaeger, H.; Janositz, A.; Knorr, D. The Maillard Reaction and Its Control during Food Processing. The Potential of Emerging Technologies. *Pathol. Biol.* **2010**, *58*, 207–213. [CrossRef] [PubMed]
82. Swanepoel, N.; Robinson, P.H.; Erasmus, L.J. Amino Acid Needs of Lactating Dairy Cows: Impact of Feeding Lysine in a Ruminally Protected Form on Productivity of Lactating Dairy Cows. *Anim. Feed. Sci. Technol.* **2010**, *157*, 79–94. [CrossRef]
83. Felfoul, I.; Jardin, J.; Gaucheron, F.; Attia, H.; Ayadi, M.A. Proteomic Profiling of Camel and Cow Milk Proteins under Heat Treatment. *Food Chem.* **2017**, *216*, 161–169. [CrossRef] [PubMed]
84. Hillier, R.M.; Lyster, R.L.J. Whey Protein Denaturation in Heated Milk and Cheese Whey. *J. Dairy Res.* **1979**, *46*, 95–102. [CrossRef]
85. Atuonwu, J.C.; Ray, J.; Stapley, A.G.F. A Kinetic Model for Whey Protein Denaturation at Different Moisture Contents and Temperatures. *Int. Dairy J.* **2017**, *75*, 41–50. [CrossRef]
86. Dannenberg, F.; Kessler, H.-G. Reaction Kinetics of the Denaturation of Whey Proteins in Milk. *J. Food Sci.* **1988**, *53*, 258–263. [CrossRef]
87. Dannenberg, F.; Kessler, H.G. Effect of Denaturation of β-Lactoglobulin on Texture Properties of Set-Style Nonfat Yoghurt. 1. Syneresis. *Milchwissenschaft* **1988**, *43*, 632–635.
88. Bulca, S.; Dumpler, J.; Kulozik, U. Kinetic Description of Heat-Induced Cross-Linking Reactions of Whey Protein-Free Casein Solutions. *Int. J. Dairy Technol.* **2016**, *69*, 489–496. [CrossRef]
89. Pitt, W.; Harden, T.; Hull, R.R. Investigation of the Antimicrobial Activity of Raw Milk against Several Foodborne Pathogens. *Milchwissenschaft* **2000**, *55*, 249–252.
90. YuQian, L.; Yousef, A.E. Characteristics of Listeria Monocytogenes Important to Food Processors. In *Listeria, Listeriosis, and Food Safety*; CRC Press: Boca Raton, FL, USA, 1999; pp. 131–224.
91. Gay, M.; Amgar, A. Factors Moderating Listeria Monocytogenes Growth in Raw Milk and in Soft Cheese Made from Raw Milk. *Lait* **2005**, *85*, 153–170. [CrossRef]
92. Dewettinck, K.; Rombaut, R.; Thienpont, N.; Le, T.T.; Messens, K.; Van Camp, J. Nutritional and Technological Aspects of Milk Fat Globule Membrane Material. *Int. Dairy J.* **2008**, *18*, 436–457. [CrossRef]
93. German, J.B.; Dillard, C.J. Composition, Structure and Absorption of Milk Lipids: A Source of Energy, Fat-Soluble Nutrients and Bioactive Molecules. *Crit. Rev. Food Sci. Nutr.* **2007**, *46*, 57–92. [CrossRef] [PubMed]
94. Spreer, E.; Mixa, A. *Milk and Dairy Product Technology*; Routledge: New York, NY, USA, 2017; pp. 1–483. [CrossRef]
95. Ceni, G.; Fernandes Silva, M.; Valério, C.; Cansian, R.L.; Oliveira, J.V.; Dalla Rosa, C.; Mazutti, M.A. Continuous Inactivation of Alkaline Phosphatase and Escherichia Coli in Milk Using Compressed Carbon Dioxide as Inactivating Agent. *J. CO_2 Util.* **2016**, *13*, 24–28. [CrossRef]
96. Wheeler, T.T.; Hodgkinson, A.J.; Prosser, C.G.; Davis, S.R. Immune Components of Colostrum and Milk—A Historical Perspective. *J. Mammary Gland. Biol. Neoplasia* **2007**, *12*, 237–247. [CrossRef]
97. Needs, E. *High Pressure Processing of Dairy Products*; Springer: Boston, MA, USA, 2001; pp. 269–296. [CrossRef]
98. Rodrigues, L.R. Milk Minor Constituents, Enzymes, Hormones, Growth Factors, and Organic Acids. In *Milk and Dairy Products in Human Nutrition: Production, Composition and Health*; Wiley-Blackwell: New York, NY, USA, 2013; pp. 220–245. [CrossRef]
99. Bogahawaththa, D.; Chandrapala, J.; Vasiljevic, T. Thermal Denaturation of Bovine Immunoglobulin G and Its Association with Other Whey Proteins. *Food Hydrocoll.* **2017**, *72*, 350–357. [CrossRef]
100. Martínez, B.; Bravo, D.; Rodríguez, A. Consequences of the Development of Nisin-Resistant Listeria Monocytogenes in Fermented Dairy Products. *J. Food Prot.* **2005**, *68*, 2383–2388. [CrossRef]

101. Li, Y.; Weng, P.; Wu, Z.; Liu, Y. Extending the Shelf Life of Raw Milk and Pasteurized Milk with Plantaricin FB-2. *Foods* **2023**, *12*, 608. [CrossRef]
102. González-Chávez, S.A.; Arévalo-Gallegos, S.; Rascón-Cruz, Q. Lactoferrin: Structure, Function and Applications. *Int. J. Antimicrob. Agents* **2009**, *33*, 301.e1–301.e8. [CrossRef]
103. Claeys, W.L.; Ludikhuyze, L.R.; Van Loey, A.M.; Hendrickx, M.E. Inactivation Kinetics of Alkaline Phosphatase and Lactoperoxidase, and Denaturation Kinetics of β-Lactoglobulin in Raw Milk under Isothermal and Dynamic Temperature Conditions. *J. Dairy Res.* **2001**, *68*, 95–107. [CrossRef]
104. Campbell, R.E.; Drake, M.A. Invited Review: The Effect of Native and Nonnative Enzymes on the Flavor of Dried Dairy Ingredients. *J. Dairy Sci.* **2013**, *96*, 4773–4783. [CrossRef]
105. Fox, P.F.; Kelly, A.L. Indigenous Enzymes in Milk: Overview and Historical Aspects—Part 2. *Int. Dairy J.* **2006**, *16*, 517–532. [CrossRef]
106. Prado, B.M.; Sombers, S.E.; Ismail, B.; Hayes, K.D. Effect of Heat Treatment on the Activity of Inhibitors of Plasmin and Plasminogen Activators in Milk. *Int. Dairy J.* **2006**, *16*, 593–599. [CrossRef]
107. Stevens, C.R.; Millar, T.M.; Clinch, J.G.; Kanczler, J.M.; Bodamyali, T.; Blake, D.R. Antibacterial Properties of Xanthine Oxidase in Human Milk. *Lancet* **2000**, *356*, 829–830. [CrossRef]
108. Alvarez, V.B. Sensory Evaluation of Milk and Milk Products. In *Dairy Processing and Quality Assurance*; Wiley-Blackwell: New York, NY, USA, 2015; pp. 467–487. [CrossRef]
109. Kokkinidou, S.; Peterson, D.G. Control of Maillard-Type off-Flavor Development in Ultrahigh-Temperature-Processed Bovine Milk by Phenolic Chemistry. *J. Agric. Food Chem.* **2014**, *62*, 8023–8033. [CrossRef]
110. Schamberger, G.P.; Labuza, T.P. Effect of Green Tea Flavonoids on Maillard Browning in UHT Milk. *LWT Food Sci. Technol.* **2007**, *40*, 1410–1417. [CrossRef]
111. Nozière, P.; Graulet, B.; Lucas, A.; Martin, B.; Grolier, P.; Doreau, M. Carotenoids for Ruminants: From Forages to Dairy Products. *Anim. Feed. Sci. Technol.* **2006**, *131*, 418–450. [CrossRef]
112. Li, Y.; Joyner, H.S.; Carter, B.G.; Drake, M.A. Effects of Fat Content, Pasteurization Method, Homogenization Pressure, and Storage Time on the Mechanical and Sensory Properties of Bovine Milk. *J. Dairy Sci.* **2018**, *101*, 2941–2955. [CrossRef]
113. Gathercole, J.; Reis, M.G.; Agnew, M.; Reis, M.M.; Humphrey, R.; Harris, P.; Clerens, S.; Haigh, B.; Dyer, J.M. Molecular Modification Associated with the Heat Treatment of Bovine Milk. *Int. Dairy J.* **2017**, *73*, 74–83. [CrossRef]
114. Datta, N.; Deeth, H.C. Age Gelation of UHT Milk—A Review. *Food Bioprod. Process.* **2001**, *79*, 197–210. [CrossRef]
115. Chojnicka-Paszun, A.; de Jongh, H.H.J.; de Kruif, C.G. Sensory Perception and Lubrication Properties of Milk: Influence of Fat Content. *Int. Dairy J.* **2012**, *26*, 15–22. [CrossRef]
116. Nguyen, P.T.M.; Bhandari, B.; Prakash, S. Tribological Method to Measure Lubricating Properties of Dairy Products. *J. Food Eng.* **2016**, *168*, 27–34. [CrossRef]
117. Clark, S.; Costello, M.; Drake, M.A.; Bodyfelt, F. The Sensory Evaluation of Dairy Products. In *The Sensory Evaluation of Dairy Products*; Van Nostrand Reinhold: New York, NY, USA, 2009; pp. 1–573. [CrossRef]
118. Reineccius, G. *Flavor Chemistry and Technology*; CRC Press: Boca Raton, FL, USA, 2005. [CrossRef]
119. Grebenteuch, S.; Kanzler, C.; Klaußnitzer, S.; Kroh, L.W.; Rohn, S. The Formation of Methyl Ketones during Lipid Oxidation at Elevated Temperatures. *Molecules* **2021**, *26*, 1104. [CrossRef] [PubMed]
120. Lee, A.P.; Barbano, D.M.; Drake, M.A. The Influence of Ultra-Pasteurization by Indirect Heating versus Direct Steam Injection on Skim and 2% Fat Milks. *J. Dairy Sci.* **2017**, *100*, 1688–1701. [CrossRef]
121. Zhao, J.; Zhou, B.; Wang, P.; Ren, F.; Mao, X. Physicochemical Properties of Fluid Milk with Different Heat Treatments and HS-GC-IMS Identification of Volatile Organic Compounds. *Int. Dairy J.* **2023**, *142*, 105654. [CrossRef]
122. Feng, D.; Wang, J.; Ji, X.; Min, W.; Yan, W. HS-GC-IMS Detection of Volatile Organic Compounds in Yak Milk Powder Processed by Different Drying Methods. *LWT* **2021**, *141*, 110855. [CrossRef]
123. Liem, D.G.; Bolhuis, D.P.; Hu, X.; Keast, R.S.J. Short Communication: Influence of Labeling on Australian and Chinese Consumers' Liking of Milk with Short (Pasteurized) and Long (UHT) Shelf Life. *J. Dairy Sci.* **2016**, *99*, 1747–1754. [CrossRef]
124. Meng, F.; Han, Z.; Zhang, Z.; Ding, H.; Lu, X.; Lu, C.; Ma, L.; Kang, Z.; Wang, B.; Li, Y. Effect of Steam Infusion and Steam Injection Ultra-High Temperature Treatment on Active Proteins and Flavor Compounds in Milk. *J. Food Sci. Technol.* **2023**, *41*, 70–80. [CrossRef]
125. Gennari, A.; Monteiro, B.; Lehn, D.; Fernanda, C.; De Souza, V. Effects of Pasteurization and Ultra-High Temperature Processes on Proximate Composition and Fatty Acid Profile in Bovine Milk. *Am. J. Food Technol.* **2015**, *10*, 265–272. [CrossRef]
126. Council Regulation (EC) No. 2597/97 Laying Down Additional Rules on the Common Organization of the Market in Milk and Milk Products for Drinking Milk. FAOLEX. Available online: https://www.fao.org/faolex/results/details/en/c/LEX-FAOC018629/ (accessed on 5 April 2025).

127. Macdonald, L.E.; Brett, J.; Kelton, D.; Majowicz, S.E.; Snedeker, K.; Sargeant, J.M. A Systematic Review and Meta-Analysis of the Effects of Pasteurization on Milk Vitamins, and Evidence for Raw Milk Consumption and Other Health-Related Outcomes. *J. Food Prot.* **2011**, *74*, 1814–1832. [CrossRef]
128. Raynal-Ljutovac, K.; Lagriffoul, G.; Paccard, P.; Guillet, I.; Chilliard, Y. Composition of Goat and Sheep Milk Products: An Update. *Small Rumin. Res.* **2008**, *79*, 57–72. [CrossRef]
129. Lyocks; Ayo, S.W.J.; Tanimu, J.; Olajide. Mineral Elements Content in Raw Milk of Grazing Cattle in Kaduna Metropolis and Environs. *Niger. J. Agric. Food Environ.* **2013**, *9*, 22–27.
130. Butler, M.I.; Bastiaanssen, T.F.S.; Long-Smith, C.; Berding, K.; Morkl, S.; Cusack, A.M.; Strain, C.; Busca, K.; Porteous-Allen, P.; Claesson, M.J.; et al. Recipe for a Healthy Gut: Intake of Unpasteurised Milk Is Associated with Increased *Lactobacillus* Abundance in the Human Gut Microbiome. *Nutrients* **2020**, *12*, 1468. [CrossRef] [PubMed]
131. Heckman, J.R. Securing Fresh Food from Fertile Soil, Challenges to the Organic and Raw Milk Movements. *Renew. Agric. Food Syst.* **2019**, *34*, 472–485. [CrossRef]
132. Schaafsma, G. Lactose and Lactose Derivatives as Bioactive Ingredients in Human Nutrition. *Int. Dairy J.* **2008**, *18*, 458–465. [CrossRef]
133. Law, D.; Conklin, J.; Pimentel, M. Erratum: Lactose Intolerance and the Role of the Lactose Breath Test. *Am. J. Gastroenterol.* **2010**, *105*, 2308. [CrossRef]
134. Claeys, W.L.; Cardoen, S.; Daube, G.; De Block, J.; Dewettinck, K.; Dierick, K.; De Zutter, L.; Huyghebaert, A.; Imberechts, H.; Thiange, P.; et al. Raw or Heated Cow Milk Consumption: Review of Risks and Benefits. *Food Control* **2013**, *31*, 251–262. [CrossRef]
135. Mummah, S.; Oelrich, B.; Hope, J.; Vu, Q.; Gardner, C.D. Effect of Raw Milk on Lactose Intolerance: A Randomized Controlled Pilot Study. *Ann. Fam. Med.* **2014**, *12*, 134–141. [CrossRef]
136. Brooks, C.; Pearce, N.; Douwes, J. The Hygiene Hypothesis in Allergy and Asthma: An Update. *Curr. Opin. Allergy Clin. Immunol.* **2013**, *13*, 70–77. [CrossRef]
137. Brick, T.; Schober, Y.; Böcking, C.; Pekkanen, J.; Genuneit, J.; Loss, G.; Dalphin, J.C.; Riedler, J.; Lauener, R.; Nockher, W.A.; et al. ω-3 Fatty Acids Contribute to the Asthma-Protective Effect of Unprocessed Cow's Milk. *J. Allergy Clin. Immunol.* **2016**, *137*, 1699–1706.e13. [CrossRef]
138. Chrischilles, E.; Ahrens, R.; Kuehl, A.; Kelly, K.; Thorne, P.; Burmeister, L.; Merchant, J. Asthma Prevalence and Morbidity among Rural Iowa Schoolchildren. *J. Allergy Clin. Immunol.* **2004**, *113*, 66–71. [CrossRef]
139. Ege, M.J.; Frei, R.; Bieli, C.; Schram-Bijkerk, D.; Waser, M.; Benz, M.R.; Weiss, G.; Nyberg, F.; van Hage, M.; Pershagen, G.; et al. Not All Farming Environments Protect against the Development of Asthma and Wheeze in Children. *J. Allergy Clin. Immunol.* **2007**, *119*, 1140–1147. [CrossRef] [PubMed]
140. Perkin, M.R.; Strachan, D.P. Which Aspects of the Farming Lifestyle Explain the Inverse Association with Childhood Allergy? *J. Allergy Clin. Immunol.* **2006**, *117*, 1374–1381. [CrossRef]
141. Riedler, J.; Braun-Fahrländer, C.; Eder, W.; Schreuer, M.; Waser, M.; Maisch, S.; Carr, D.; Schierl, R.; Nowak, D.; Von Mutius, E. Exposure to Farming in Early Life and Development of Asthma and Allergy: A Cross-Sectional Survey. *Lancet* **2001**, *358*, 1129–1133. [CrossRef]
142. Tse, K.; Horner, A.A. Allergen Tolerance versus the Allergic March: The Hygiene Hypothesis Revisited. *Curr. Allergy Asthma Rep.* **2008**, *8*, 475–483. [CrossRef]
143. Von Mutius, E. The Environmental Predictors of Allergic Disease. *J. Allergy Clin. Immunol.* **2000**, *105*, 9–19. [CrossRef]
144. Von Mutius, E.; Vercelli, D. Farm Living: Effects on Childhood Asthma and Allergy. *Nat. Rev. Immunol.* **2010**, *10*, 861–868. [CrossRef]
145. Waser, M.; Michels, K.B.; Bieli, C.; Flöistrup, H.; Pershagen, G.; Von Mutius, E.; Ege, M.; Riedler, J.; Schram-Bijkerk, D.; Brunekreef, B.; et al. Inverse Association of Farm Milk Consumption with Asthma and Allergy in Rural and Suburban Populations across Europe. *Clin. Exp. Allergy* **2007**, *37*, 661–670. [CrossRef] [PubMed]
146. Wickens, K.; Lane, J.M.; Fitzharris, P.; Siebers, R.; Riley, G.; Douwes, J.; Smith, T.; Crane, J. Farm Residence and Exposures and the Risk of Allergic Diseases in New Zealand Children. *Allergy* **2002**, *57*, 1171–1179. [CrossRef] [PubMed]
147. Rosenlund, H. *Protective Factors in Childhood Allergy Related to Diet and Lifestyle*; Karolinska Institutet: Stockholm, Sweden, 2010.
148. Loss, G.; Depner, M.; Ulfman, L.H.; Van Neerven, R.J.J.; Hose, A.J.; Genuneit, J.; Karvonen, A.M.; Hyvärinen, A.; Kaulek, V.; Roduit, C.; et al. Consumption of Unprocessed Cow's Milk Protects Infants from Common Respiratory Infections. *J. Allergy Clin. Immunol.* **2015**, *135*, 56–62.e2. [CrossRef]
149. Strachan, D.P. Hay Fever, Hygiene, and Household Size. *BMJ* **1989**, *299*, 1259–1260. [CrossRef]
150. Morin, S.; Bernard, H.; Przybylski-Nicaise, L.; Corthier, G.; Rabot, S.; Wal, J.M.; Hazebrouck, S. Allergenic and Immunogenic Potential of Cow's Milk β-Lactoglobulin and Caseins Evidenced Without Adjuvant in Germ-Free Mice. *Mol. Nutr. Food Res.* **2011**, *55*, 1700–1707. [CrossRef] [PubMed]

151. Barnes, M.; Cullinan, P.; Athanasaki, P.; MacNeill, S.; Hole, A.M.; Harris, J.; Kalogeraki, S.; Chatzinikolaou, M.; Drakonakis, N.; Bibaki-Liakou, V.; et al. Crete: Does Farming Explain Urban and Rural Differences in Atopy? *Clin. Exp. Allergy* **2001**, *31*, 1822–1828. [CrossRef]
152. Remes, S.T.; Iivanainen, K.; Koskela, H.; Pekkanen, J. Which Factors Explain the Lower Prevalence of Atopy Amongst Farmers' Children? *Clin. Exp. Allergy* **2003**, *33*, 427–434. [CrossRef]
153. Radon, K.; Windstetter, D.; Eckart, J.; Dressel, H.; Leitritz, L.; Reichert, J.; Schmid, M.; Praml, G.; Schosser, M.; Von Mutius, E.; et al. Farming Exposure in Childhood, Exposure to Markers of Infections and the Development of Atopy in Rural Subjects. *Clin. Exp. Allergy* **2004**, *34*, 1178–1183. [CrossRef] [PubMed]
154. Bieli, C.; Eder, W.; Frei, R.; Braun-Fahrländer, C.; Klimecki, W.; Waser, M.; Riedler, J.; von Mutius, E.; Scheynius, A.; Pershagen, G.; et al. A Polymorphism in CD14 Modifies the Effect of Farm Milk Consumption on Allergic Diseases and CD14 Gene Expression. *J. Allergy Clin. Immunol.* **2007**, *120*, 1308–1315. [CrossRef]
155. Pfefferle, P.I.; Büchele, G.; Blümer, N.; Roponen, M.; Ege, M.J.; Krauss-Etschmann, S.; Genuneit, J.; Hyvärinen, A.; Hirvonen, M.R.; Lauener, R.; et al. Cord Blood Cytokines Are Modulated by Maternal Farming Activities and Consumption of Farm Dairy Products during Pregnancy: The PASTURE Study. *J. Allergy Clin. Immunol.* **2010**, *125*, 108–115.e3. [CrossRef] [PubMed]
156. Khaliq, A.; Rabbani, A. Exploring Commercially Available Camel Milk Dairy Products: Medicinal, Nutritional and Antimicrobial Properties. *Res. Innov. Food Sci. Technol.* **2025**, *14*, 37–48. [CrossRef]
157. Mainer, G.; Sánchez, L.; Ena, J.M.; Calvo, M. Kinetic and Thermodynamic Parameters for Heat Denaturation of Bovine Milk IgG, IgA and IgM. *J. Food Sci.* **1997**, *62*, 1034–1038. [CrossRef]
158. Kulczycki, A. Bovine IgG Can Aggregate at Conditions Simulating Pasteurization and Binds to Some Human Fcγ Receptors. *Mol. Immunol.* **1987**, *24*, 259–266. [CrossRef]
159. Donnet-Hughes, A.; Duc, N.; Serrant, P.; Vidal, K.; Schiffrin, E.J. Bioactive Molecules in Milk and Their Role in Health and Disease: The Role of Transforming Growth Factor-β. *Immunol. Cell Biol.* **2000**, *78*, 74–79. [CrossRef]
160. Van Neerven, R.J.J.; Knol, E.F.; Heck, J.M.L.; Savelkoul, H.F.J. Which Factors in Raw Cow's Milk Contribute to Protection Against Allergies? *J. Allergy Clin. Immunol.* **2012**, *130*, 853–858. [CrossRef] [PubMed]
161. Blaser, M.J.; Duncan, D.J.; Osterholm, M.T.; Istre, G.R.; Wang, W.L. Serologic Study of Two Clusters of Infection Due to Campylobacter Jejuni. *J. Infect. Dis.* **1983**, *147*, 820–823. [CrossRef]
162. Capuano, E.; Ferrigno, A.; Acampa, I.; Ait-Ameur, L.; Fogliano, V. Characterization of the Maillard Reaction in Bread Crisps. *Eur. Food Res. Technol.* **2008**, *228*, 311–319. [CrossRef]
163. Masatcioglu, M.T.; Gokmen, V.; Ng, P.K.W.; Koksel, H. Effects of Formulation, Extrusion Cooking Conditions, and CO_2 Injection on the Formation of Acrylamide in Corn Extrudates. *J. Sci. Food Agric.* **2014**, *94*, 2562–2568. [CrossRef]
164. Melnik, B.C.; Schmitz, G. Exosomes of Pasteurized Milk: Potential Pathogens of Western Diseases. *J. Transl. Med.* **2019**, *17*, 3. [CrossRef] [PubMed]
165. Khaliq, A.; Mishra, A.K.; Shaukat, M.N.; Rabbani, A. Plant-Based Dairy Substitutes: Current Scenario and Future Prospects. *Innov. Agric.* **2023**, *6*, e32880. [CrossRef]
166. Badawy, S.; Liu, Y.; Guo, M.; Liu, Z.; Xie, C.; Marawan, M.A.; Ares, I.; Lopez-Torres, B.; Martínez, M.; Maximiliano, J.E.; et al. Conjugated Linoleic Acid (CLA) as a Functional Food: Is It Beneficial or Not? *Food Res. Int.* **2023**, *172*, 113158. [CrossRef]
167. Ali, M.A.; Kamal, M.M.; Rahman, M.H.; Siddiqui, M.N.; Haque, M.A.; Saha, K.K.; Rahman, M.A. Functional Dairy Products as a Source of Bioactive Peptides and Probiotics: Current Trends and Future Prospectives. *J. Food Sci. Technol.* **2021**, *59*, 1263–1279. [CrossRef]
168. Siddiqui, S.A.; Khan, S.; Bahmid, N.A.; Nagdalian, A.A.; Jafari, S.M.; Castro-Muñoz, R. Impact of High-Pressure Processing on the Bioactive Compounds of Milk—A Comprehensive Review. *J. Food Sci. Technol.* **2024**, *61*, 1632–1651. [CrossRef]
169. Smithers, G.; Versteeg, C.; Augustin, M.; Kelly, A.; Kothari, K.; Simpson, A.; Mulvihill, D.; Kelly, P. Symposium: Dairy Foods: Emerging Non-Thermal Food Processing Technologies-Their Potential in Dairy Systems. Available online: https://www.jtmtg.org/JAM/2008/abstracts/0553.PDF (accessed on 5 March 2025).
170. Pitino, M.A.; Unger, S.; Doyen, A.; Pouliot, Y.; Aufreiter, S.; Stone, D.; Kiss, A.; O'Connor, D.L. High Hydrostatic Pressure Processing Better Preserves the Nutrient and Bioactive Compound Composition of Human Donor Milk. *J. Nutr.* **2019**, *149*, 497–504. [CrossRef]
171. Chawla, R.; Patil, G.R.; Singh, A.K. High Hydrostatic Pressure Technology in Dairy Processing: A Review. *J. Food Sci. Technol.* **2011**, *48*, 260. [CrossRef] [PubMed]
172. Rademacher, B.; Hinrichs, J. Effects of High Pressure Treatment on Indigenous Enzymes in Bovine Milk: Reaction Kinetics, Inactivation and Potential Application. *Int. Dairy J.* **2006**, *16*, 655–661. [CrossRef]
173. Pegu, K.; Arya, S.S. Non-Thermal Processing of Milk: Principles, Mechanisms and Effect on Milk Components. *J. Agric. Food Res.* **2023**, *14*, 100730. [CrossRef]

174. Farkas, D.F.; Hoover, D.G. High Pressure Processing. *J. Food Sci.* **2000**, *65*, 47–64. [CrossRef]
175. Hite, B.H. The Effect of Pressure in the Preservation of Milk: A Preliminary Report. In *West Virginia Agricultural and Forestry Experiment Station Bulletins*; Generic: Flemington, Australia, 1899. [CrossRef]
176. Sehrawat, R.; Kaur, B.P.; Nema, P.K.; Tewari, S.; Kumar, L. Microbial Inactivation by High Pressure Processing: Principle, Mechanism and Factors Responsible. *Food Sci. Biotechnol.* **2021**, *30*, 19–35. [CrossRef]
177. Ribeiro, L.R.; Magalhães, I.S.; Tribst, A.A.L.; de Castro Leite Júnior, B.R. Effects of High-Pressure Processing on Enzyme Activity in Milk and Dairy Products: Science and Applications. In *Foundations and Frontiers in Enzymology*; Academic Press: Cambridge, MA, USA, 2023; pp. 169–193. [CrossRef]
178. Borda, D.; Van Loey, A.; Smout, C.; Hendrickx, M. Mathematical Models for Combined High Pressure and Thermal Plasmin Inactivation Kinetics in Two Model Systems. *J. Dairy Sci.* **2004**, *87*, 4042–4049. [CrossRef]
179. Huppertz, T.; Fox, P.F.; Kelly, A.L. Plasmin Activity and Proteolysis in High Pressure-Treated Bovine Milk. *Lait* **2004**, *84*, 297–304. [CrossRef]
180. Yu, T.; Zhang, X.; Feng, R.; Wang, C.; Wang, X.; Wang, Y. Comparison of the Effects of High Hydrostatic Pressure and Pasteurization on Quality of Milk during Storage. *Foods* **2022**, *11*, 2837. [CrossRef]
181. Ravichandran, C.; Jayachandran, L.E.; Kothakota, A.; Pandiselvam, R.; Balasubramaniam, V.M. Influence of High Pressure Pasteurization on Nutritional, Functional and Rheological Characteristics of Fruit and Vegetable Juices and Purees-an Updated Review. *Food Control* **2023**, *146*, 109516. [CrossRef]
182. Akdeniz, V.; Akalın, A.S. Recent Advances in Dual Effect of Power Ultrasound to Microorganisms in Dairy Industry: Activation or Inactivation. *Crit. Rev. Food Sci. Nutr.* **2022**, *62*, 889–904. [CrossRef]
183. Bermudez-Aguirre, D.; Niemira, B.A. Pasteurization of Foods with Ultrasound: The Present and the Future. *Appl. Sci.* **2022**, *12*, 10416. [CrossRef]
184. Islam, M.N.; Zhang, M.; Adhikari, B. The Inactivation of Enzymes by Ultrasound—A Review of Potential Mechanisms. *Food Rev. Int.* **2014**, *30*, 1–21. [CrossRef]
185. Villamiel, M.; De Jong, P. Influence of High-Intensity Ultrasound and Heat Treatment in Continuous Flow on Fat, Proteins, and Native Enzymes of Milk. *J. Agric. Food Chem.* **2000**, *48*, 472–478. [CrossRef] [PubMed]
186. Chiozzi, V.; Agriopoulou, S.; Varzakas, T. Advances, Applications, and Comparison of Thermal (Pasteurization, Sterilization, and Aseptic Packaging) Against Non-Thermal (Ultrasounds, UV Radiation, Ozonation, High Hydrostatic Pressure) Technologies in Food Processing. *Appl. Sci.* **2022**, *12*, 2202. [CrossRef]
187. Xu, S.; Qiao, Y.; Liu, X.; Church, C.C.; Wan, M. Fundamentals of Cavitation. In *Cavitation in Biomedicine*; Springer: Boston, MA, USA, 2015; pp. 1–46. [CrossRef]
188. Chauhan, O.P.; Sayanfar, S.; Toepfl, S. Effect of Pulsed Electric Field on Texture and Drying Time of Apple Slices. *J. Food Sci. Technol.* **2018**, *55*, 2251–2258. [CrossRef]
189. Wouters, P.C.; Alvarez, I.; Raso, J. Critical Factors Determining Inactivation Kinetics by Pulsed Electric Field Food Processing. *Trends Food Sci. Technol.* **2001**, *12*, 112–121. [CrossRef]
190. Roobab, U.; Abida, A.; Chacha, J.S.; Athar, A.; Madni, G.M.; Ranjha, M.M.A.N.; Rusu, A.V.; Zeng, X.A.; Aadil, R.M.; Trif, M. Applications of Innovative Non-Thermal Pulsed Electric Field Technology in Developing Safer and Healthier Fruit Juices. *Molecules* **2022**, *27*, 4031. [CrossRef]
191. Michalac, S.; Alvarez, V.; Ji, T.; Zhang, Q.H. Inactivation of Selected Microorganisms and Properties of Pulsed Electric Field Processed Milk. *J. Food Process Preserv.* **2003**, *27*, 137–151. [CrossRef]
192. Sampedro, F.; Rodrigo, M.; Martínez, A.; Rodrigo, D.; Barbosa-Cánovas, G.V. Quality and Safety Aspects of PEF Application in Milk and Milk Products. *Crit. Rev. Food Sci. Nutr.* **2007**, *45*, 25–47. [CrossRef]
193. Sepulveda, D.R.; Góngora-Nieto, M.M.; Guerrero, J.A.; Barbosa-Cánovas, G.V. Production of Extended-Shelf Life Milk by Processing Pasteurized Milk with Pulsed Electric Fields. *J. Food Eng.* **2005**, *67*, 81–86. [CrossRef]
194. Šalaševičius, A.; Uždavinytė, D.; Visockis, M.; Ruzgys, P.; Šatkauskas, S. Comparative Analysis of Pulsed Electric Fields (PEF) and Traditional Pasteurization Techniques: Comparative Effects on Nutritional Attributes and Bacterial Viability in Milk and Whey Products. *Appl. Sci.* **2023**, *13*, 12127. [CrossRef]
195. Su, W.; Wang, Q.; Li, J.; Qiu, Z.; Qiu, Y. Effects of Pulsed Electric Field Technology on the Nutritional Value and Biological Function of Plant Food. *Front. Sustain. Food Syst.* **2024**, *8*, 1385533. [CrossRef]
196. Dhillon, P.K.; Singla, P. Advances in Non-Thermal Membrane Processing for Nutrient Recovery in the Dairy Industry: A Comprehensive Review. In *Non-Thermal Processing of Functional Foods*; CRC Press: Boca Raton, FL, USA, 2024; pp. 223–245. [CrossRef]
197. Rabbani, A.; Srikumar, S.; Niroula, A.; Khan, M.K.I.; Nazir, A. Stability and Shelf-Life Modeling of Lemongrass Essential Oil-in-Water Nanoemulsions. *Eur. J. Lipid Sci. Technol.* **2025**, *127*, e202400096. [CrossRef]

198. Zhang, S.; Liu, L.; Pang, X.; Lu, J.; Kong, F.; Lv, J. Use of Microfiltration to Improve Quality and Shelf Life of Ultra-High Temperature Milk. *J. Food Process Preserv.* **2016**, *40*, 707–714. [CrossRef]
199. Zhang, W.; Liu, Y.; Li, Z.; Xu, S.; Zhang, J.; Hettinga, K.; Zhou, P. Effects of Microfiltration Combined with Ultrasonication on Shelf Life and Bioactive Protein of Skim Milk. *Ultrason. Sonochem* **2021**, *77*, 105668. [CrossRef] [PubMed]
200. Khatkar, S.K.; Dudi, K.; Lonkar, S.A.; Sidhu, K.S.; Khatkar, A.B.; Chandla, N.K.; Panghal, A. An Overview of Membrane Technology in Dairy & Food Industry. In *Novel Technologies in Food Science*; Wiley-Scrivener: Austin, TX, USA, 2022; pp. 65–108. [CrossRef]
201. Żbik, K.; Onopiuk, A.; Górska-Horczyczak, E.; Wierzbicka, A. Trends and Opportunities in the Dairy Industry: A2 Milk and Processing Methods. *Appl. Sci.* **2024**, *14*, 6513. [CrossRef]
202. Becker, K.M.; Parsons, R.L.; Kolodinsky, J.; Matiru, G.N. A Cost and Returns Evaluation of Alternative Dairy Products to Determine Capital Investment and Operational Feasibility of a Small-Scale Dairy Processing Facility. *J. Dairy Sci.* **2007**, *90*, 2506–2516. [CrossRef]
203. Lau, S.; Wiedmann, M.; Adalja, A. Economic and Environmental Analysis of Processing Plant Interventions to Reduce Fluid Milk Waste. *J. Dairy Sci.* **2023**, *106*, 4773–4784. [CrossRef]
204. Merlino, V.M.; Massaglia, S.; Borra, D.; Mimosi, A.; Cornale, P. Which Factors Drive Consumer Decisions During Milk Purchase? New Individuals' Profiles Considering Fresh Pasteurized and Uht Treated Milk. *Foods* **2022**, *11*, 77. [CrossRef]

MDPI AG
Grosspeteranlage 5
4052 Basel
Switzerland
Tel.: +41 61 683 77 34

Foods Editorial Office
E-mail: foods@mdpi.com
www.mdpi.com/journal/foods

www.ingramcontent.com/pod-product-compliance
Lightning Source LLC
Chambersburg PA
CBHW040816250926
51850CB00007BA/1179